# Lecture Notes in Physics

## Bisher erschienen/Already published

Vol. 1: J. C. Erdmann. Wärmeleitung in Kristallen, theoretische Grundlagen und fortgeschrittene experimentelle Methoden. II, 283 Seiten. 1969.

Vol. 2: K. Hepp, Théorie de la renormalisation. III, 215 pages. 1969.

Vol. 3: A. Martin, Scattering Theory: Unitarity, Analyticity and Crossing. IV, 125 pages. 1969.

Vol. 4: G. Ludwig, Deutung des Begriffs „physikalische Theorie" und axiomatische Grundlegung der Hilbertraumstruktur der Quantenmechanik durch Hauptsätze des Messens. 1970. Vergriffen.

Vol. 5: Schaaf, The Reduction of the Product of Two Irreducible Unitary Representations of the Proper Orthochronous Quantummechanical Poincare Group. IV, 120 pages. 1970.

Vol. 6: Group Representations in Mathematics and Physics. Edited by V. Bargmann. V, 340 pages. 1970.

Vol. 7: R. Balescu, J. L. Lebowitz, I. Prigogine, P. Résibois, Z. W. Salsburg, Lectures in Statistical Physics. V, 181 pages. 1971.

Vol. 8: Proceedings of the Second International Conference on Numerical Methods in Fluid Dynamics. Edited by M. Holt. 1971. Out of print.

Vol. 9: D. W. Robinson, The Thermodynamic Pressure in Quantum Statistical Mechanics. V, 115 pages. 1971.

Vol. 10: J. M. Stewart, Non-Equilibrium-Relativistic Kinetic Theory. III, 113 pages. 1971.

Vol. 11: O. Steinmann, Pertubation Expansions in Axiomatic Field Theory. III, 126 pages. 1976.

Vol. 12: Statistical Models and Turbulence. Edited by C. Van Atta and M. Rosenblatt. Reprint of the First Edition. VIII, 492 pages. 1975.

Vol. 13: M. Ryan, Hamiltonian Cosmology. VII, 169 pages. 1972.

Vol. 14: Methods of Local and Global Differential Geometry in General Relativity. Edited by D. Farnsworth, J. Fink, J. Porter, and A. Thompson. V, 188 pages.

Vol. 15: M. Fierz, Vorlesungen zur Entwicklungsgeschichte der Mechanik. V, 97 Seiten. 1972.

Vol. 16: H.-O. Georgii, Phasenübergang 1. Art bei Gittergasmodellen. IX, 167 Seiten. 1972.

Vol. 17: Strong Interaction Physics. Edited by W. Rühl and A. Vancura. V, 405 pages. 1973.

Vol. 18: Proceedings of the Third International Conference on Numerical Methods in Fluid Mechanics, Vol. I. Edited by H. Cabannes and R. Temam. VII, 186 pages. 1973.

Vol. 19: Proceedings of the Third International Conference on Numerical Methods in Fluid Mechanics, Vol. II. Edited by H. Cabannes and R. Temam. VII, 275 pages. 1973.

Vol. 20: Statistical Mechanics and Mathematical Problems. Edited by A. Lenard. VIII, 247 pages. 1973.

Vol. 21: Optimization and Stability Problems in Continuum Mechanics. Edited by P. K. C. Wang. V, 94 pages. 1973.

Vol. 22: Proceedings of the Europhysics Study Conference on Intermediate Processes in Nuclear Reactions. Edited by N. Cindro, P. Kulišic and Th. Mayer-Kuckuk. XIV, 329 pages. 1973.

Vol. 23: Nuclear Structure Physics. Proceedings 1973. Edited by U. Smilansky, I. Talmi, and H. A. Weidenmüller. XII, 296 pages. 1973.

Vol. 24: R. F. Snipes, Statistical Mechanical Theory of the Electrolytic Transport of Nonelectrolytes. V, 210 pages. 1973.

Vol. 25: Constructive Quantum Field Theory. The 1973 "Ettore Majorana" International School of Mathematical Physics. Edited by G. Velo and A. Wightman. III, 331 pages. 1973.

Vol. 26: A. Hubert, Theorie der Domänenwände in geordneten Medien. XII, 377 Seiten. 1974.

Vol. 27: R. K. Zeytounian, Notes sur les Ecoulements Rotationnels de Fluides Parfaits. XIII, 407 pages. 1974.

Vol. 28: Lectures in Statistical Physics. Edited by W. C. Schieve and J. S. Turner. V, 342 pages. 1974.

Vol. 29: Foundations of Quantum Mechanics and Ordered Linear Spaces. Advanced Study Institute, Marburg 1973. Edited by A. Hartkämper and H. Neumann. VI, 355 pages. 1974.

Vol. 30: Polarization Nuclear Physics. Proceedings 1973. Edited by D. Fick. IX, 292 pages. 1974.

Vol. 31: Transport Phenomena. Sitges International Schools of Statistical Mechanics, June 1974. Edited by G. Kirczenow and J. Marro. XIV, 517 pages. 1974.

Vol. 32: Particles, Quantum Fields and Statistical Mechanics. Proceedings 1973. Edited by M. Alexanian and A. Zepeda. V, 132 pages. 1975.

Vol. 33: Classical and Quantum Mechanical Aspects of Heavy Ion Collisions. Proceedings 1974. Edited by H. L. Harney, P. Braun-Munzinger, and C. K. Gelbke. VII, 311 pages. 1975.

Vol. 34: One-Dimensional Conductors GPS Summer School Proceedings, 1974. Edited by H. G. Schuster. VII, 371 pages. 1975.

Vol. 35: Proceedings of the Fourth International Conference on Numerical Methods in Fluid Dynamics, 1974. Edited by R. D. Richtmyer. V, 457 pages. 1975.

Vol. 36: R. Gatignol, Théorie Cinétique des Gaz à Répartition Discrète de Vitesses. II, 219 pages. 1975.

Vol. 37: Trends in Elementary Particle Theory. Proceedings 1974. Edited by H. Rollnik and K. Dietz. V, 472 pages. 1975.

Vol. 38: Dynamical Systems, Theory and Applications. Proceedings 1974. Edited by J. Moser. VI, 624 pages. 1975.

Vol. 39: International Symposium on Mathematical Problems in Theoretical Physics. Proceedings 1975. Edited by H. Araki. XII, 562 pages. 1975.

Vol. 40: Effective Interactions and Operators in Nuclei. Proceedings 1975. Edited by B. R. Barrett. XII, 339 pages. 1975.

Vol. 41: Progress in Numerical Fluid Dynamics. Proceedings 1974. Edited by H. J. Wirz. V, 471 pages. 1975.

Vol. 42: H II Regions and Related Topics. Proceedings 1975. Edited by D. Downes and T. L. Wilson. XII, 488 pages. 1975.

Vol. 43: Laser Spectroscopy. Proceedings 1975. Edited by S. Haroche, J. C. Pebay-Peyroula, T. W. Hänsch, and S. E. Harris. X, 466 pages. 1975.

# Lecture Notes in Physics

Edited by J. Ehlers, München, K. Hepp, Zürich,
R. Kippenhahn, München, H. A. Weidenmüller, Heidelberg,
and J. Zittartz, Köln
Managing Editor: W. Beiglböck, Heidelberg

## 62

# Photonuclear Reactions II

International School on Electro- and
Photonuclear Reactions,
Erice, Italy 1976
Edited by S. Costa and C. Schaerf

Springer-Verlag Berlin Heidelberg GmbH

**Editors**
Prof. Sergio Costa
Instituto di Fisica, Università di Torino
Corso M. d'Azeglio, 46
10125 Torino, Italia

Prof. Carlo Schaerf
Instituto di Fisica, Università di Roma
Piazzale delle Scienze, 5
00185 Roma, Italia

Library of Congress Cataloging in Publication Data

International School on Electro and Photonuclear Reactions, Erice, Italy, 1976.
   Photonuclear reactions.

   (Lecture notes in physics ; 61-62)
   1.  Photonuclear reactions--Congresses.  2.  Electro-
magnetic interactions--Congresses.  I. Costa, Sergio.
II.  Schaerf, Carlo, 1935-       III.  Ettore Majorana
Centre for Scientific Culture.  IV.  Title.  V.  Series.
QC794.8.P4I55   1976      539.7'56        77-4456

ISBN 978-3-540-08140-1        ISBN 978-3-540-37387-2 (eBook)
DOI 10.1007/978-3-540-37387-2

Originally published by Springer-Verlag Berlin Heidelberg New York in 1977.

2153/3140-543210

# P R E F A C E

Our understanding of the nature of fundamental interactions has been greatly helped by experiments involving electrons and photons. These tools had, and still have, indeed, a relevant impact on the investigation of atoms and molecules, as well as nuclear structure.

In the case of nuclear systems held together by forces not yet completely understood, the electromagnetic probes, whose interaction with the nucleons in the nucleus is basically well known, provided us with matrix elements giving direct information on the nuclear wave functions.

The study of the electromagnetic porperties of nuclear states has been crucial for the development of nuclear models and the good knowledge of the quantum numbers associated with specific multipoles has notably simplified the analysis of the fundamental types of nuclear motion.

A large number of laboratories are currently carrying out experiments using both electron and real photon beams and, in many of these, a noticeable effort is being made to improve the quality of the existing instrumentation, as well as to set up new facilities.

The purpose of the School on Electro- and Photonuclear Reactions is, therefore, to review and discuss the most significant achievements in the study of nuclear properties using electromagnetic probes with special emphasis on the most recent experimental and theoretical results obtained in this field.

Because of the vastness of the subject, the first course of the School was essentially devoted to photonuclear interactions from few MeV up to intermediate energy.

Ten series of lectures (collected in Volume I) and a number of seminars were given. Most of the seminars were presented in the form of status reports from important laboratories, and these reports are collected in Volume II.

In the lecture sessions, the classic topic of the giant dipole resonance was first reviewed and the phenomenology discussed together with the mechanisms leading to the excitation of the giant states. Collective and microscopic models were proposed in order to push the theoretical description closer to the experimental results. Isotopic spin effects and sum rules completed the study of the doorway states through which the G.D.R. is formed.

The competition between the decay channels from the G.D.R. was analysed and the existing knowledge of M1 and E2 resonances, both isoscalar and isovector, was surveyed.

Nuclear elastic photon scattering was discussed in detail, including the contributions of Thomson, Rayleigh and Delbruck scattering. Suggestions for future experimentation, mainly with polarized photons, were also made.

A new way of calculating the transition matrix at intermediate energy was presented, underlining the relevance of gauge terms and describing the direct coupling of the photon to nucleon-nucleon correlations.

Exchange-current phenomena were also discussed in connection with the integrated photo-absorption cross section, which is shown to contain information on mesonic degrees of freedom in nuclei. The effects of explicitly introducing isobars on nuclear constitutents were explored, in particular, in the case of electromagnetic interactions in the two-nucleon system.

Finally the few-body systems and their interaction with real and virtual photons and with hadrons were discussed as a check of our present understanding of nuclear properties in terms of the basic n-n force.

During the course special theoretical topics were introduced by some participants: these are also summarized in Volume II.

The course could have not been so rewarding as it was without the enthusiastic collaboration of all the lecturers and participants and the entire staff of the Centro di Cultura Scientifica " Ettore Majorana ". It is a pleasure to thank in particular Prof. A. Zichichi, Director of the Centre, Dr. S.A. Gabriele, Miss P. Savalli and Miss M. Zaini for their generous help.

The organization of the course was made possible by the financial contributions of the National Research Council (CNR) and the National Institute of Nuclear Physics (INFN). We wish to express our gratitude to Prof. E. Amaldi, President of the National Committee for the Physical Sciences of CNR and Prof. A. Gigli, President of INFN.

TABLE OF CONTENTS

INVITED SEMINARS

BERTOZZI, W.      :   Recent Developments at M.I.T. ........................... 1

CATILLON, Ph.    :     News from Saclay ...................................... 47

DE VRIES, C.     :   Electron Scattering Work at Amsterdam - past, present
                     and future activities ................................... 62

DRECHSEL, D.     :   Electronuclear Sum Rules ................................ 92

LINDGREN, K.     :   A Review of Present Photonuclear Research at Lund and
                     future Accelerator Plans ................................ 132

MATONE, G.    :   A Monochromatic and Polarized Photon Beam for Photo-
                  nuclear Reactions, The Ladon Project at Frascati .......... 149

RICHTER, A.      :   Latest from Dalinac ..................................... 165

SANZONE, M.      :   Preliminary Results on the Annihilation Photon Beam at
                     the Frascati Linac Laboratory ........................... 199

SOLODUKHOV, G.V.    :   Some Experimental Results on the Measurement of the
                        the total Photoabsorption Cross Sections ............. 216

TORIZUKA, Y.    :   Electroexcitation of Giant Multipole Resonances ......... 258

SUMMARIES OF CONTRIBUTED SEMINARS

BOHIGAS, O.    :   Description of Isoscalar Resonances. A Sum-Rule Approach .. 294

DELSANTO, P.P.    :    The Center of Mass Problems in Continuum ............... 295

GIANNINI, M.    :   A Simple Model for Resonance Shifts ..................... 297

PROSPERI, D.    :   Nucleon Polarizabilities and Deep Inelastic Electron
                    Scattering .............................................. 298

CHRISTILLIN, P. and ROSA-CLOT, M.   :    Exchange Effects in Photon
                                         Scattering in Nuclei ................ 301

TABLE OF CONTENTS
(to Volume I)

LECTURES

BERGERE, R.    :   Features of the Giant EI Resonance ......................... 1

RICCO, G.    :   Photonuclear Reactions above the Giant Dipole Resonance.
A Survey ..................................................... 223

HANNA, S.    :   Giant Multipole Resonances ................................. 275

HAYWARD, E.    :    Photon Scattering in the Energy Range 5-30 MeV .......... 340

HEBACH, H.    :   Mechanisms of Photonuclear Reactions at Intermediate
Energies (40 - 140 MeV) ................................. 407

BOSCO, B.    :   Real and Virtual Photons ................................... 461

WEISE, W.    :   Sum Rules in Photonuclear Physics ......................... 484

LEONARDI, R.    :    Isospin Structure of the Dipole Giant Resonance ........ 501

CIOFI DEGLI ATTI, C.    :    Electromagnetic and Hadronic Interactions with
the Few-Body Systems at Intermediate Energies ... 521

ARENHÖVEL, H.    :   Bayron Resonances in Nuclei ........................... 586

# List of Participants

ARENHOEVEL, H.,  Mainz

BERGERE, R.,  Bures sur Yvette

BERTOZZI, W.,  Cambridge (USA)

BOHIGAS, O.,  Orsay

BOSCO, B.,  Arcetri

CALOI, R.,  Roma

CARCHON, R.,  Gent

CATILLON, P.,  Gif sur Yvette

CHEW, S.H.,  Birmingham

CIOFI DEGLI ATTI, C.,  Roma

COSTA, S.,  Torino

DELSANTO, P.P.,  Cagliari

D'ERASMO, G.,  Bari

DEVOS, J.,  Gent

DE VRIES, C.,  Amsterdam

DRECHSEL, D.,  Mainz

EPPEL, D.,  Hamburg

FABRE DE LA RIPELLE, M.,  Orsay

FINDLAY, D.J.S.,  Glasgow

GIANNINI, M.,  Genova

GIUSTI, C.,  Pavia

GOERINGER, H.,  Mainz

HANNA, S.S.,  Stanford (USA)

HAYWARD, E.,  Washington, DC

HEBACH, H.,  Bochum

JOHNSSON, B.,  Lund

LAUTERBACH, C.,  Garching

LEONARDI, R.,  Bologna

LEPRETRE, A.,  Gif sur Yvette

LINDGREN, K.,  Lund

LIPPARINI, E.,  Trento

MATONE, G.,  Frascati

MATTHEWS, J.L.,  Cambridge (USA)

MECKING, B.,  Bonn

PANTALEO, A.,  Bari

PROSPERI, D.,  Frascati

RICCO, G.,  Genova

RICHTER, A.,  Darmstadt

ROSA-CLOT, M.,  Geneva

ROSS, C.,  Ottawa

SANZONE, M.,  Genova

SCHAERF, C.,  Roma

SOLODUKHOV, G.V.,  Moscow

STRANGIO, C.,  Roma

TERRANOVA, M.L.,  Roma

TORIZUKA, Y.,  Sendai

TRAINI, M.,  Trento

VAN CAMP, E.,  Gent

WEISE, W.,  Erlangen

WOLYNEC, E.,  Sao Paulo

W. Bertozzi

Department of Physics and Laboratory for Nuclear Science

Cambridge, Mass., U.S.A.

## I.  Introduction; Facility Status

The experimental program at MIT has been in operation since early
1975.  There have been three major areas of activity; threshold photo-
pion studies; photoproton studies; and electron scattering.  Electron
scattering has come to use the major share of beam time because of the
uniqueness of the high resolution energy-loss spectrometer.  Also, it
is our only major experimental instrument and it occupies our single
experimental hall.  The $(\gamma,p)$ experiments also use the spectrometer
while the $(\gamma,\pi)$ work is carried out in a small room originally intended
to house part of the beam transport for a second experimental hall that
was not funded in our original construction.  The Energy Research and
Development Agency (ERDA), the supporting agency for medium energy
physics at MIT is in the process of seeking funds for a second experimental area that is shown in Figure 1, which also indicates some
of the physics we are planning.

The Linac is designed to operate at a duty ratio of 1.8% up to
400 MeV and a duty ratio of 5.6% up to 200 MeV.  Because of funding
limitations, the operation has been conservative to date extending only to
0.8% duty.  The main concern has been to carry on research and at the
same time to gain operating experience on reliability without exhausting
a very limited inventory of spare parts.  The spectrum from the
accelerator is excellent with about 80% of the beam in a momentum region
less than 0.3%.  The momentum spread is made up of two components.  For
short time periods, less than one second, the spread is about 0.1%.  This
narrow beam then wanders in momentum filling out the 0.3% bin.  We have
used average currents up to 70 μA in experiments, the limitation having
been radiation levels and the lack of protective devices for the
accelerator and switchyard elements.  This situation has been improved

recently so that average currents of 150 μA are now possible.

We are proposing to recirculate the beam through the accelerator for a second pass, to raise the energy to about 700 MeV. Our studies indicate that several passes are feasible to achieve even higher energies. However, one turn will yield an important increment in energy and can be achieved at moderate cost with only a small modification of the existing physical plant. The beam switchyard and the energy-loss spectrometer in particular have been designed to perform at momenta in excess of 900 MeV/c and are therefore ready in their present state for experiments in electron scattering at the higher energy. The high resolution $\frac{\Delta p}{p} = 10^{-4}$ or about 70 keV at 700 MeV would produce a research facility capable of exploring very fully the small scale features of nuclear structure. In addition, many recent questions concerning meson degrees of freedom and relativistic effects could be studied extensively. These effects are known to become detectable in the two body system at low q but dominate the scattering at $q \sim 4\,\mathrm{fm}^{-1}$. The higher q behavior is at present only a matter of conjecture.

## II.   Recent Results in Electron Scattering from MIT

The first results with the high resolution, energy-loss spectrometer at MIT have been very exciting and clearly it is a pleasure for me personally to report them at this meeting. Even though we have not had time to produce a complete analysis of the spectra, I believe you will at least be impressed by some of the clear implications for the physics.

The MIT group was the first to recognize the potential of energy-loss spectroscopy in electron scattering and proposed in 1966 to construct a rather extensive spectroscopic system based on this idea, with a resolution approaching one part in $10^4$. The details of the design and motivations were reported in the "1967 MIT Summer Study" [1] and at the "Sardina Conference on Electron Scattering" in 1970 [2]. Since then, the technique has come to be used at Darmstadt and there are at least a handful of other laboratories planning to implement such a system. In hadron scattering, the technique was first used by Bernie Cohen in the late 1950's. Because you are all by now very familiar with the technique, I will not dwell on the spectrometer design but go directly to our results.

    We use the spectrometer with a beam energy spread on the target of
$\Delta E/E = 3 \times 10^{-3}$, corresponding to 80% of the beam from the accelerator.
In Figure 2, I show the results for the scattering of 200 MeV electrons
from $^{19}F$ (teflon in transmission mode) using a vertical aperture of
± 70 mr and a horizontal aperture (scattering angle) of ± 5.0 mr to
define the spectrometer solid angle. The $1/2^+$ ground state, the $1/2^-$
state at 110 keV and the $5/2^+$ state at 197 keV are completely resolved.
The widths of the prominent $1/2^+$, $5/2^+$ states along with the spacings
point out that our resolution (FWHM) is about 20 KeV and that our
fractional resolution is one part in $10^4$. Clearly, the spectrometer
is functioning as well as expected, if not better.

    We have recently expanded the focal plane to accept a 4.2 msr solid
angle (± 70 mr x ± 15 mr) with a unique counter and readout system. The
system employs a standard multiwire proportion chamber. A measurement
of the drift times of ions to adjacent wires defines the electron track
location with a resolution of 130μm and the track angle with a
resolution of less than 15 mr. A second chamber defines the track
location perpendicular to the momentum direction. These parameters
provide the information needed to correct for focal plane curvature
and spectrometer aberrations. The beam switchyard-spectrometer system
has a resolution of $\Delta p/p \sim 10^{-4}$ over approximately a 1.5% range in
momentum and a resolution of about $1.5 \times 10^{-4}$ over the 6% range of
its present instrumentation. Our plans are to extend the instrumentation to a range of about 11% in momentum.

    Experiments have been performed on the following nuclei: $^{16}O$, $^{12}C$,
$^{13}C$, $^{19}F$, $^{27}Al$, $^{24}Mg$, $^{25}Mg$, $^{150}Nd$, $^{166}Er$, $^{176}Yb$, $^{156}Gd$ and $^{181}Ta$. In
the following I display some sample spectra and point out some of the
qualitative physics that can be observed in the data.

    A.) $\underline{^{16}O}$:  The $0^+$ state at 6 MeV has been well resolved from the
$3^-$ state separated by 80 keV. The form factor for the $0^+$ is shown in
Figure 3 along with theoretical calculations [3]. The extended range
in momentum transfer of the new data provides a more stringent test of
the theory.

B.) $\underline{^{19}F}$: There is considerable experimental evidence that the nucleus $^{19}F$ is a strongly deformed system. In particular, the ground state in the Nilsson classification should have the configuration $1/2^+$ [220], the negative parity level at 0.110 MeV should have the configuration $1/2^-$ [101], and the positive parity level at 3.907 MeV should have the configuration $3/2^+$ [211]. Practically all the observed levels below 5 MeV excitation energy can be ascribed to rotational bands built on these levels.

The E2 transitions to the $3/2^+$ and $5/2^+$ members of the $1/2^+$ [220] band were studied in an earlier electron scattering experiment [4] where the ratios of the $3/2^+$ to $5/2^+$ form factors were found to be very nearly 2/3 as predicted by a pure rotational model. A subsequent experiment [5] observed a strong E4 transition to the $9/2^+$ (2.780 MeV) member of this rotational band, but did not observe the $7/2^+$ (4.378 MeV) level ascribed to this band. The pure rotational model predicts a ratio of 4/5 for the $7/2^+$ to $9/2^+$ form factors. This region of the excitation spectrum as seen by 243.4 MeV electrons scattered through 70° is shown in Figure 4. It is clear that the intensity of the $7/2^+$ (4.378 MeV) level is at least an order of magnitude smaller than theory.

It has been suggested that the 5.464 MeV level is in fact the $7/2^+$ member of the ground-state band. However, this level fits very well the energetics of a rotational band built on the $3/2^+$ (3.907 MeV) level. Even with the excellent resolution available, it is difficult to extract the form factors for this $7/2^+$ level because of the nearby $7/2^-$ (5.428 MeV) level. However, a preliminary analysis indicates that the ratio of the strengths of the $7/2^+$ (5.464 MeV) level to the $9/2^+$ (2.780 MeV) level is consistent with 4/5 (with rather large errors). Thus, it is likely that either the 5.464 MeV level actually belongs to the ground-state rotational band, as has been suggested previously, or there is strong Coriolis mixing between the $1/2^+$ [220] and $3/2$ [211] bands.

The $1/2^-$ (0.110 MeV) and $3/2^-$ (1.458 MeV) members of the $1/2^-$ [101] rotational band have never previously been resolved in electron scattering. As can be seen in Figure 5, the excellent resolution of the MIT-Bates Linac Facility allows a detailed study of the form factors of these levels. Comparison of the form factors of the $1/2^+$ [200] rotational band with the form factors for the $1/2^-$ and $3/2^-$ members of the $1/2^-$ [101]

band indicate that the latter require a much smaller transition charge radius. This is not completely unexpected since the Nilsson configurations for the 1/2$^-$ [101] band are 1p while those for the 1/2$^+$ [200] band are 2s and 1d: 1p particles tend to reside more in the interior of the nucleus.

Note in Figure 5 the clean separation of the one percent isotope $^{13}$C from $^{12}$C by the differing kinematic recoils. The background level in the region of $^{13}$C is essentially due to the radiative tail from the $^{19}$F scattering. One dividend of high resolution is very dramatically illustrated in this figure. If the resolution were to have been ~ 10$^{-3}$ instead of ~10$^{-4}$, the radiative background would have registered about ten times higher in intensity. By the same token the $^{13}$C line would have registered with a 10-fold decrease in intensity and would have effectively vanished. The effective signal to noise ratio is improved by a factor of ~ 100 by this 10-fold improvement in resolution.

C.) $^{27}$Al: In Figure 6 we show the spectrum of electrons scattered from $^{27}$Al at 70° with 248 MeV incident energy. An experiment performed at Tohoku [6] with a resolution of $\Delta p/p$ ~ 10$^{-3}$ is shown for comparison. The spectrum at MIT was taken in about one hour using one setting of the spectrometer. We have noticed in these results that the radii of the transition densities of different states are considerably different as with $^{19}$F above.

D.) Nuclear Deformations - I wish now to turn our attention to deformed even-even nuclei and our efforts to measure their deformations or shapes. These nuclei have zero spin and their ground states are spectroscopically spherical; that is, there can be no expectation value for any multipole other than the charge monopole. Nevertheless, because these nuclei have well developed rotation spectra, 0$^+$, 2$^+$, 4$^+$ ... with energies $E_I = E_o I(I + 1)$, it is assumed that they possess a well deformed and stable intrinsic state $\emptyset_{\underline{int}}$. It is also assumed that the various rotational states of these nuclei can be described by the factorized wave function

$$\Psi_I(\vec{r}_1 \ldots \vec{r}_A) = \emptyset_{\underline{int}}(r_1 \ldots r_A - 3) D_I(\Omega)$$

where the coordinates have been written to denote that three degrees of freedom are included in the rotational wave function $D_I(\Omega)$, describing the orientation of the body fixed axis with respect to the laboratory. Using this wave function we can write

$$\rho_{\underline{int}}\,(r,\theta,\phi) = |\emptyset_{\underline{int}}(\vec{r})|^2$$

where the coordinates $r,\theta,\phi$ are measured with respect to the body axes which are fixed to the intrinsic shape. Elastic scattering samples the spherical part of $\rho_{\underline{int}}\,(r,\theta,\phi)$ and the inelastic scattering samples the charge densities responsible for the higher multipoles according to the expressions,

$$0^+ \rightarrow 0^+: \quad \rho_0(r) = \int \rho_{\underline{int}}(r,\theta)\ Y_0^0(\Omega)\,d\Omega$$

$$0^+ \rightarrow 2^+: \quad \rho_2(r) = \int \rho_{\underline{int}}(r,\theta)\ Y_2^0(\Omega)\,d\Omega$$

$$0^+ \rightarrow I^+: \quad \rho_I(r) = \int \rho_{\underline{int}}(r,\theta)\ Y_I^0(\Omega)\,d\Omega$$

This notation assumes that the nuclei have axial symmetry and require only the $m = 0$ part of the Legendre expansion in multipoles. Clearly the intrinsic charge distribution is given by

$$\rho_{\underline{int}}(r,\theta) = \sum_{\substack{L \\ (\text{even})}} \rho_L(r)\ Y_L^0(\Omega).$$

Each of the $\rho_L(r)$ is derived from the form factors $F_L(q)$ for the various transitions by a distorted wave analysis.

In this case of spin-zero nuclei, nature has conveniently provided the equivalent of an aligned system to separate the various multipoles. If the same intrinsic state is common to each of the rotational states as assumed, then the rotational separations in energy are unimportant. They merely provide a way of separating the different multipole components of this intrinsic shape if one has a sufficiently high experimental resolution. The problem may be compared to studying the structure of a proton in a box. The energy levels of the box, if sufficiently small, do not affect the internal structure of the proton that is common to all of the levels.

There are many potential problems with this simplified approximation
of a factorizable nuclear wave function. One must keep in mind that
a microscopic theory to back-up this assumption does not exist. The
only justification is the empirical analogy with the predictions of
the quantum mechanics of a simple axially symmetic rotor - the $j(j + 1)$
spacing of the energy levels. One understands the high precision of this
model in molecular physics through its numerous applications but with
nuclei experience is limited. These reservations are amplified through
examples that illustrate possible ambiguities. In the s-d shell nuclei,
the shell model with SU3-symmetry itself leads to an energy level
spacing that is close to that of a rotor. On the other hand, in the
rare-earth and actinide region, the regularity and constancy of the
rotational feature and the deformations is too strong to be considered
ambiguous as with s-d shell nuclei. Thus, for these heavier nuclei, one
proceeds with this simple view with increased confidence. Nevertheless,
this discussion is continued to point out more problems.

The very first difficulty encountered is that the moments of
inertia of nuclei, even good rotators according to the energy spacings,
are always less than the expected rigid value, $\vartheta$ $<0.6\vartheta_R$. Clearly, the
semiclassical picture is not correct. It seems that some of the nucleus
does not participate in the rotation. This however, may not be an
important difficulty for our interpretations. In particular, assume
that deformed nuclei can be visualized as having two components, a
spherical superfluid part that does not participate in the rotation and
a deformed component that is responsible for the rotational spectrum.
In this case, the factorizable wave function is still possible for part
of the system, but a second, inert component is added. The spherical or
elastic scattering would have contributions from each part which combine
to give one $\rho_o(r)$ as before. Only the deformed part would contribute to
the inelastic scattering and one would work with the $\rho_L(r)(L>0)$ as before.

These factorized models are useful in providing simple ways of
treating the data and in allowing for idealized pictures of these nuclei
in the absence of a more fundamental theory. On the other hand, it is
not clear that it is correct to treat these excitations as resulting
from a static charge distribution that is the $L^{th}$ Legendre projection
of $\rho_{\underline{int}}$. It is not absolutely certain that the above prescription really

indicates what these nuclei look like. The idea of an intrinsic state with a stable shape may be too approximate a description. Perhaps the energetic situation does not have a very sharp minimum as a function of deformation and the nucleus is allowed to sample many different shapes. The approximation may have meaning for the E2 deformations which have large amounts of matter involved. On the other hand, the higher multipoles which involve smaller amounts of matter may not be attributable to permanent shape deformations in the intrinsic state. In this connection, it should be remembered that inelastic scattering measures a transition charge that is the overlap of the ground state wave function with the wave function of an excited state. Clearly, states of angular momentum should be projected from the intrinsic state and transition matrix elements calculated for comparison to experiment. These transition densities will be the same as the Legendre components of the intrinsic density only if the deformations are large and energetically dominate the nuclear Hamiltonian and if the wave function factorizes as discussed above.

Other problems we should discuss relate to the validity of the distorted wave Born approximation. We assume this approximation for inelastic scattering since the electromagnetic interaction is weak. We hope that second order processes such as $0^+ \to 2^+ \to 4^+$ are small enough to be neglected compared to the direct process $0^+ \to 4^+$. Ravenhall and others [7,8] have developed a coupled channel calculation involving the $0^+$, $2^+$ and $4^+$ states assuming a rotational nuclear wave function and they have shown that the $0^+$ and $2^+$ scattering has a contribution that depends on the relative size of $\rho_4(r)$ and amounts to ~ 5% in the cross section at $q \sim 1$ fm$^{-1}$ in the cases of $^{152}$Sm and $^{154}$Sm. This field needs considerably more work, in terms of dispersive effects and, in particular, with respect to the theory underlying our treatment of rotating systems.

1.) <u>Results on Nuclear Deformations at Low Momentum Transfer</u>
<u>from MIT-NBS</u> - Recently, electron scattering experiments have been performed by the MIT group in collaboration with the group at the National Bureau of Standards on the deformed nuclei $^{152}$Sm [9], $^{154}$Sm, $^{166}$Er, $^{176}$Yb, $^{232}$Th and $^{238}$U [10]. In line with the above discussion, these authors assumed that these nuclei are described by an axially

symmetric, deformed intrinsic state that contains all the degrees of
freedom of the nucleus except for the rotational degrees described in
the usual way by the $D_{MK}^{J}$ $(\alpha, \beta, \gamma)$ functions. The transition charge
densities are related to the reduced transition probabilities and intrinsic
multipole moments by the relations

$$[B(EL\uparrow)]^{1/2} = <f|EL|i> = \int_{0}^{\infty} \rho_{L}^{tr}(r) r^{L+2} \, dr.$$

Using the B(E2) values from Coulomb excitation and life time measurements,
and the rms radii from $\mu$-mesic x-ray data, these authors proceed to
determine the $\rho_{L}^{tr}(r)$ from the cross sections for excitation of the $0^{+}$,
$2^{+}$ and $4^{+}$ states. In the cases of $^{152}Sm$, $^{154}Sm$, $^{166}Er$ and $^{176}Yb$, the
three rotational states are resolved. In the case of $^{232}Th$ and $^{238}U$, the
$0^{+}$, $2^{+}$ are measured together and only the $4^{+}$ is resolved.

$^{152}Sm$ is well described by a deformed Fermi shape

$$\rho(r, \theta) = \bar{\rho} \, [1 + \exp \, (\frac{r-R(\theta)}{t})]^{-1}$$

$$R(\theta) = R_{o} \, [1 + \beta_{2}Y_{2}^{o} + \beta_{4}Y_{4}^{o} + \ldots]$$

The shape is characterized by a constant skin thickness that is not an
assumption and is required by the data. Allowing for a $\theta$ dependence in
skin thickness of the form $t \, (\theta) = t_{o}(1 + \gamma_{2}Y_{2}^{o})$ makes it possible to
determine $\gamma_{2} = 0.035 \pm 0.035$. This is consistent with a skin thickness
that varies by less than 5%. A similar result is obtained with $^{154}Sm$.
The final sets of parameters for the deformed Fermi shape where $\beta_{6}$ is
taken from $\alpha$-scattering results [11] tell us that $\beta_{4}$ is approximately
50% larger than that derived from $\alpha$-particle work [11] and is much
closer to the results derived from Coulomb scattering.

For $^{166}Er$ and $^{176}Yb$ the data cannot be fitted with the deformed
Fermi distribution used for other nuclei. To generalize the shapes
for these nuclei, each of the $\rho_{L}^{tr}(r)$ is parametrized separately by
the form

$$\rho_{L}^{tr}(r) = \int \bar{\rho} \, [1 + \exp \, (\frac{r-R_{L}(\theta)}{t_{L}})]^{-1} Y_{L}^{o}(\theta) \, d\Omega$$

with $R_{L}(\theta)$ defined as before. The resulting charge distributions of $^{166}Er$,

$^{176}$Yb and $^{154}$Sm are compared in Figure 7. The shapes are very different and reflect a decrease in the transition charge radius for the quadrupole distribution from $^{154}$Sm to $^{166}$Er and $^{176}$Yb.

These results were based on experiments at $q \lesssim 1.1$ fm$^{-1}$. The highest electron energy used was about 110 MeV. Recently, the $0^+$, $2^+$ and $4^+$ states of $^{152}$Sm have been reexamined at Saclay using 252 MeV electrons, and extending the range of momentum transfer to $> 2$ fm$^{-1}$. The cross sections are displayed in Figure 8 along with the predictions derived from the parameters described above [12]. The new data is in excellent agreement with these parameters in the case of the $0^+$ and $2^+$ scattering. The $4^+$ scattering seems to result in a substantially larger cross section than predicted by these parameters. The effect is not understood. It could be spurious but perhaps it is due to dispersive effects being more important at higher energies.

2.) Recent MIT Results on $^{150}$Nd, $^{156}$Gd, $^{166}$Er, $^{176}$Yb: In Figures 9 - 12 we show assorted spectra of electrons scattered from these rare earth nuclei. We have data ranging from $q = 0.5$ fm$^{-1}$ to $q = 2.25$ fm$^{-1}$. The rotational levels $0^+$, $2^+$, $4^+$ and $6^+$ are clearly resolved and observed in these nuclei. The $2^+$ γ-vibration state is also observed, often along with the $4^+$ rotational member of this vibration band. Many other states are also observed with spins ranging up to 5 (possibly 8) and many are observed for which we have no identification as yet from the standard tables. We verify in this data that the radii of the transition charges $\rho_2$ and $\rho_4$ are smaller in $^{176}$Yb and $^{166}$Er than in $^{150}$Nd and $^{156}$Gd (near $^{152}$Sm) and we notice that the $6^+$ state has characteristics that are different in each nucleus.

In Figures 13 - 16 note the experimentally determined form factors for the $0^+$, $2^+$, $4^+$ and $6^+$ states of these rotational nuclei. Also shown on these graphs are predictions based on a density dependent Hartree-Fock theory by Rinker and Negele [13]. These predictions assume that $\rho_L$ (r) is given by the $L^{th}$ term of the Legendre expansion of the intrinsic density. In the case of $^{176}$Yb the Hartree-Fock density is shown in the insert of Figure 16. The agreement for the $0^+$ and $2^+$ states is particularly significant, since the Legendre expansion should be reasonably accurate for these states and the theory is based on a realistic interaction the only adjustable parameters of which were previously determined from spherical nuclei. The discrepancies in $4^+$ and $6^+$ transitions can be attributed to the fact that the wave functions have not been properly projected and thus indicate the limitations of the description in terms of a single intrinsic state density distribution. Indeed, Negele and co-workers are just now calculating these effects and they have shown that for the $4^+$ and $6^+$ states, there are substantially corrections to the transition density derived by simply projecting the intrinsic density. On the other hand, it is interesting also to note that the worst disagreement for the $4^+$ excitation occurs for $^{150}$Nd. Since $^{148}$Nd is more a vibrator than a rotator, this disagreement may be related to a possible lack of rigidity of the higher order deformations in this transition region.

The experimental form factor for the $2^+$ γ-vibration in $^{166}$Er is shown in Figure 17. This is compared to the form factor derived from the intrinsic state density of the above density dependent Hartree-Fock theory by a vibrational model normalized to give the experimental BE2 shown. The disagreement may be a significant deficiency of the vibrational model as we apply it or perhaps it is related to problems related to the projection of wave functions discussed above.

It is clear that we are able to explore these deformed nuclei in great detail. Hopefully we will also come to a better understanding of the concept of an intrinsic state and its range of validity.

E.) $^{181}$TA: In Figure 18 we show the spectrum of electrons scattered from $^{181}$Ta at 70° using an incident energy of 260 MeV. The spectrum shows that the scattering from the ground state rotational band [$7/2^+$, $9/2^+$, $11/2^+$, $13/2^+$, $15/2^+$] is clearly resolved and that all members are observed. The data span a range of momentum transfer from 0.5 fm$^{-1}$ to 1.8 fm$^{-1}$. The $9/2^+$ and $11/2^+$ states are predominantly excited by E2 transitions. However, in the diffraction minima we observe that the ratio of the $9/2^+$ and $11/2^+$ form factors deviate from that expected for a rotator and the contaminant appears to be E4. The $13/2^+$ and $15/2^+$ states appear to be predominantly E4 at the lower momentum transfers consistent with the admixture required in the $9/2^+$ and $11/2^+$ states. At the higher momentum transfer, the $13/2^+$ and $15/2^+$ states appear to require an additional component that could be either E6 or a high magnetic multipole.

It is interesting to note that a measurement of the $\rho_L$ (r) along with the magnetization density of these odd-even deformed nuclei might provide us with insight into the velocity profile of the deformed core and the development of the moment of inertia. Towards this end we have examined the scattering from $^{181}$Ta at angles of 160° and 120° to better understand the nature of the magnetism of this nucleus. Figure 19 shows the Coulomb cross section curve for CØ + C2 elastic scattering from $^{181}$Ta computed using a deformed Fermi distribution which was fit to the $^{181}$Ta data taken at 70°. Also shown are the cross sections for M5 and M7 elastic scattering assuming a single particle transition using spherical SHO wave functions for the $g_{7/2}$ odd proton. The oscillator parameter is b = 2.1 fm.

The M5 and M7 cross sections are normalized such that their sum plus the Coulomb scattering correctly predicts the 150 and 185 MeV 160° data points. This required approximately a factor of two reduction from the single particle value. The Coulomb CØ + C2 curve was later checked by going to 120° where the transverse scattering is down by a factor of ~ 10 from the 160° data. Points were taken with the same $q_{eff}$ as the 150 MeV and the 215 MeV at 160°. This data verified the validity of the Coulomb curve.

The sum of these normalized M5 + M7 cross sections plus the cross section for CØ + C2 was used to predict points at 210 and 215 MeV. As can be seen, the data shows that the transverse transitions must have a different q dependence than that predicted by this simple model. At the very least one is able to conclude that the M7 form factor may be significantly suppressed (x 1/3) over the single particle model in contrast to recent results on $^{93}$Nb [14], $^{51}$V [15], and $^{209}$Bi [16] where a SHO prediction fits the data reasonably well for the highest magnetic multipole. Perhaps one shouldn't really expect the spherical H. O. to give good results because $^{181}$Ta is permanently deformed, thus radically changing the form of the transition operators.

F.) $\underline{^{25}Mg}$: We have also begun to examine the magnetic scattering from $^{25}$Mg in collaboration with the group from the University of Massachu-

setts and the group from the University of Sao Paolo, Brazil. Some
very recent results for the magnetic form factor are shown in Figure 20.
Although the statistical errors on the points are fairly small, the
large error bars come from uncertainties in subtraction of Coulomb
multipole $C\emptyset$, $C2$ and $C4$. The curve is a single particle $(d_{5/2})$ prediction for the $M5$ moment corrected for the nucleon form factor and
assuming the SHO parameter is $b = 1.35$ fm. It appears that the simple
model has an incorrect shape and it is possible to interpret the data
as indicating a reduction in the peak of the form factor by a factor
of about 2 - 3 over the SHO predictions. Perhaps this supression of
the highest moment in $^{25}$Mg is caused by the strong deformations of
the even-even core as in the case of $^{181}$Ta.

G.) $^{24}$Mg: I report here the results of some experiments at 160°
and 120° on the inelastic scattering from $^{24}$Mg in collaboration with
the group from the University of Toronto and the group from the Naval
Research Laboratory in Washington, D. C.. We have seen many states, but
I mention explicitly the discovery of a very strong $6^-$ state at 15.1 MeV
of excitation. From the q - dependence we know that L = 6 and from the
angular dependence we know it to be transverse, hence, the assignment
$6^-$. The state appears to carry much of the $(d^{-1}_{5/2} f^1_{7/2})$ particle-hole
strength with a form factor that is very reasonable and in agreement with
the predictions of this simple configuration.

H.) $^{207}$Pb: In collaboration with the group from the University of
Virginia and the group from Chalk River, we have begun a series of experiments on the isotopes of Lead. As a first result we have examined
the excitation of the two states in $^{207}$Pb that are formed by coupling a
$P_{1/2}$ neutron hole to the $3^-$ state in $^{208}$Pb. The data presently extend
up to q ~ 1.8 fm$^{-1}$. The form factors sum up to the value of the $3^-$ form
factor in $^{208}$Pb and the branching ratio seems to follow the $2j + 1$
intensity rule expected for a weak coupling model.

III. Photo-Pion Studies at MIT-Bates Linac

The first experiments at the new accelerator involved photo-meson
production near the meson threshold from nuclei giving radioactive

daughters which could be examined after the beam burst. These beta active nuclei have lifetimes from milliseconds; $^{12}C$ $(\gamma, \pi^-)^{12}N$, to days; $^7Li$ $(\gamma, \pi^-)$ $^7Be$. Although extensive measurements of this type had been carried out before, particularly at Lund and at Urbana, they were done at energies well above threshold, generally with interest center-ing on the T = 3/2, J = 3/2 resonance. The aim was to obtain data at such low energies that the contribution to the total cross-section from individual levels in the daughter nucleus could be identified, whereas previous work had involved only a sum over all bound states of the daughter nucleus. The theoretical interpretation also is cleaner at very low pion energies because for the transitions corres-ponding to pure Gamow-Teller cases in beta decay, such as $^{12}N \rightarrow ^{12}C$, only the $\vec{\sigma} \cdot \vec{\epsilon}$ term remains in the interaction. Moreover, the effect of final state interactions is small at low outgoing pion energies where the pion-nucleus interaction is known to be small. The hope was to pick cases which were known from beta decay to have large $<f|\vec{\sigma} \cdot \vec{\epsilon}|i>$ and which were already well studied by electron scattering and by $\pi^-$ and $\mu^-$ capture so as to remove the nuclear structure uncertainties and allow the examina-tion of the pion production mechanism. These considerations restrict one to the lightest nuclei.

A.) $\underline{^{12}C~(\gamma, \pi^-)^{12}N}$: The yield and cross section for the $^{12}C$ $(\gamma, \pi^-)^{12}N$ reaction is shown in Figure 21a, along with the theoretical predictions. The calculation of Koch [17,18] uses the $\vec{\sigma} \cdot \vec{\epsilon}$ interaction with a coupling strength adjusted to fit the $(\gamma, \pi^-)$ production on the neutron. Only s-wave pions, distorted in a Kisslinger type optical potential [19] are considered. Because of these approximations, this calculation should be accurate only near the threshold region. The nuclear transition density [20] is obtained from the electron scattering data to the 15.1 MeV, $1^+$ state of $^{12}C$ which is the isospin analogue of the ground state of $^{12}N$; this has been parameterized with shell model wave functions. The calcula-tion of Koch [17,18] gives results which are in agreement with experiment up to 4 MeV above threshold, but are too small above that energy.

The calculation of Nagl and Uberall [21] uses the full interaction Hamiltonian [22] and includes s, p and d pion waves distorted in a Kisslinger type optical potential. The transition density is obtained

from the electron scattering data to the $1^+$ state of $^{12}$C and are parameterized by the Helm model [21]. The differences between the curves in
Figure 21 are due to different choices for the pion momentum in the
coefficients of the interaction Hamiltonian. In curve 1, the asymptotic
pion momentum is used. In curve 2, the local pion momentum is used. In
the curves shown, a correction for the Lorentz–Lorentz effect has been
made. If this correction is omitted the results lie between curves 1
and 2.

A two parameter "best fit" to the data was made assuming a step
at the threshold, which is the effect of the final state Coulomb
interaction in the $(\gamma, \pi^-)$ reaction [23], and a linear rise above
threshold. This curve is shown in Figure 21. This does not imply that
the data excludes curvature in the cross section, but because of the
limited number of data points, a fit with more unknown parameters is
not justified.

The cross sections versus energy are shown in Figure 21b. The best
fit extracted for the cross section at threshold is 2.9 ±1.1 μb, and is
consistent with all of the theoretical calculations. In the radiative
pion capture by $^{12}$C the theoretical predictions [24] (with errors of
approximately 30%) are in agreement with experiment [25]. The $(\gamma, \pi)$
calculations of Nagl and Uberall [21] have a larger slope than that of
Koch [18] because p wave pions and momentum dependent terms in the
Hamiltonian were included. However, it can be seen that the slope of
the best fit cross section is somewhat greater than any of the theoretical
curves. Further experimental and theoretical work on this reaction is
needed.

B.) <u>Threshold Photoproduction of $\pi^-$ Mesons in Deuterium</u>: The theory
of pion photoproduction in complex nuclei is based on the physical picture
of a photon interacting with an individual nucleon, the interaction am-
litude being identical with the free nucleon case. This is the essential
assumption of the impulse approximation which is the basis for most
calculations reported thus far [26].

The validity of the impulse approximation is difficult to establish
in complex nuclei because calculations involve assumptions about nuclear
wave functions and final state interactions which inevitably introduce

theoretical uncertainties. These are minimal for pion production from the deuteron particularly near threshold, where the dominant interaction amplitude is well known $(\vec{\sigma} \cdot \vec{\epsilon})$ [22].

The excitation function for $\pi^+$ production from the deuteron was obtained by detecting positrons emitted in the $\pi\mu e$ decay. The 1.6 μsec half life of the muon permitted detection of the positrons after the short (2 μsec) beam pulses.

In this experiment the measured quantity is

$$Y(E_o) = \int_{E_T}^{E_o} N(E, E_o) \sigma_d(E) dE,$$

where $N(E, E_o)$ is the bremsstrahlung shape function, $E_T$ is the threshold energy and $\sigma_d(E)$ is the total deuteron photoproduction cross section. In order to normalize the measurements, comparison is made to measurements of $\gamma(p,n)\pi^+$. The fit to the experimental points for $\gamma(p,n)\pi^+$ involved a scaling of the "theoretical" excitation function, and shifting the energy scale. Thus the absolute energy scale was ultimately determined by fitting the proton excitation function.

Figure 22 shows the experimental excitation function for the deuteron. The solid curves were obtained from calculations of $\sigma_d(E)$ carried out by Tzara [27] representing three values of the neutron-neutron scattering length, $A_n$. They have been scaled by a factor determined by the proton fit, and are corrected for the difference in H and D target thickness. It can be seen that there is good agreement with the data for $A_n > 16$ fm; $A_n \stackrel{\sim}{\sim} 18$ fm gives perhaps the best fit. Since the currently accepted value of $A_n = 17 \pm 1.5$ fm, this experiment qualitatively supports Tzara's calculation. Small target thickness corrections are still to be made.

C.) <u>Other Photo-Pion Work</u>: In order to examine the differential cross section for $(\gamma, \pi^+)$ in the range of $E_\pi \sim 5 - 50$ MeV, a low energy spectrometer is under construction which will be ready in the Fall of 1976. It will yield data at $\theta = 90°$ with resolution $\Delta E/E \sim 1\%$. Other specifications are: Solid angle = 15 msr; Momentum range = 20%. The program will include an examination of $\pi^-/\pi^+$ ratios, isoboric analog states and spin-isospin collective resonances.

# IV. Photoproton Studies: $^{16}O(\gamma,p)^{15}N$

The tantalizing hope offered by these reactions is best understood in terms of the somewhat oversimplified description provided by the impulse approximation. In this picture, a photon of energy $E_o$ interacts with a proton of momentum $\vec{p}$ and energy $-E$ leading to a final detected proton of momentum $\vec{p}_f$. From a measurement of $E_o$ and $p_f$, $p$ can be determined as well as $E$. Hence, in impulse approximation the cross section specifies the spectral function

$$S(p,E) = \Sigma |<A-1^*|\phi(p)|A>|^2$$

which is the probability of finding a proton in a nucleus with momentum $p$ and separation energy $E$. Compared to $(e,e'p)$ studies $(\gamma,p)$ offers the kinematic advantage that even a modest energy of 100 MeV probes $p$ on the order of 300 MeV/c. Since $(\gamma,p)$ does not require a coincidence measurement, it is immune to accidental counting problems and can probe the low amplitudes at high momenta using the limited duty ratio of present facilities.

The first experiment at MIT has examined the $^{16}O\ (\gamma,p)^{15}N$ reaction leading to the $^{15}N$ ground state and the $3/2^-$ state of $^{15}N$ at about 6 MeV excitation. The experiments have been performed using photons as high as 280 MeV and for protons emitted at 43° and 90° relative to the photon beam. The results, which are still preliminary, are shown in Figures 23 and 24. The data join on the earlier results at Glasgow that stop at about $E_\gamma = 100$ MeV. The interesting feature is the change in character that occurs at about 180 MeV at 43° and 150 MeV at 90°. The explanation of this feature is not clear. Perhaps it is a reflection of the actual single particle momentum distribution of the p-shell. On the other hand, a Woods-Saxon type potential seems to be deficient by about one order of magnitude in these high momentum regions [29]. Recently, Walker, Londergan and Nixon have derived some preliminary results including the two step process whereby the excitation of the $\Delta$ is taken into account [28]. In the region $E_\gamma = 100 - 280$ MeV, this idea seems promising. If the $\Delta$ plays an important role, the $(\gamma,p)$ process changes its character from the simple one described above to a more complicated one whereby the mesonic degrees of freedom begin to be important even at rather low photon energies. The results from Mainz on the $(\gamma,n)$ process at $E_\gamma \sim 150$ MeV also support this view [30]. The $(\gamma,n)$ cross section is comparable to the $(\gamma,p)$ requiring in the dipole approximation an effective charge for the neutron comparable to that of the proton. Clearly, this implies that the effective nuclear sub-unit in the $(\gamma,n)$ process is a shorter range system with higher momentum states than that provided by the typical average shell model potential.

Another unexplained feature of the data is the ratio of the cross section for the excited state $(3/2^-)$ of $^{15}N$ to the ground state $(1/2^-)$ of $^{15}N$. The experimental ratio is about six. The simple impulse approximation would predict a ratio of two based on the $2j + 1$ populations of the p shell components.

These $(\gamma,p)$ results point to the need for some rather detailed and new theoretical work.

## ACKNOWLEDGEMENTS

I am grateful to C. Williamson for providing the material on $^{19}F$. I also thank my colleagues at MIT for many useful discussions and, in particular, J. W. Negele for discussions and the material on the few-body problems. My experimental colleagues, who have contributed to the energy-loss spectrometer and the new MIT data include: C. Creswell, J. Heisenberg, A. Hirsch, S. Kowalski, M. V. Hynes, C. Rad, C. P. Sargent and W. Turchinetz. I am grateful to E. Booth, B. Chasan and A. Bernstein for the material on photopion production and I am also grateful to J. Matthews, R. Owens and D. Findlay for material and discussions about the $(\gamma,p)$ studies. I must also thank Mr. M. V. Hynes, whose editorial comments and assistance in preparing the many figures was invaluable. I am indebted to Mrs. Natalie Alger for her patient and rapid typing of the manuscript.

## REFERENCES

1.  "Medium Energy Nuclear Physics with Electron Accelerators", USAEC, TID Report #24667, Edited by W. Bertozzi and S. Kowalski.

2.  W. Bertozzi, Lectures at the International Summer School on Electron Scattering, Sardinia, Italy 1970, Edited by B. Bosco. (To be published.)

3.  T. Erikson, Nucl. Phys. $\underline{A170}$, 513 (1971)

4.  P. L. Hallowell et al., Phys. Rev. $\underline{C7}$, 1396 (1973).

5.  M. Oymada et al., Phys. Rev. $\underline{C11}$, 1578 (1975).

6.  Research Report of the Laboratory of Nuclear Science, Tohoku University Vol. 4, No. 2, p. 28, Feb. 1972.

7.  R. L. Mercer and D. G. Ravenhall, Phys. Rev. $\underline{C10}$, 2002 (1974).

8.  F. Hackenberg and R. Rosenfelder, Preprint "Eikonal Expansion in Electron Scattering, II. Inelastic Scattering", Institute for Theoretical Physics, Heidelberg Report 26/75 .

9.  W. Bertozzi et al., Phys. Rev. Lett. $\underline{28}$, 1711 (1972).

10. T. Cooper et al , Phys. Rev. $\underline{C13}$, 1083 (1976).

11. D. L. Hendrie Phys. Rev. Lett. $\underline{36}$, 571 (1973) and D. L. Hendrie et al., Phys. Lett. $\underline{26B}$, 127 (1968).

12. J. Alster et al., private communication .

13. G. Rinker and J .Negele. (To be published.)

14. P. K. A. DeWitt Huberts et al., Phys. Lett. $\underline{60B}$, 157 (1976).

15. J. C. Nascimento et al., Phys. Lett. $\underline{53B}$, 168 (1974).

16. J. R. Moreira et al., Phys. Rev. Lett. $\underline{36}$, 566 (1976).

17. J. H. Koch and T. W. Donnelly, Nucl. Phys. $\underline{B64}$, 478 (1973); Phys. Rev. $\underline{10C}$, 2618 (1974).

18. J. H. Koch, private communication.

19. L. Tauscher and W. Schneider, Z. Physik, 271, 409 (1974).

20. J. S. O'Connell, T. W. Donnelly and J. D. Walecka, Phys. Rev. $\underline{C6}$, 719 (1972).

21. A. Nagl and H. Uberall, Private Communication and to be published, A. Nagl, F. Cannata and H. Uberall, Phys. Rev. $\underline{C12}$, 1586 (1975).

22. G. F. Chew, M. L. Goldberger, F. E. Low and Y Nambu, Phys. Rev. $\underline{106}$ 1345 (1957), F. A. Berends, A. Donnachie and D. L. Weaver, Nucl. Phys. $\underline{B4}$, 1 (1967).

23.  C. Tzara, Nucl. Phys. B18, 246 (1970)

24.  W. Maguire and C. Wentz, Nucl. Phys. A205, 211 (1973).

25.  H. W. Baer et al, Phys. Rev. C12, 921 (1975) H. W. Baer and
     K. M. Crowe, Conf. on Photonuclear Reactions and Applications,
     Asilomar, Calif., March 1973, B. L. Burman, editor, and
     references quoted there.

26.  F. J. Kelly, L. S. McDonald and H. Uberall, Nucl. Phys. A139,
     329 (1969).

     J. H. Koch and T. W. Donnelly, Nucl. Phys. B64, 478 (1973).

27.  C. Tzara, Nucl. Phys. A256 381 (1976).

28.  R. Owens, private communication.

29.  Walker, Londergan and Nixon, private communication.

30.  B. Schoch, private communication

FIGURE CAPTIONS

Figure 1.  New Experimental Area

Figure 2.  Spectrum of 199 MeV electrons scattered from $^{19}$F (teflon) at 70°.

Figure 3.  Form factor for the $0^+$ state at 6 MeV in $^{16}$O.

Figure 4.  A portion of the spectrum of 243.4 MeV electrons scattered from $^{19}$F (teflon) at 70°.

Figure 5.  Spectrum of 203 MeV electrons scattered at 70° from $^{19}$F, $^{13}$C and $^{12}$C in a teflon target.

Figure 6.  Spectrum of 248 MeV electrons scattered from $^{27}$Al at 70°. The dashed curve shows a sketch of the data from Tohoku [5] for 250 MeV electrons scattered from $^{27}$Al at 60°.

Figure 7.  Constant charge density contours for $^{154}$Sm, $^{166}$Er, and $^{176}$Yb based on best fit Fermi charge distributions.

Figure 8.  Differential cross sections (mb/sr) for the $0^+$, $2^+$ and $4^+$ states of $^{152}$Sm from Saclay []2]. The curves assume a rotational model and the parameters of [9].

Figure 9.  Spectrum of 256 MeV electrons scattered from $^{166}$Er at 90°.

Figure 10.  Spectrum of 319.4 MeV electrons scattered from $^{150}$Nd at 90°.

Figure 11.  Spectra of 256 MeV electrons scattered from $^{150}$Nd, $^{166}$Er and $^{176}$Yb at 90°.

Figure 12.  Spectra of electrons scattered from $^{166}$Er at 90° for three different incident electron energies, 202 MeV, 160 MeV and 123 MeV.

Figure 13.  Form factors for rotational states of $^{150}$Nd.

Figure 14.  Form factors for the rotational states of $^{156}$Gd.

Figure 15.  Form factors for the rotational states of $^{166}$Er.

Figure 16.  Form factors for the rotational states of $^{176}$Yb. The insert shows the constant density contours of the intrinsic density from the Hartree-Fock theory of [38].

Figure 17.  Form factors for the $2^+$ γ-vibration in $^{166}$Er.

Figure 18.  Spectrum of 260 MeV electrons scattered from $^{181}$Ta at 70°.

Figure 19.  Coulomb cross section (CØ + C2) for $^{181}$Ta along with magnetic scattering predictions and experimental cross sections.

Figure 20. Magnetic scattering from $^{25}$Mg.

Figure 21. a)  Yield and b) Cross section versus energy above threshold
for $^{12}$C $(\gamma,\pi^-)$ $^{12}$N.  The various curves are:
K:  Koch [18]
1 and 2:  Nagl and Uberall [20]
The "Best-Fit" is a two parameter fit described in the text.

Figure 22. The yield $(\gamma,\pi^+)$ from the proton and from the deuteron as
a function of the difference between the bremstrahlung end-
point energy E and threshold energy $E_+$.

Figure 23. The $^{16}$O$(\gamma,p)^{15}$N cross section at $\theta_p = 43°$.

Figure 24. The $^{16}$O$(\gamma,p)^{15}$N cross section at $\theta_p = 90°$.

23

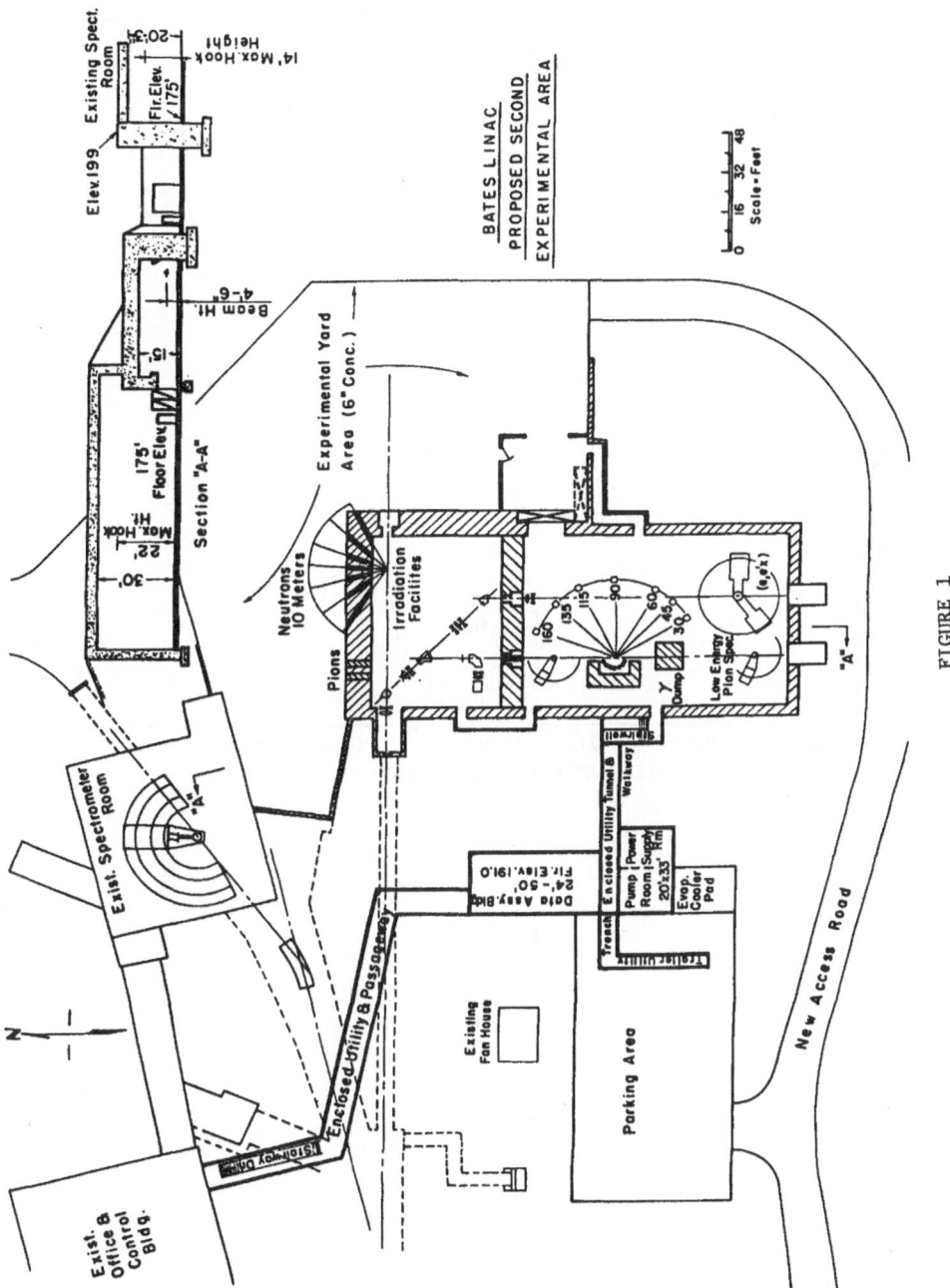

FIGURE 1

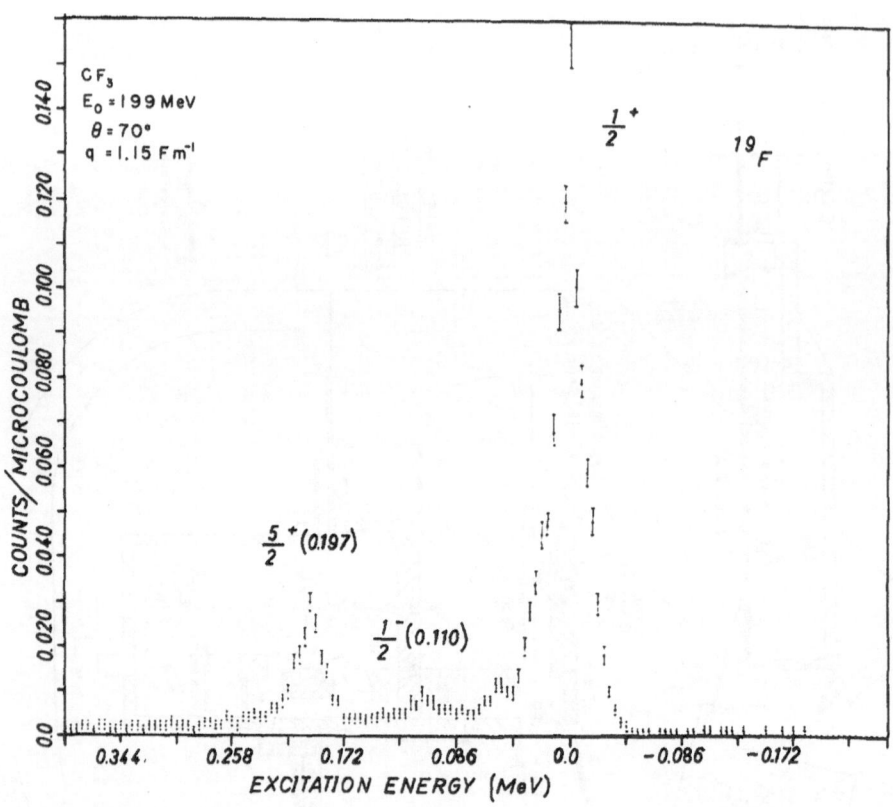

FIGURE 2

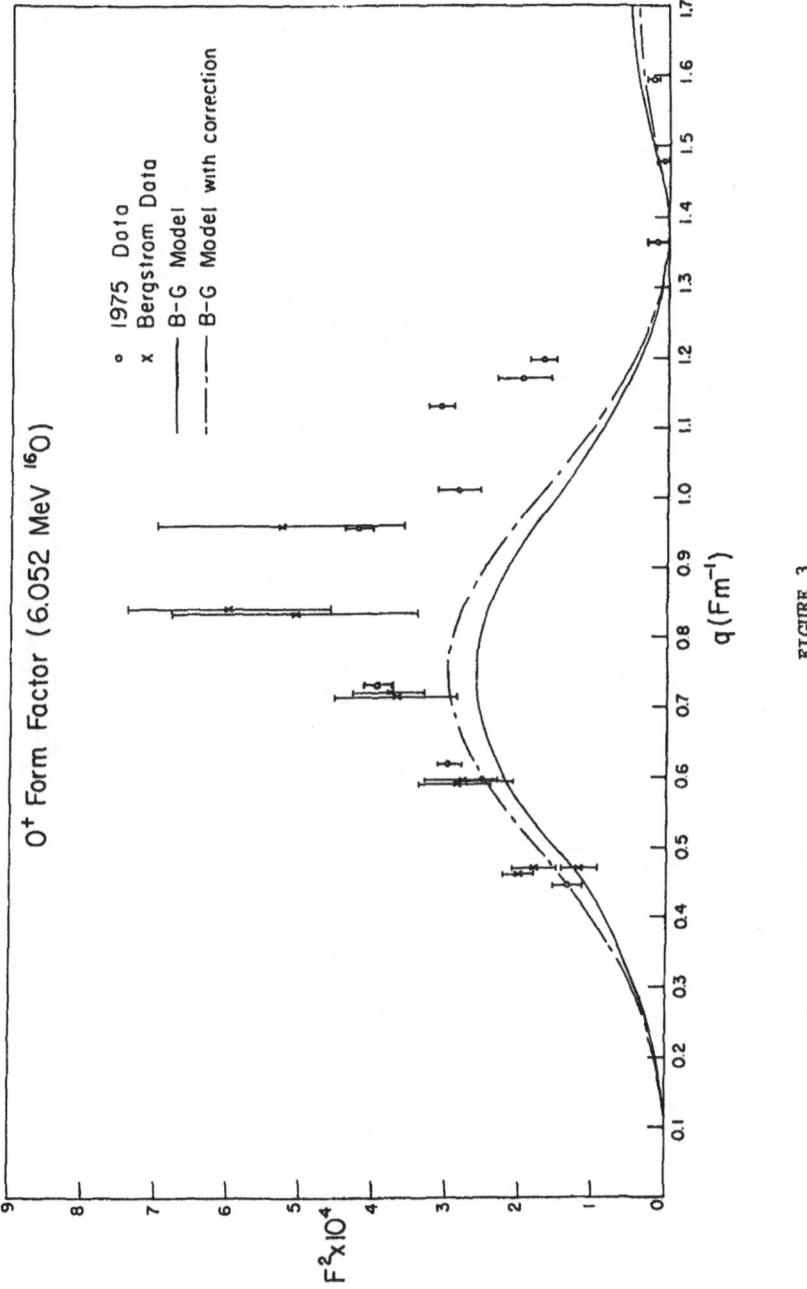

O⁺ Form Factor (6.052 MeV $^{16}$O)

○  1975 Data
×  Bergstrom Data
—— B-G Model
—·— B-G Model with correction

FIGURE 3

26

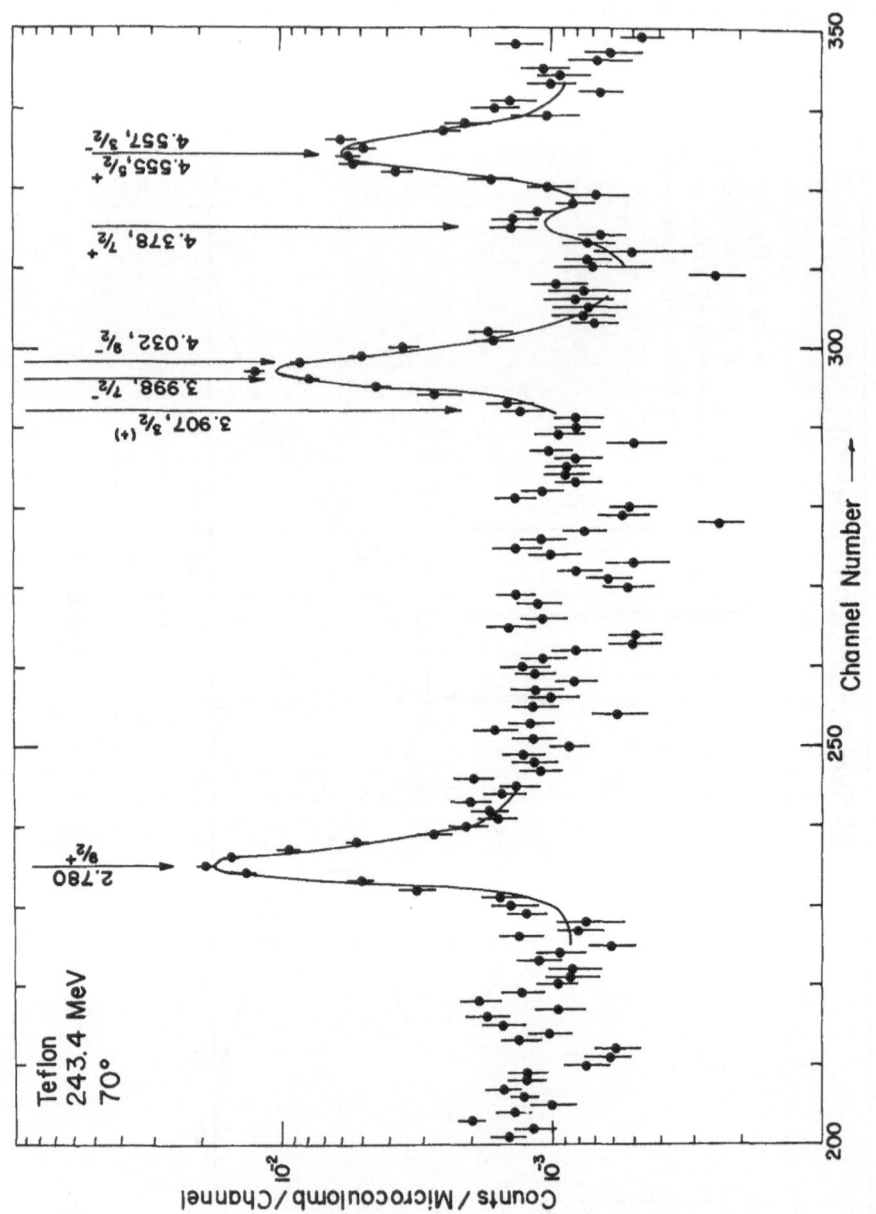

FIGURE 4

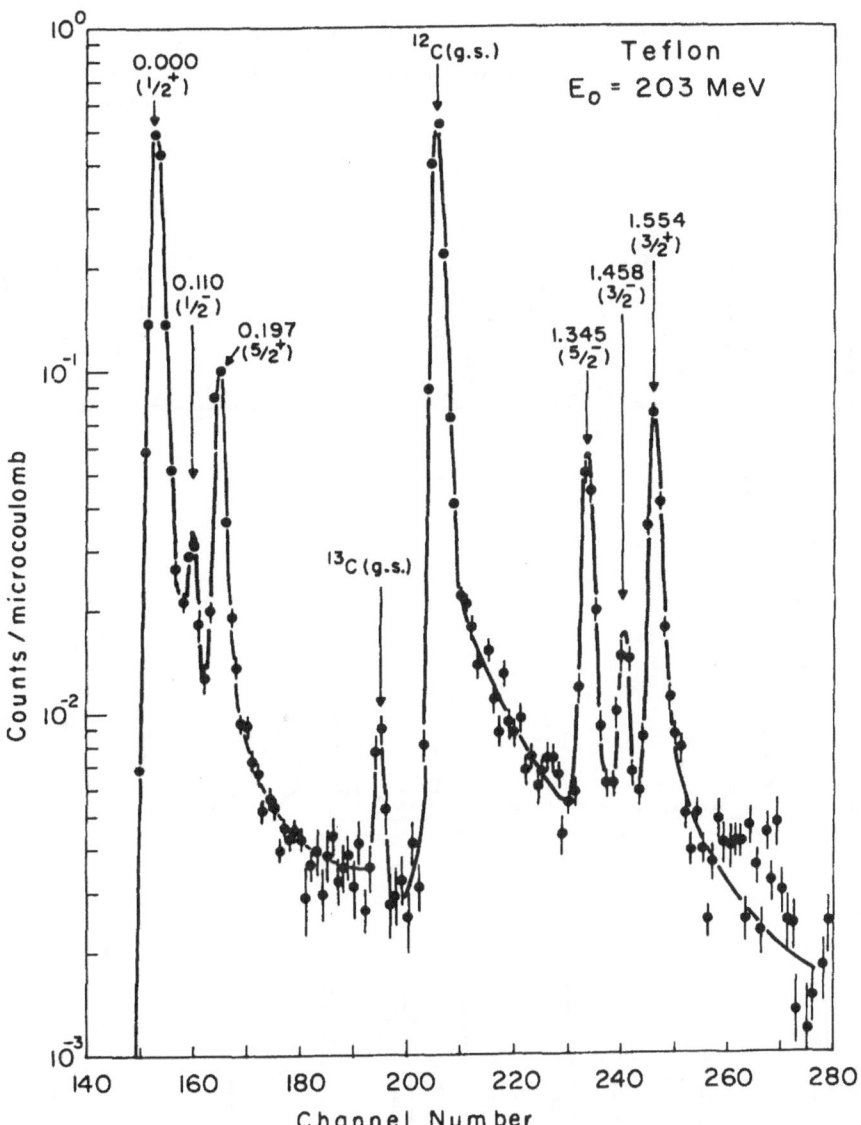

FIGURE 5

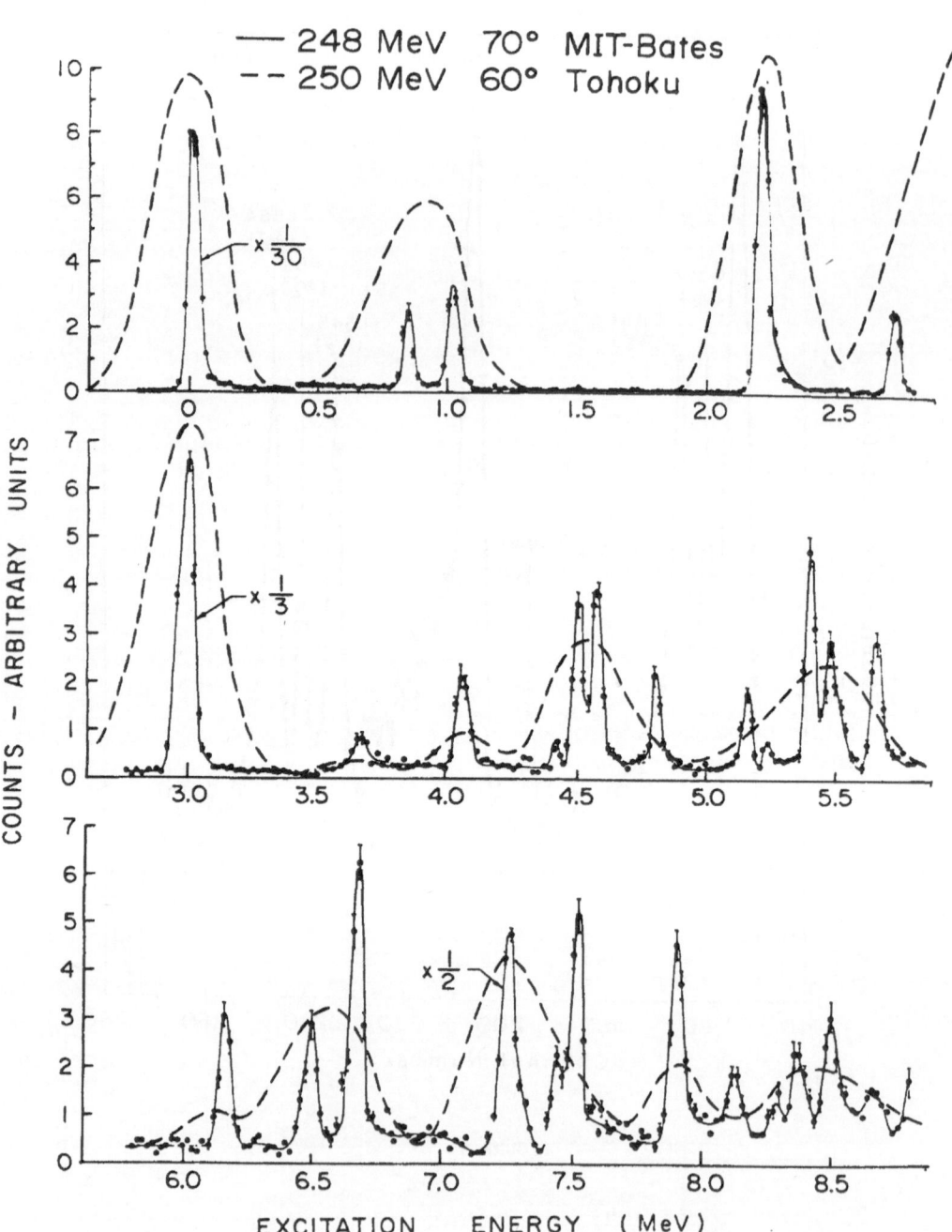

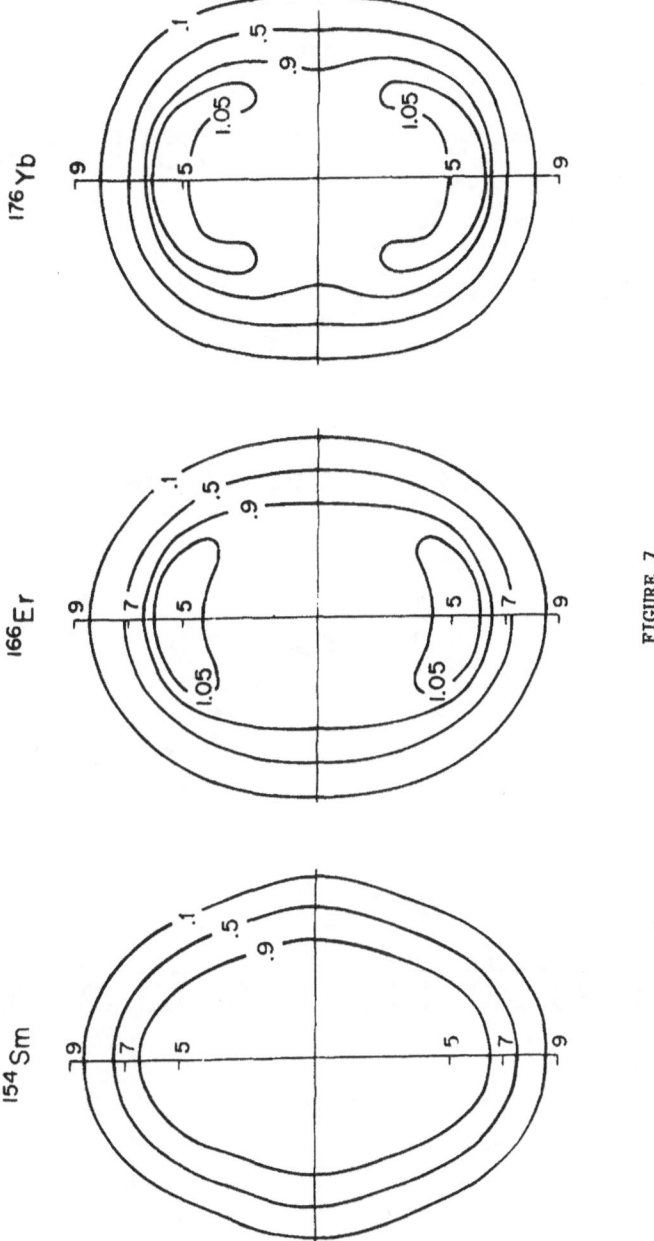

FIGURE 7

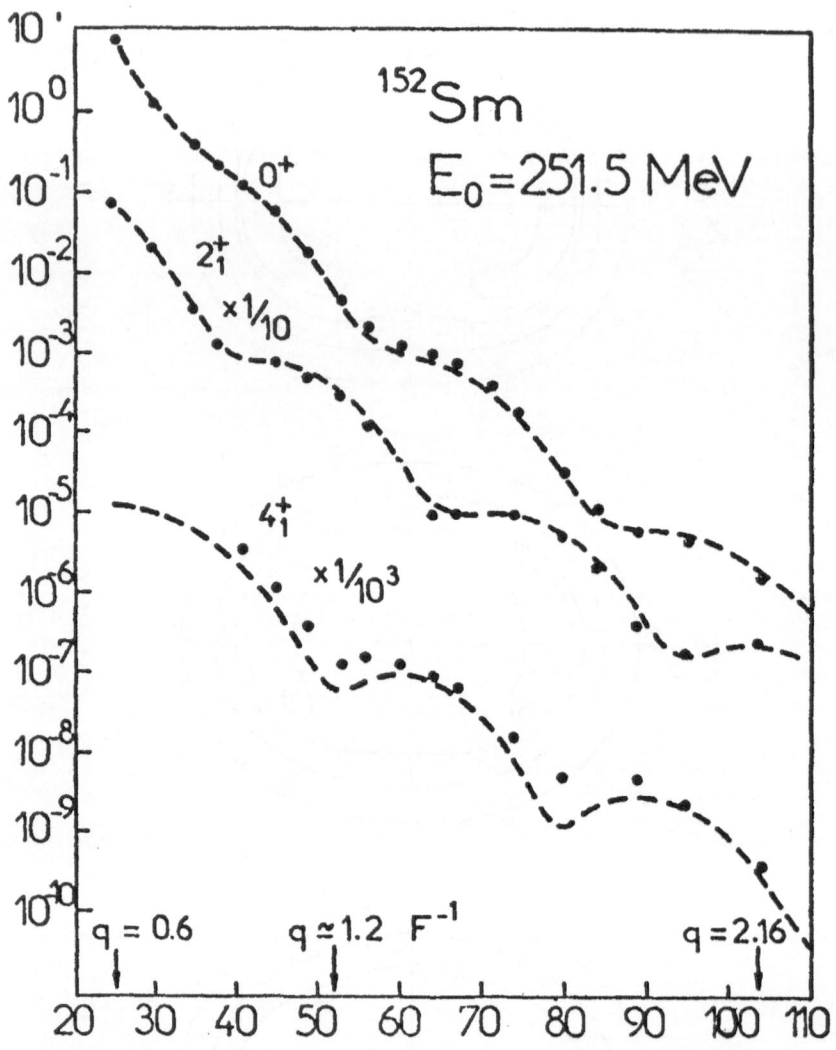

FIGURE 8

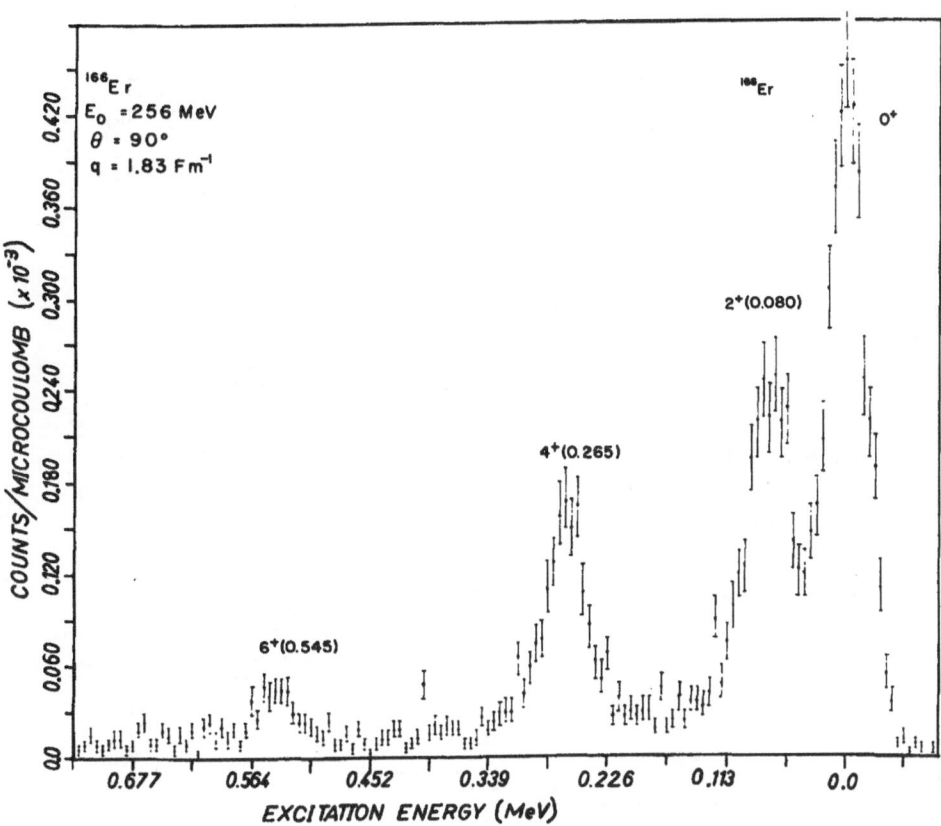

FIGURE 9

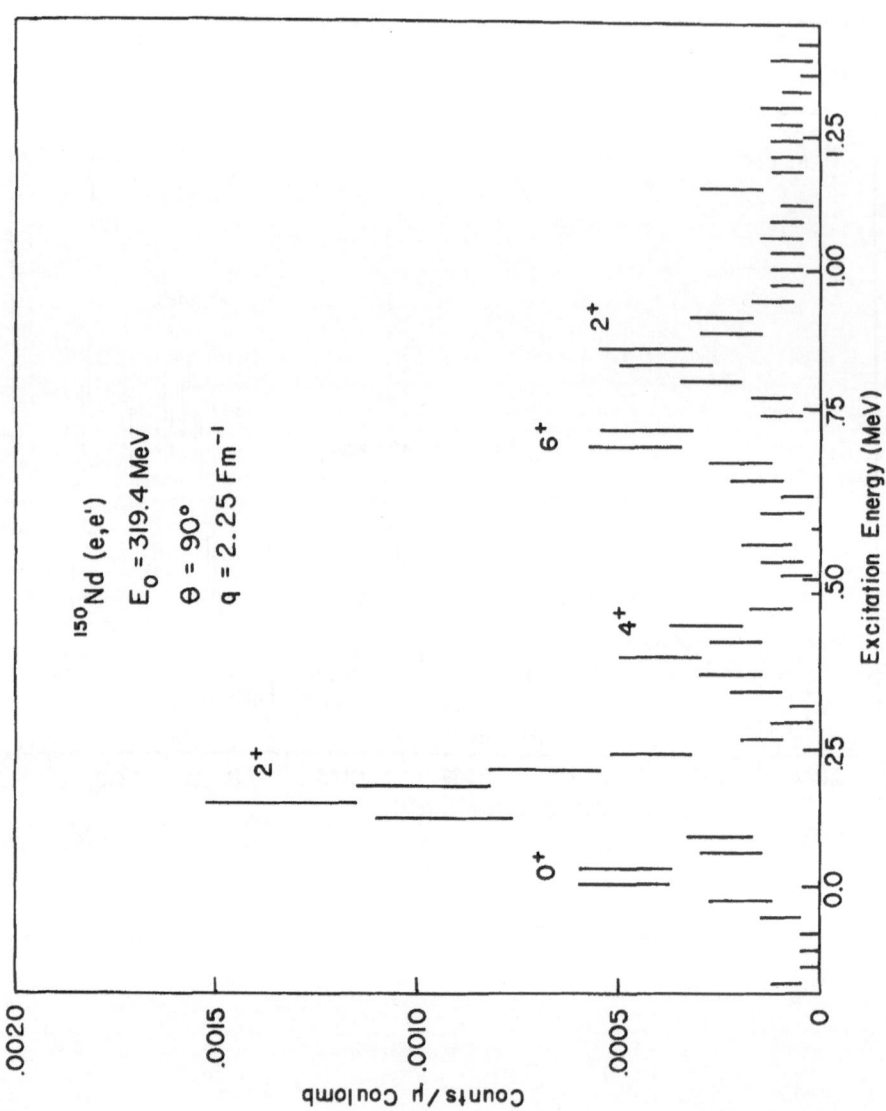

FIGURE 10

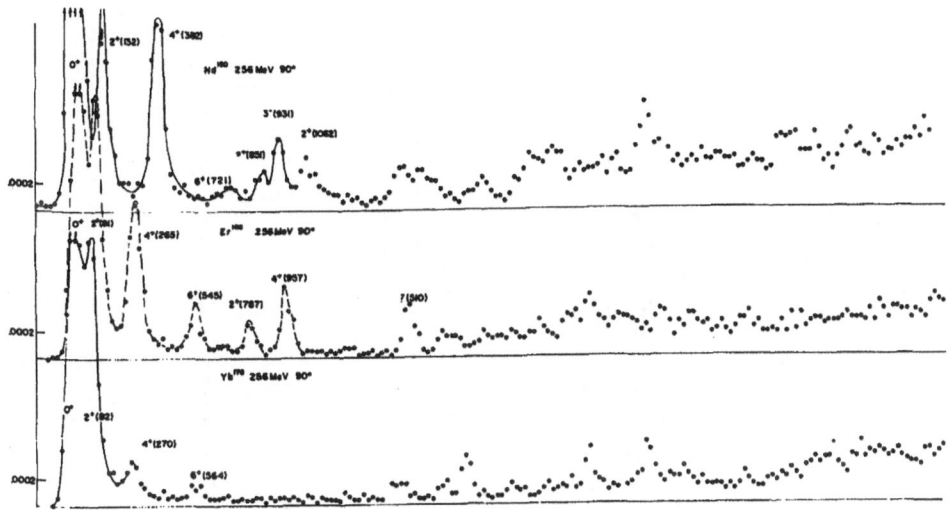

FIGURE 11

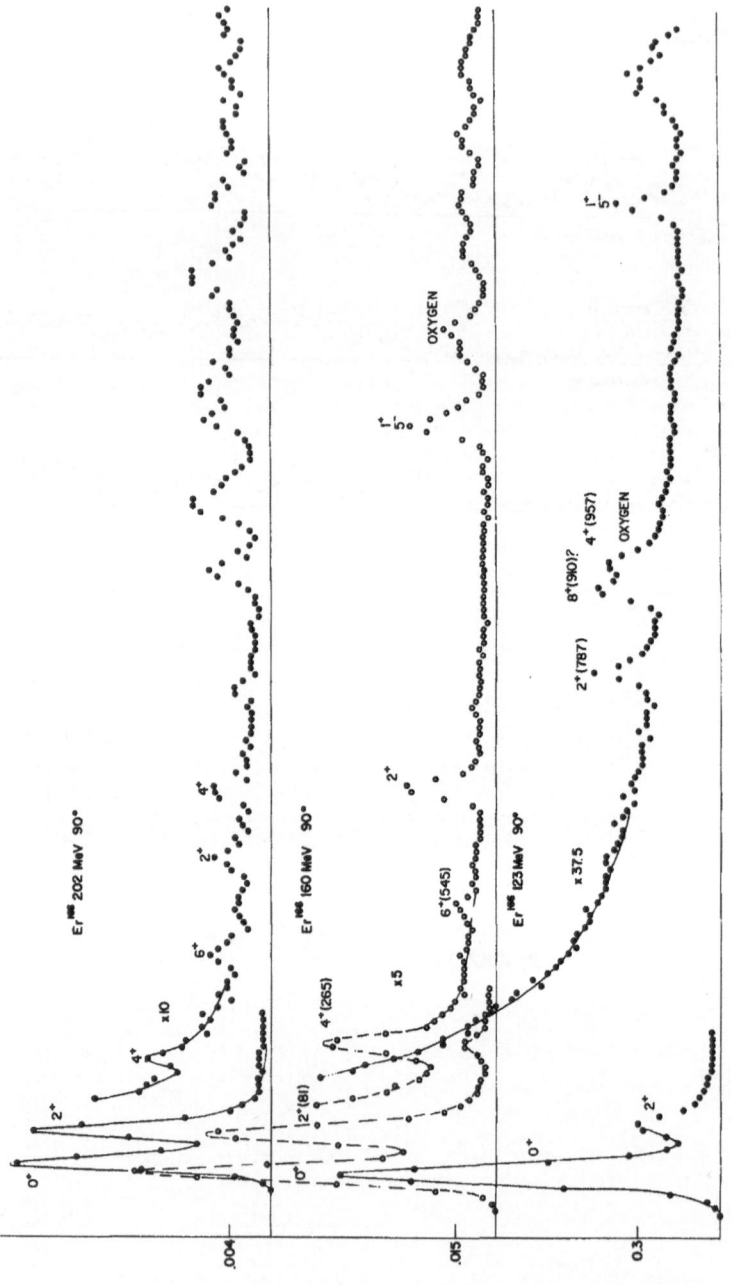

FIGURE 12

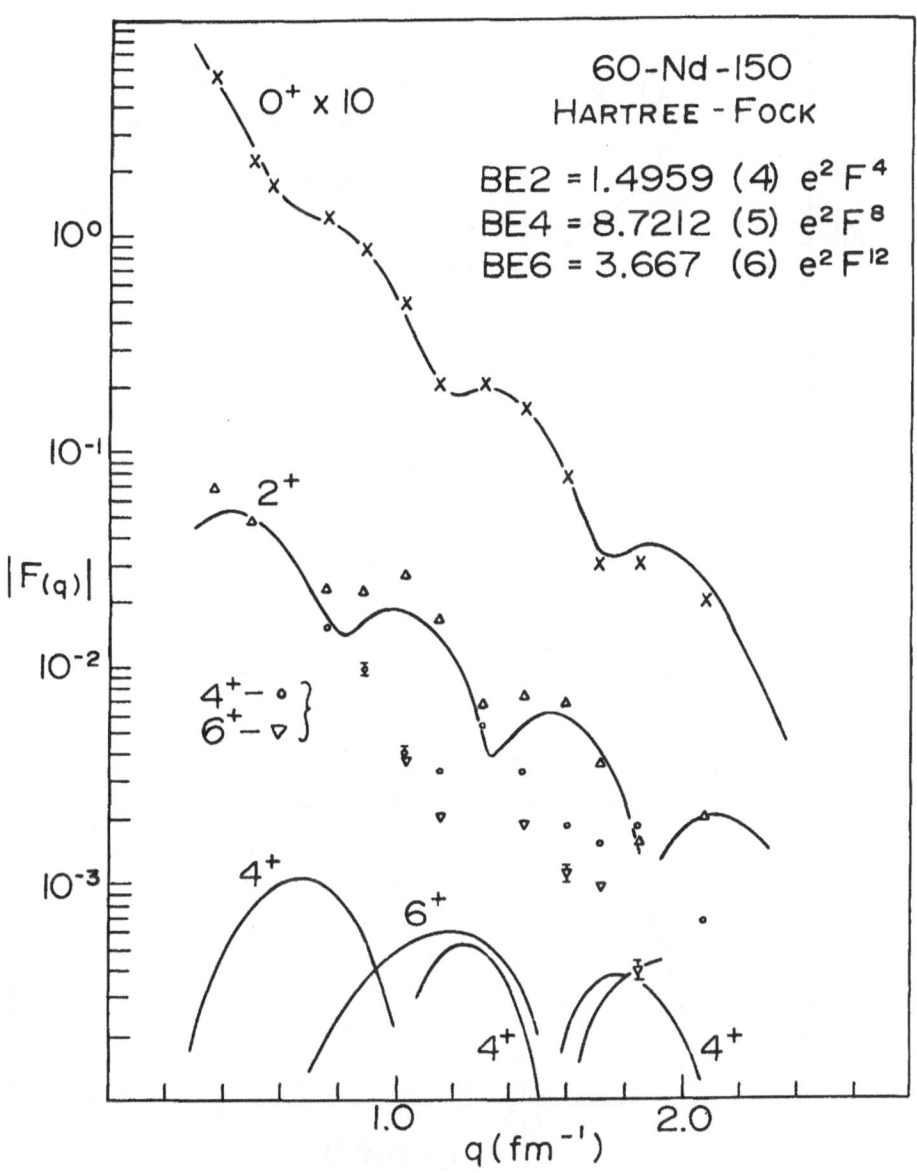

FIGURE 13

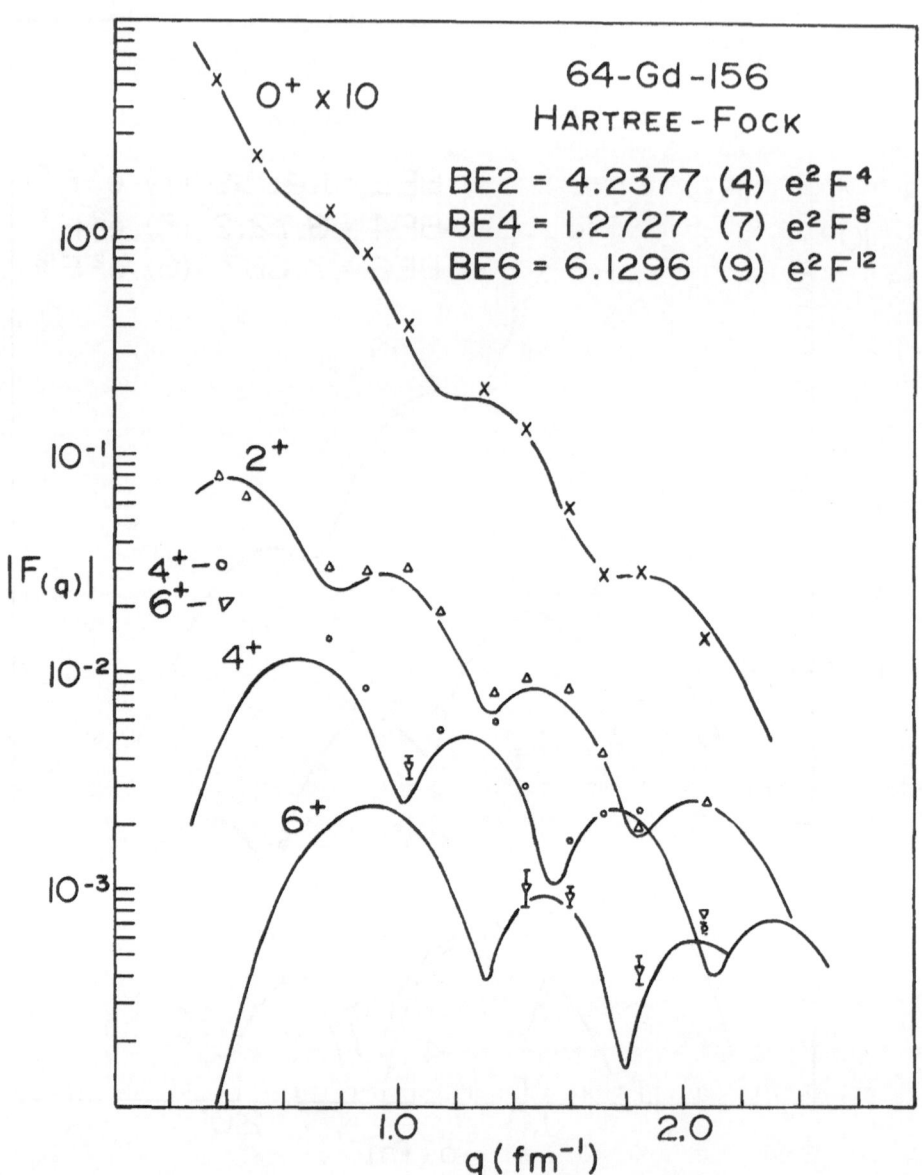

FIGURE 14

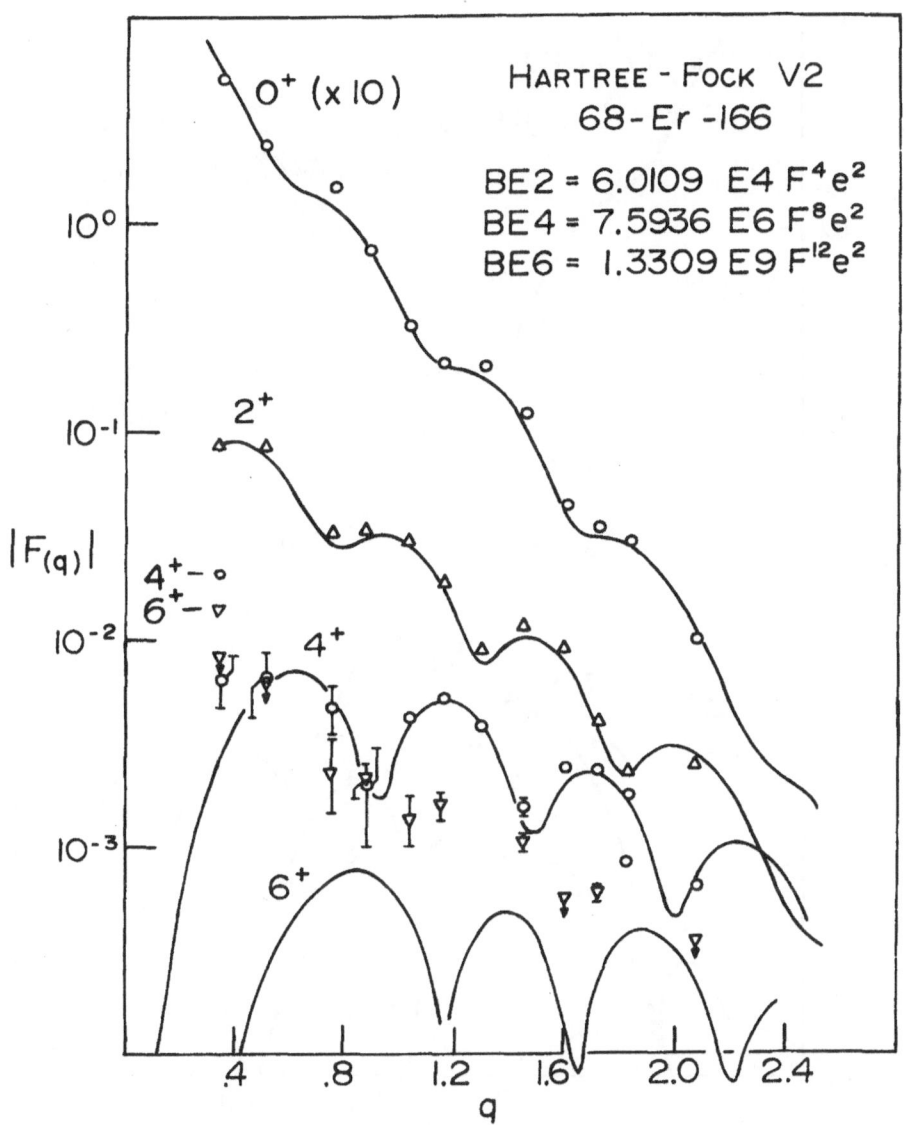

FIGURE 15

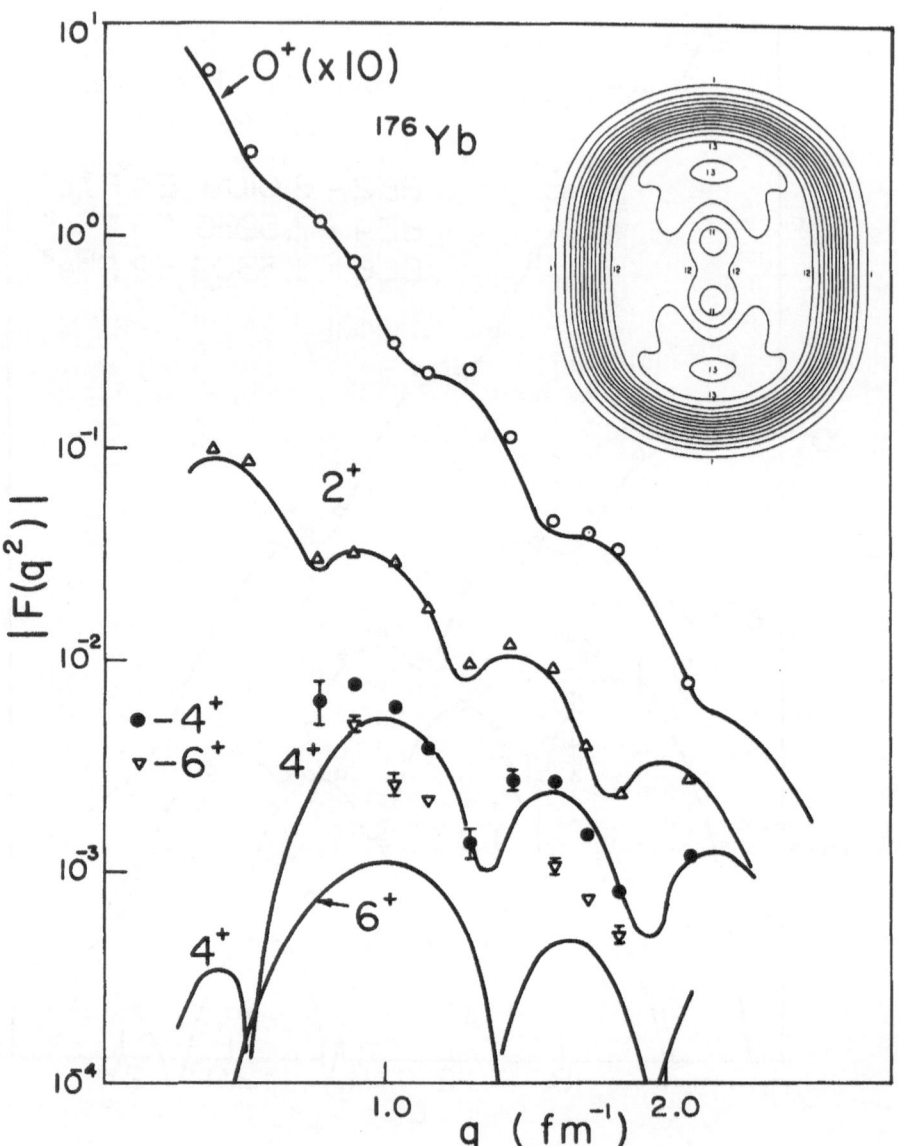

FIGURE 16

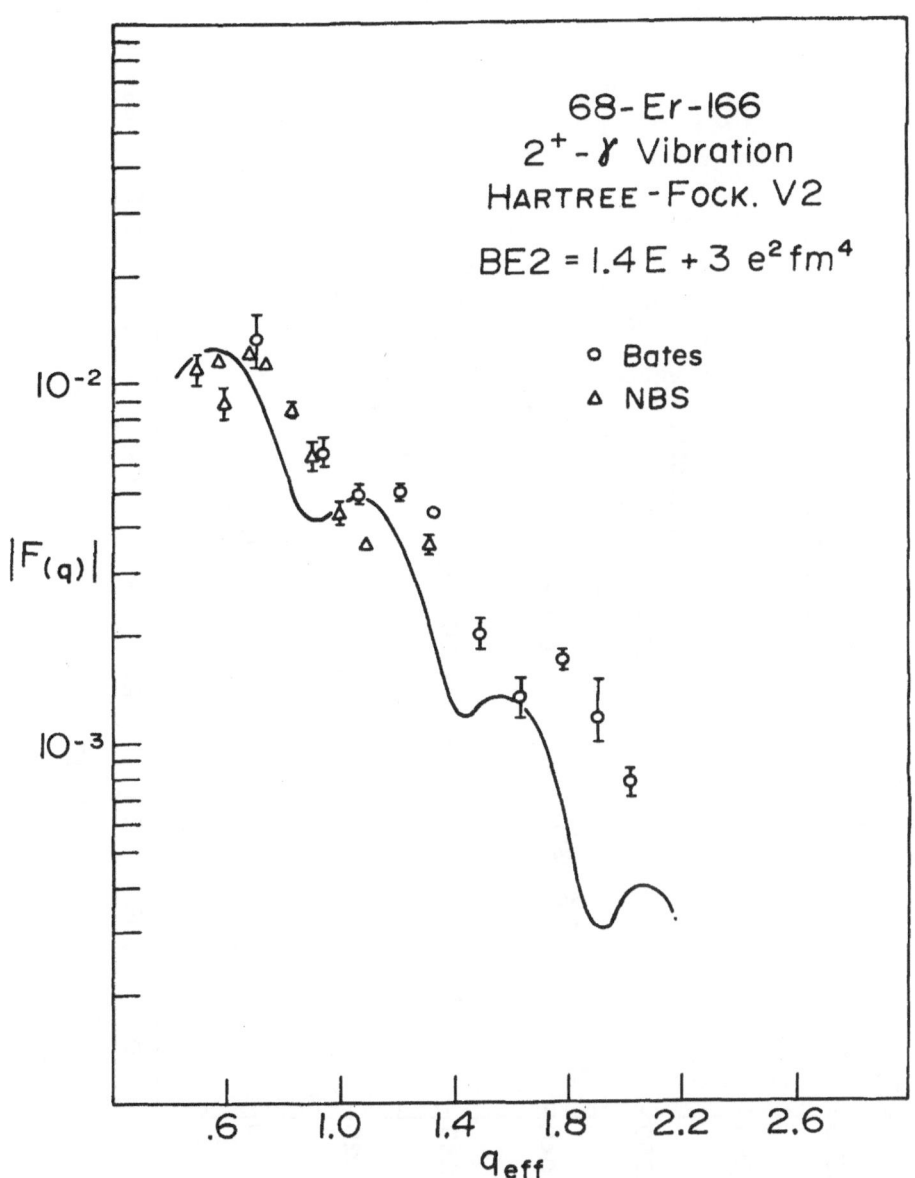

FIGURE 17

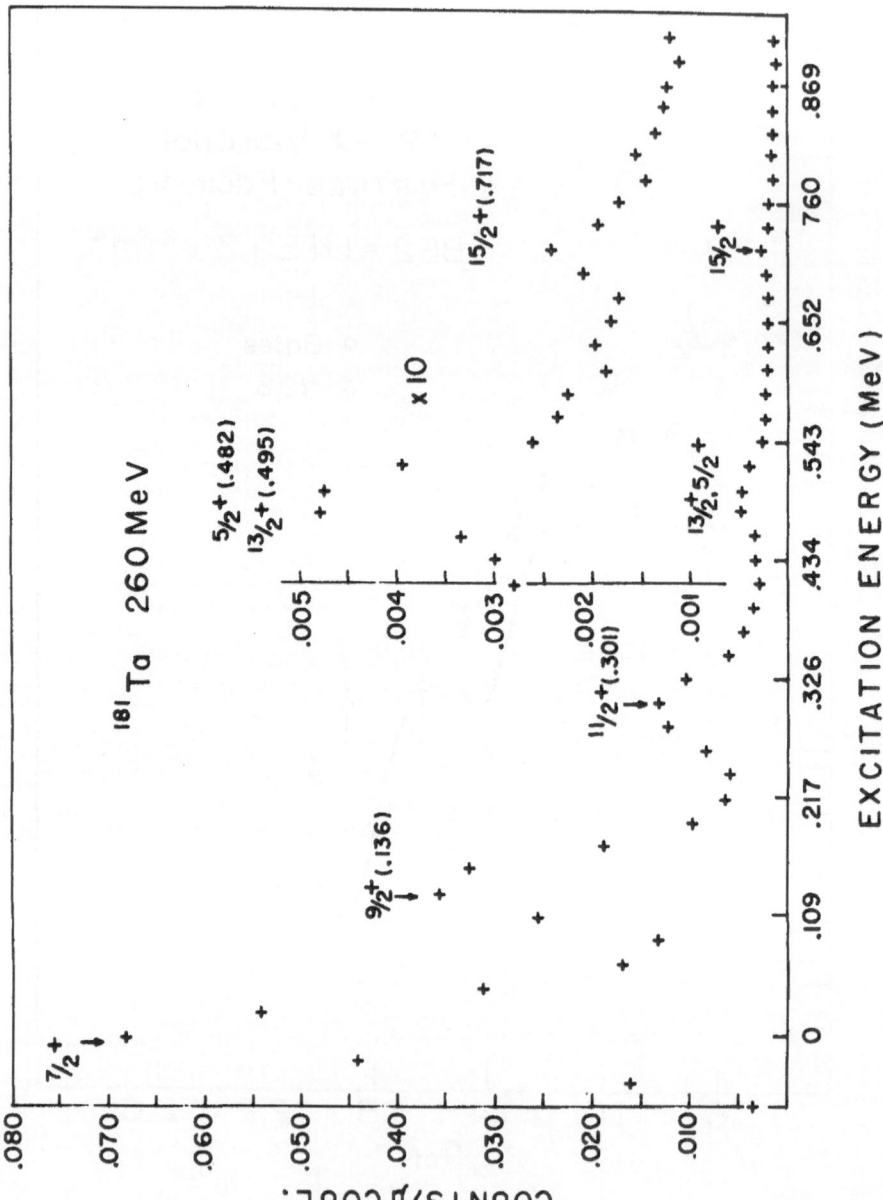

FIGURE 18

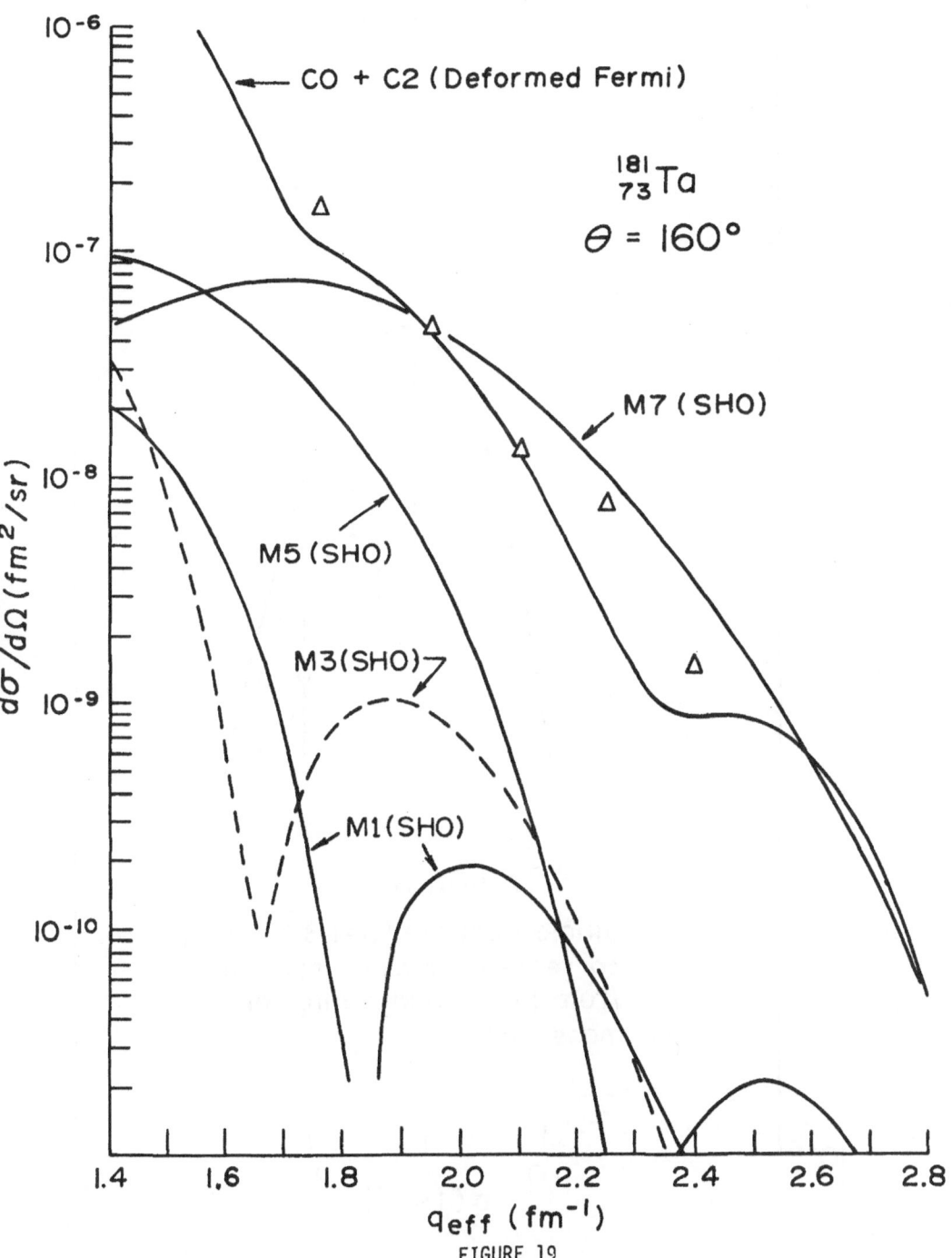

FIGURE 19

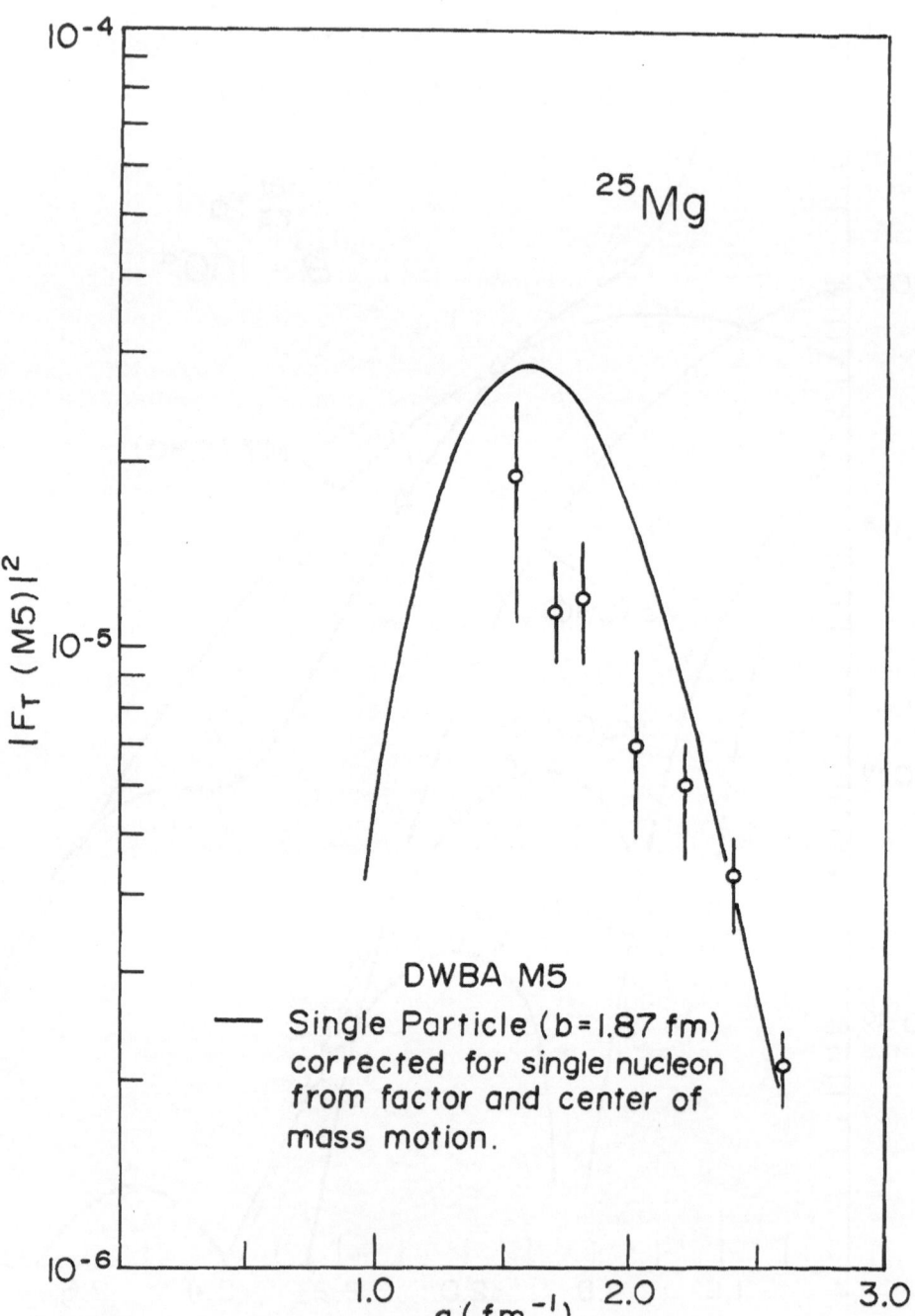

FIGURE 20

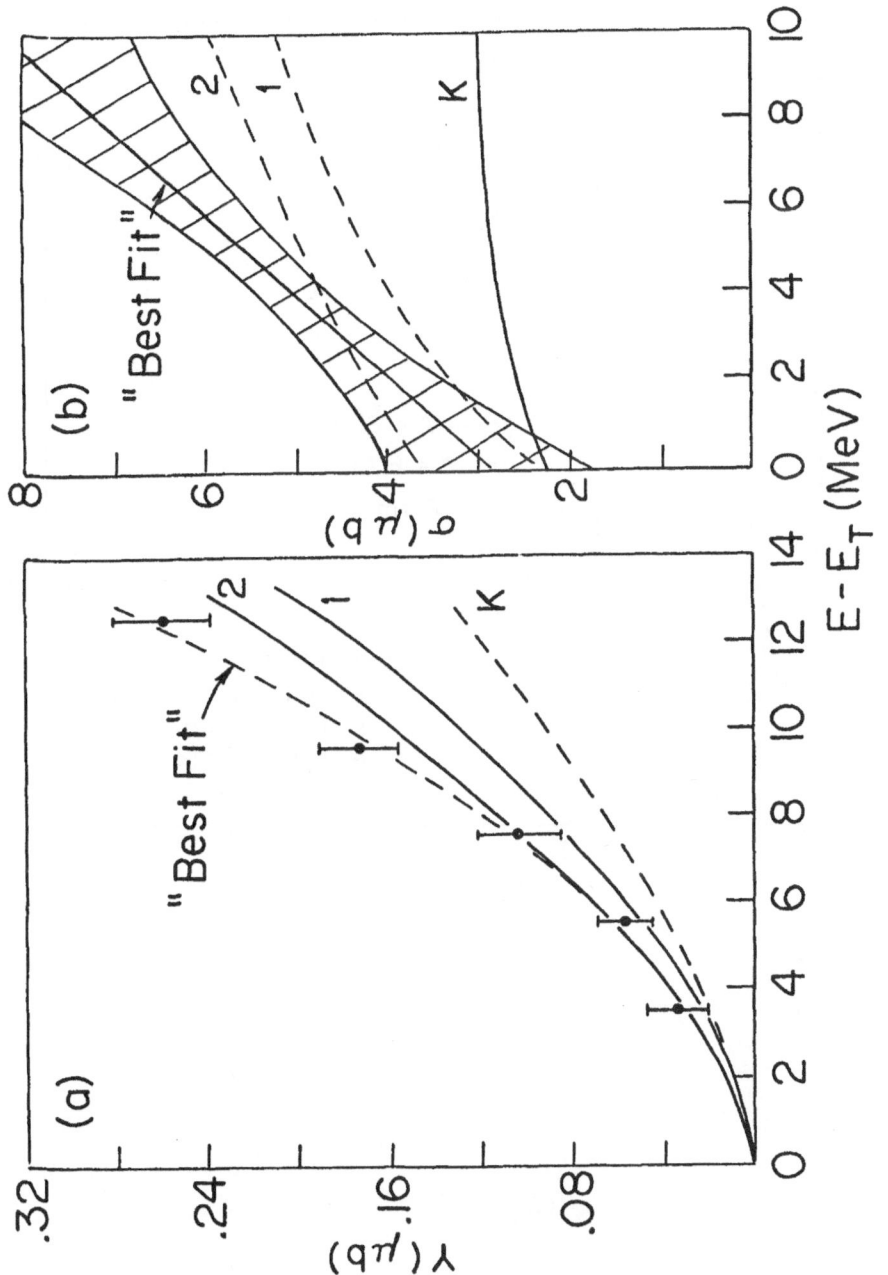

FIGURE 21 a,b

44

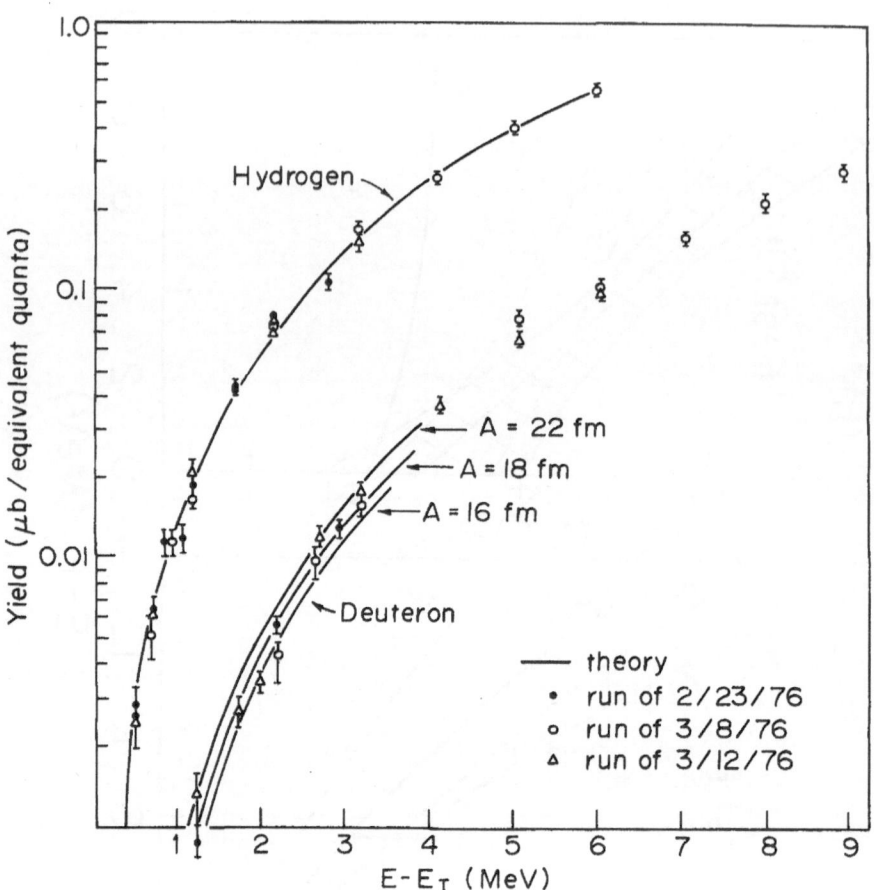

FIGURE 22

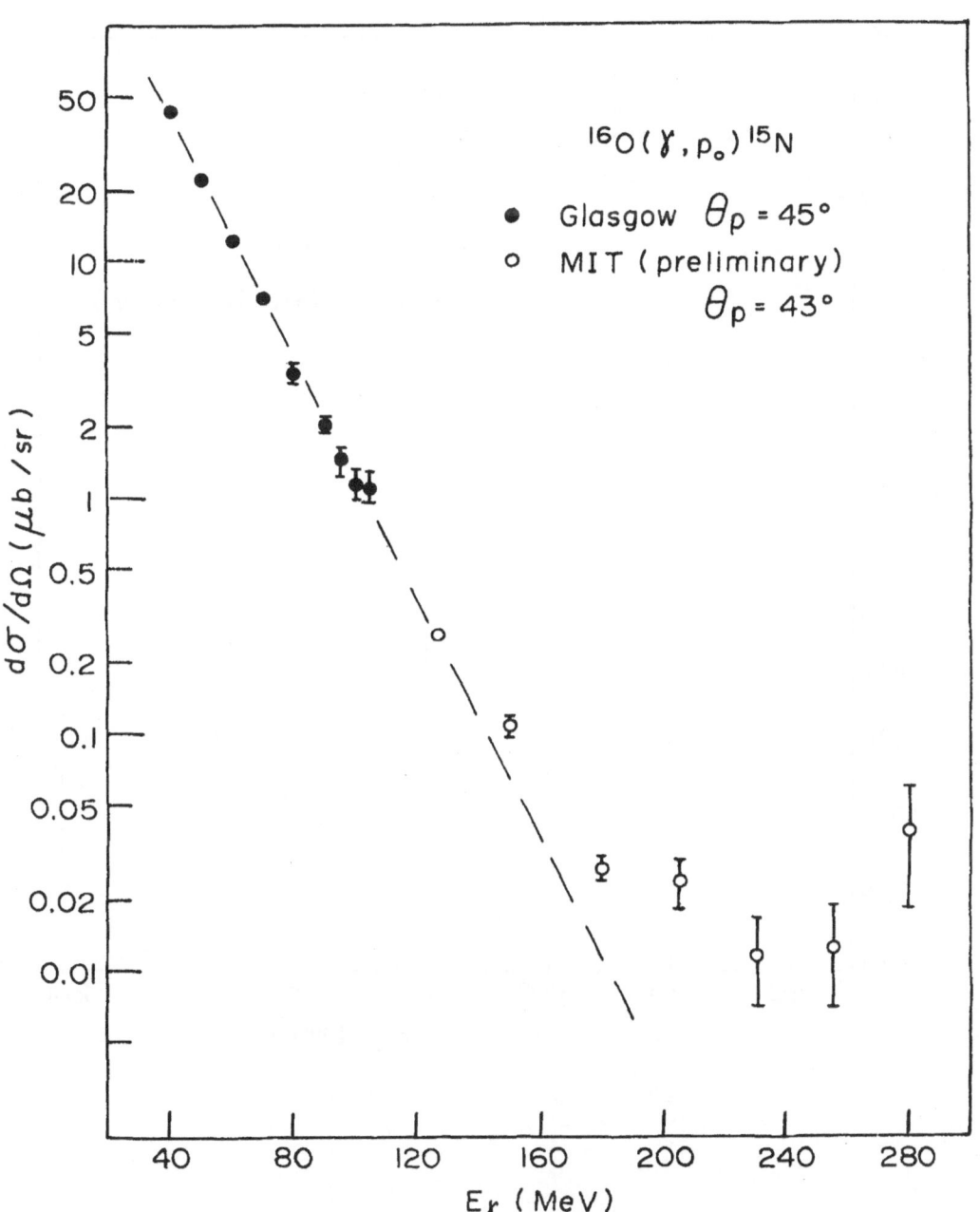

FIGURE 23

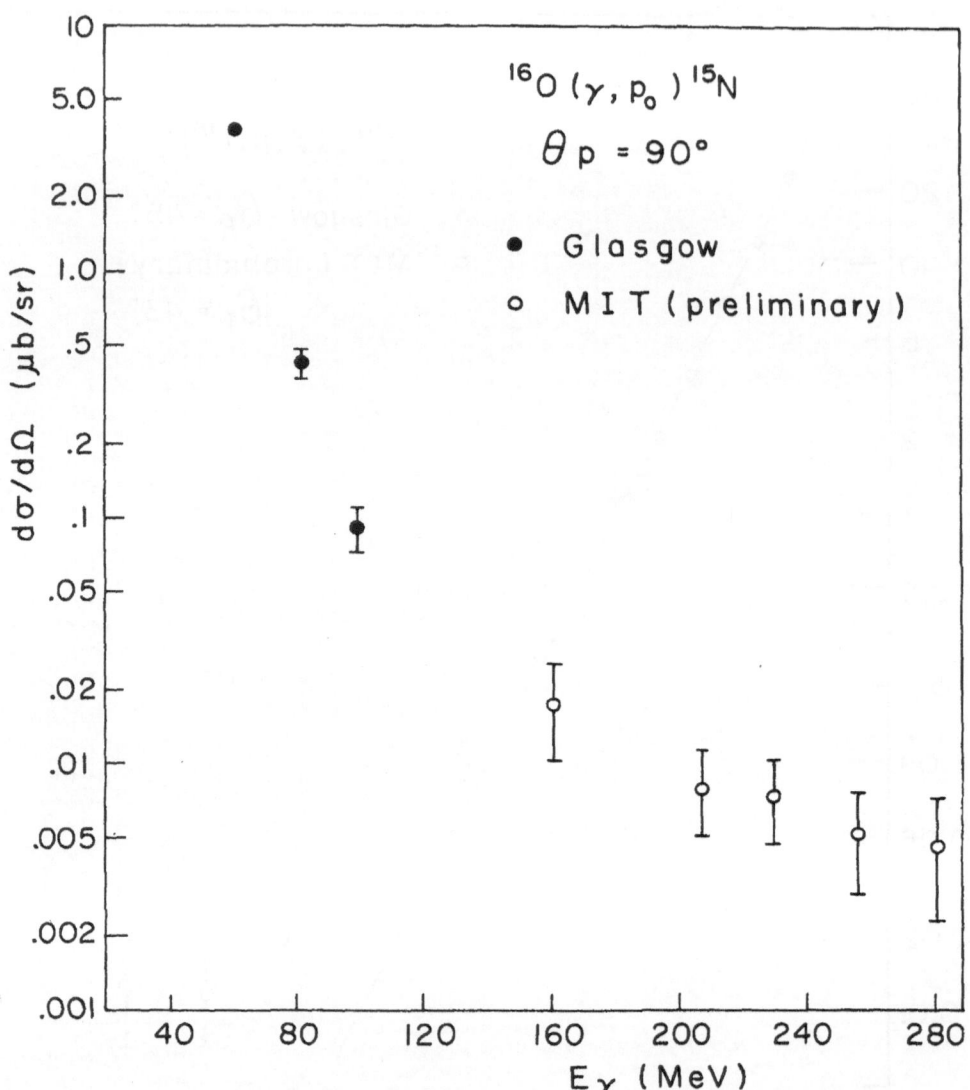

FIGURE 24

# NEWS FROM SACLAY

Ph. Catillon

DPh-N/HE, CEN SACLAY,
BP 2, 91190 Gif-sur-Yvette, France

## Introduction

The experimental programme of the Saclay electron linac (A.L.S.) is a direct consequence of the main characteristics of the accelerator. It can be seen on fig.1 that from the electron linacs built during the years 60's, the improvements have been made towards higher energy (Karkov, Orsay, Stanford) or higher duty cycle (Illinois or Stanford superconducting linacs). Between these two directions three accelerators are working or in completion with quite similar energies (400 to 600 MeV) and duty cycles (1 to 10 %) at Saclay (A.L.S), Bates (MIT) and Amsterdam (IKO). From their characteristics it is easy to deduce the general trends of new research :

. high momentum transfer reactions
. photonuclear reactions up to the (3,3) resonance region
. coincidence experiments
. physics with secondary beams of pions and muons.

I shall restrict myself to the limits of this School, that is the electro and photonuclear reactions but fig. 2 gives an idea of the complete field of interest studied with the A.L.S. in nuclear physics and particle physics using strong, e.m. or weak interactions.

I shall restrict this paper to the electron scattering experiments and to the photoproduction of pions at threshold. Informations concerning the experiments not given here can be found in the Annual Report [1].

## A. Electron scattering

### 1° Charge density

The cross-section for elastic scattering of electrons by a nucleus is given by

$$d\sigma/d\Omega = \frac{Z^2 e^4 \cos^2\theta/2}{4E^2 \sin^4\theta/2} \left[ F_L^2(q) + (\tfrac{1}{2} + tg^2\theta/2) F_T^2(q) \right]$$

If the spin of the nucleus is zero the transverse form factor cancels and from the longitudinal form factor $F_o(q)$ it is possible to deduce the charge density

$$\rho(r) = Ze \int_o^\infty e^{i\vec{q}\cdot\vec{r}} F(q)\, d^3\vec{q}$$

In fact the integral is taken from o to q max and through a model independent analysis [2] the larger is the maximum momentum transfert $q_{max}$ the more accurate is the density in particular the central part.

The elastic scattering of electrons has been performed at 450 MeV on $^{58}$Ni up to 3.9 fm$^{-1}$ [3]. The resulting density is shown on fig. 3 with two recent calculations made by Gogny [4]. DDHFB-D1 uses a density dependent finite range interaction in a Hartree-Fock Bogoliubov calculation. DYN-CALC is a "dynamic" calculation integrating over different deformations made by the same group [5].

Such a calculation is especially interesting since the nickel isotopes (even though nearly spherical) are fairly soft to deformation [5].

A similar experiment has been done on $^{208}$Pb at 500 MeV up to 3.7 fm$^{-1}$. It is shown on fig. 4 a fit made to the previous Stanford data extending up to 2.7 fm$^{-1}$. The analysis giving $\rho(r)$ is presently in progress [6] and gives a density exhibiting much less structure than expected theoretically.

### 2° Magnetization distribution

The cross section for elastic electron scattering from a non-zero spin ground state consists of incoherent contributions of electric and magnetic multipoles. The magnetic scattering can be measured at 180° or deduced from two experiments at different energies and angles for the same momentum transfert. The magnetic form factor is given by [7]

$$F_T^2(q) = \sum_{\text{odd } J} \frac{1}{2J_o+1} |<J_o||\hat{T}_J^{magn}(q)||J_o>|^2$$

In a stretched configuration ($J_o = \ell + \frac{1}{2}$) only the intrinsic magnetization contributes to the highest multipole and comes from the unpaired nucleon.

Experiments have been done with different targets ($^{51}V$, $^{59}Co$, $^{93}Nb$, $^{209}Bi$) in the transfer momentum range corresponding to the highest multipole. The results on $^{93}Nb$ and $^{59}Co$ are shown on figs. 5 and 6.

The analysis are carried out in the framework of the single particle shell model with harmonic oscillator or Woods-Saxon radial wave functions.

The oscillator lengths b have been respectively found equal to $2.04 \pm 0.04$ fm for $^{93}Nb$ [8] and $1.84 \pm 0.05$ fm for $^{59}Co$ [9].

3° Inelastic scattering

Recently inelastic scattering of electrons has been done at 250 MeV in a momentum transfer range from 0.6 to 2.2 $fm^{-1}$. The rotational states $2^+ 4^+ 6^+$ have been analysed [10] and compared with an Hartree-Fock Bogoliubov calculation [11] as shown on fig. 7 for $^{152}Sm$.

4° Quasi elastic scattering

In order to study the mechanism of the (e,e'p) reaction, the cross-section of the $^{12}C$(e,e'p) reaction has been measured in different kinematical conditions. It will be possible to test the factorization of the cross-section and the validity of usual treatments of the distortion. The data are under analysis.

The electrodisintegration of deuterium has been performed up to 500 MeV for a momentum transfer of 2.2 $fm^{-1}$. The distribution in momentum of the recoiling neutron gives, in PWIA, the wave function of the deuterium in momentum space. Fig. 8 shows the preliminary result for $\sigma($ e,e'p/$C\sigma_{e\ell})$ where $\sigma_{e\ell}$ is the cross-section for hydrogen and C a phase space factor .

B. Pion photoproduction at threshold

Pion photoproduction near threshold on nuclei $\gamma + i \rightarrow \pi + f$ can be viewed as a transition, from the ground-state $|i>$ to a very peculiar state in which in addition to the nuclear state $|f>$ a pion is present. In the vicinity of threshold, the pion interacts quite weakly with the nucleus so that the transition amplitude will be, with a good approximation

$$\phi\pi \ (o) < f \ | \ e^{i\vec{k}\vec{r}} \ \theta \ |i>;$$

$\phi\pi(o)$ is the value at origin of the pion wave function which is a constant in the nuclear volume, $\vec{k}$ is the photon momentum and $\theta$ the photoproduction operator. In a microscopic approach $\theta$ is dominated by the axial term

$$E_{o^+} \sum_{i=1}^{A} \vec{\sigma}_i \cdot \hat{\varepsilon}$$

$\hat{\varepsilon}$ is the photon polarization and $E_{o^+}$ the threshold value of the electric dipole amplitude in the photoproduction on the nucleon.

For $\pi^+$ photoproduction the experimental method consists in a measurement relative to the proton case ; bremsstrahlung is the photon source and detection of the $\pi^+$ is made by observation of the $e^+$ from $\pi\mu e$ decay chain in Cerenkov counters.

The cross section of the process can be parametrized easily by the expression

$$\sigma = a \frac{q}{k} \frac{2\pi\gamma}{e^{2\pi\gamma}-1}$$

q and k are the pion photon momenta in the center of mass referential, $\gamma = Z \frac{e^2}{\hbar v}$ the Sommerfeld factor describing the Coulomb interaction between the $\pi$ and the final nucleus. The parameter a countains the nuclear physics information and the effect of the $\pi$-nucleus strong interaction. The result of a relative measurement can be expressed in terms of $a_{i \to f}/a_{proton}$ where

$$a_{proton} = 4\pi \, |E_{o_+}^{(\pi^+)}|^2 = (193.5 \pm 6.7)\mu b$$

1). A first measurement on the $^6Li \to {}^6He$ g.s. case, yielded a result in strong disagreement with existing theoretical estimations. This discrepancy prompted theoretical efforts which through reexamination of $^6Li$ electron inelastic scattering data lead to a decrease of the ratio $a_{^6Li \to {}^6He}/a_7$. On the other hand, an improved experiment has shown that some systematic errors in the first measurement lead to an underestimation of the $^6Li$ rate. The new value $a_{^6Li \to {}^6He}/a_7 = 0.098 \pm 0.004$ is in reasonable agreement with the most recent calculations.

2). $\pi^+$ photoproduction on deuterium near threshold is dominated by the neutron-neutron scattering length $a_{nn}$ ; using the recommended value $a_{nn} = -16.4 \pm 0.9$ fm we should be able to test our overall understanding of the process. Preliminary results are shown on fig. 9, where the experimental deuterium yielded has been filled to a theoretical one obtained in folding Tzara PWIA Zero-range potential estimation [14] with a bremsstrahlung spectrum ; in the fitting procedure we allowed for a normalization factor of .73. The proton yield is used for calibration. A more refined calculation [15] using realistic wave functions yields a cross section lower than Tzara estimation by 20 % giving almost agreement with experiment.

$3°$). The neutral pion photoproduction amplitude at threshold on the proton is one order of magnitude less than the charged pion one. In the case of the neutron, there is no experimental value of $\dot{E}_{o+}^{(\pi°n)}$ but it is expected to be less than $E_{o+}^{(\pi°p)}$. Other multipoles, especially $M_{1+}$, become very important even in the vicinity of threshold. In the nuclear case one expects large pion rescattering effects : a charged pion photoproducted on a nucleon undergoes charge exchange on another nucleon.

A preliminary experiment has shown that the measurement is feasible in d and $^3$He, 2 MeV above threshold using Čerenkov counters with lead converters in front, to detect in coïncidence the two photons of the $\pi°$ decay. An estimation of the rescattering by Koch and Wolshyn [16] predicts drastic effects which should be easy to bring into light.

In conclusion we can say that $\pi^+$ photoproduction constitutes a tool to investigate very precisely nuclear structure whereas $\pi°$ photoproduction will shed light on the reaction mechanism.

## References

[1] Departement de Physique Nucléaire. Compte rendu d'activité. Note CEA-N 1861.
[2] I. Sick, Nucl. Phys. A218 (1974) 509.
[3] I. Sick, J.B. Bellicard, M. Bernheim, B. Frois, M. Huet, Ph. Leconte, J. Mougey, Phan Xuan-Ho, D. Royer and S. Turk, Phys. Rev. Lett. 35 (1975) 910.
[4] D. Gogny , Nucl. self consistent fields (Ed. Ripka, Porneuf, 1975).
[5] G.A. Rinker and J.W. Negele, Santa-Fe Abstracts II.28 page 148.
[6] B. Frois, J.B. Bellicard, J.M. Cavedon, M. Huet, P. Leconte, A. Nakada, Phan Xuan Ho and I. Sick, to be published.
[7] T.W. Donnelly, J.D. Walecka, Nucl. Phys. A201 (1973) 81.
[8] P.K.A. de Witt Huberts et al. Phys. Lett. 60B (1976) 157.
[9] P.K.A. de Witt Huberts et al. Crakow Conf. june1976 "Radial Shape of Nuclei".
[10] A. Nakada et al., to be published.
[11] J. Decharge, M. Girod and D. Gogny, Phys. Lett. 55B (1975) 361.
[12] J. Deutsch et al. Phys. Rev. Lett. 33, (1974) 316.
[13] J.L. Bergström, J.P. Auer and R.S. Hickes, Nucl. Phys. A251, (1975) 401.
J.B. Laummarata and T.W. Donnelly, submitted to Nucl. Phys. A (1976).
J. Delorme and A. Figureau (private communication).
[14] C. Tzara, Nucl. Phys. A256 (1976) 381.
[15] J. O'Connell and L. Maximon (private communication).
[16] J.H. Koch and R.M. Woloshyn, Phys. Lett. 60B, (1976) 221.

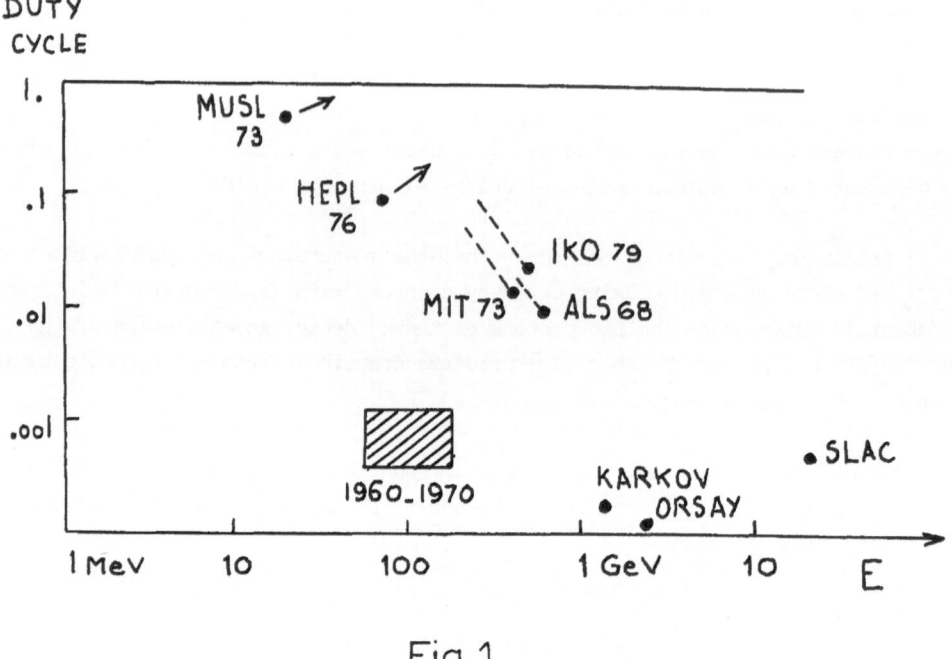

Fig.1

| | STRONG | E.M. | WEAK |
|---|---|---|---|
| PARTICLE | $\pi$-nucleon interaction<br><br>$\pi^\pm p \longrightarrow \pi^\pm$<br><br>$\pi^- p \rightarrow \pi^0 n$<br>$\quad\quad \rightarrow \pi^- p$<br>$\quad\quad \rightarrow n\,\gamma$ | • Electrophotoproduction of pion on hydrogen<br><br>$H(e,e'\pi^+)n$<br>$\quad\quad\longrightarrow \langle r_\pi^2 \rangle$ | • $\mu^+$ life time<br>• $\mu^-$ capture in hydrogen |
| NUCLEAR | $\pi$-nucleus interaction | • Electron scattering<br>• $(e,e'p)$ reaction<br>• Photonuclear reactions<br>• Photoproduction of pions<br>• Muonic X rays | • $\mu^-$ capture in nuclei |

Fig. 2

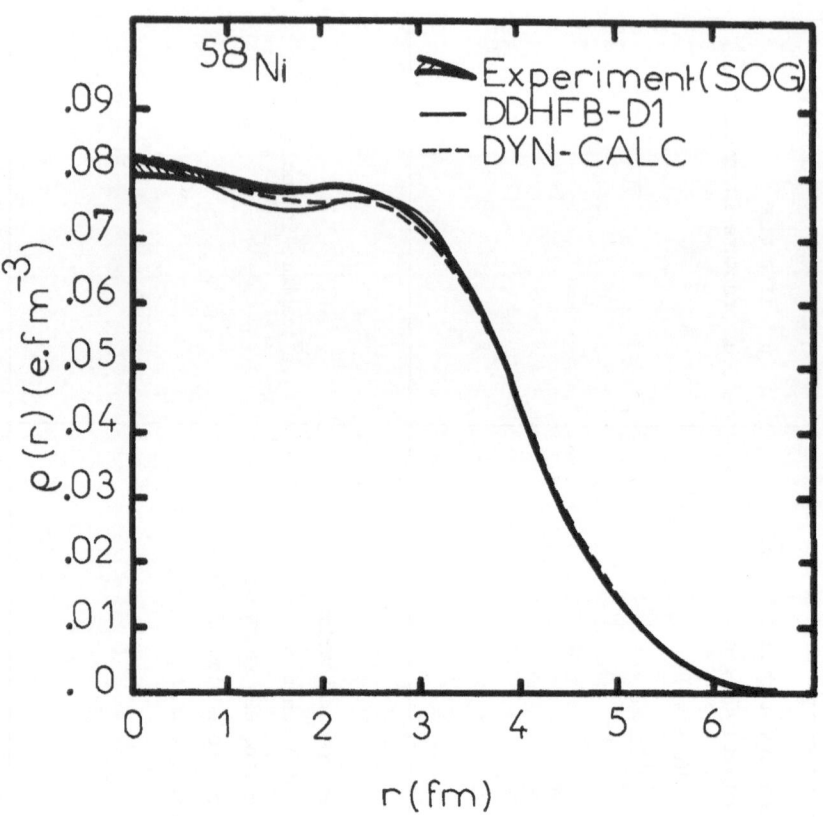

Fig 3

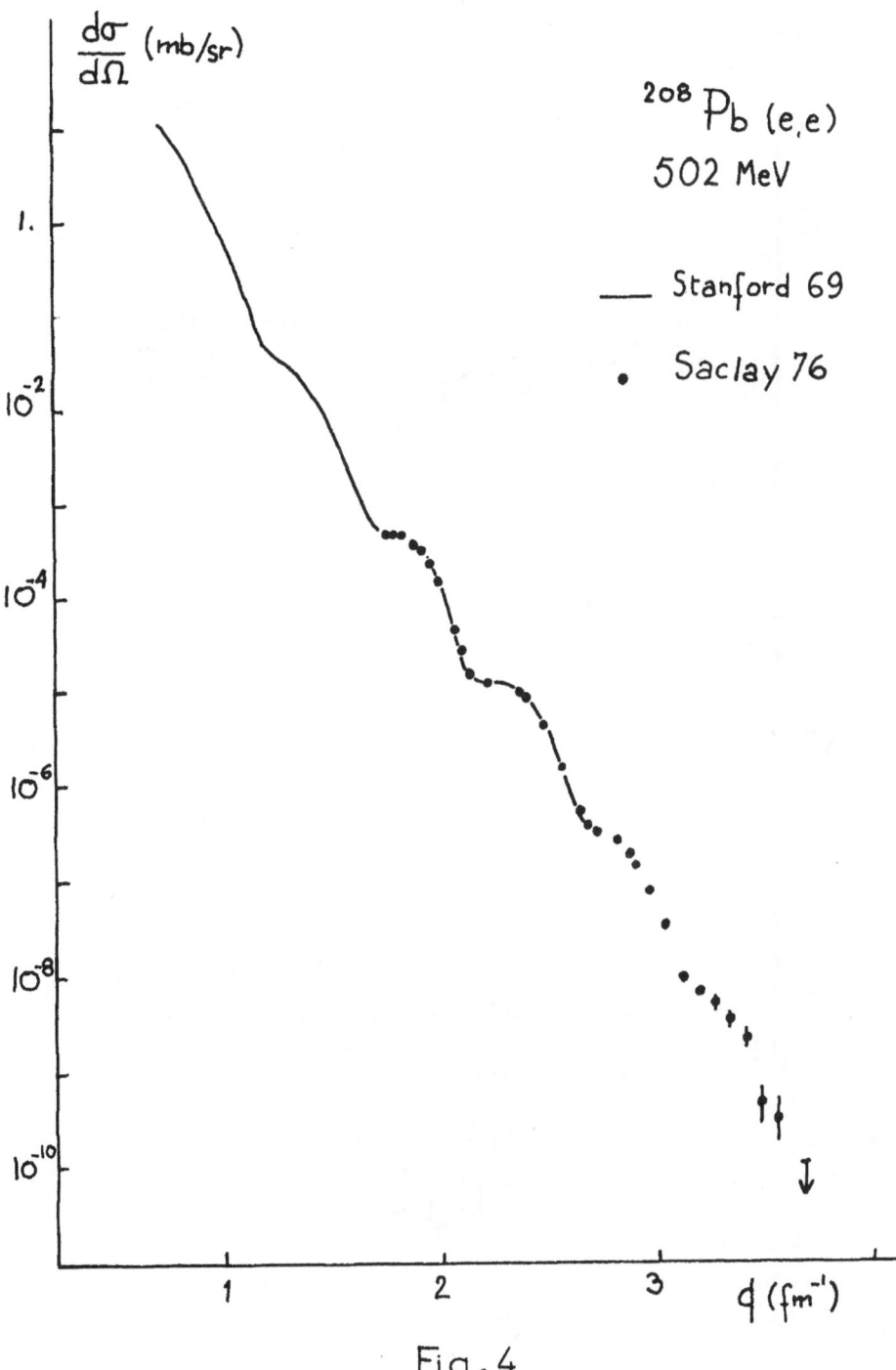

Fig.4

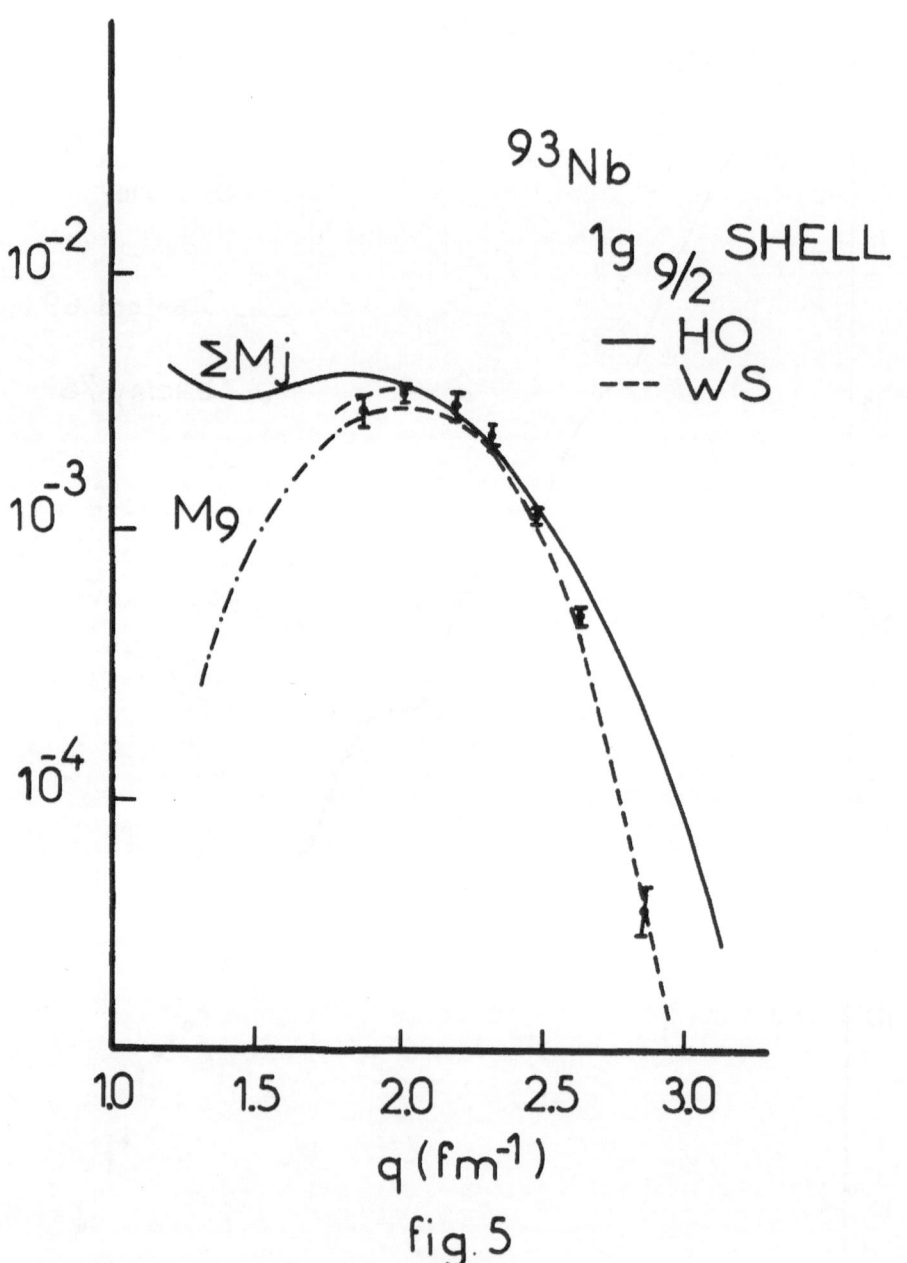

fig. 5

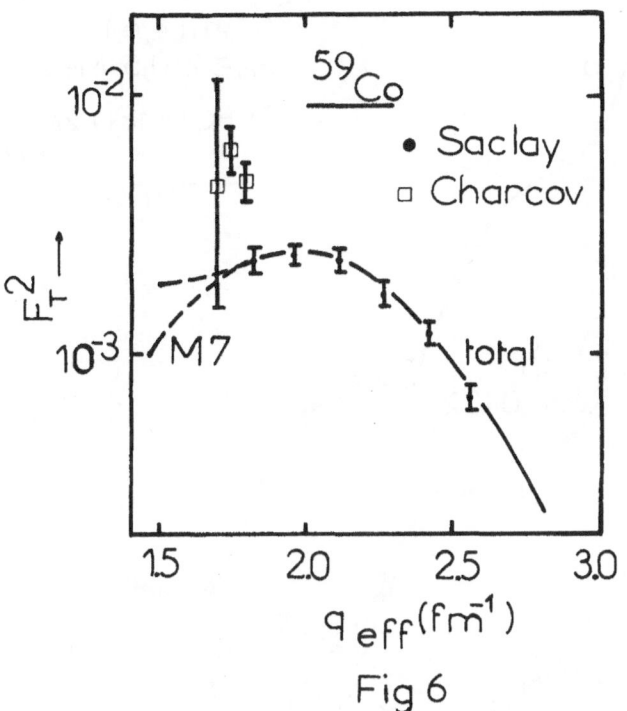

Fig 6

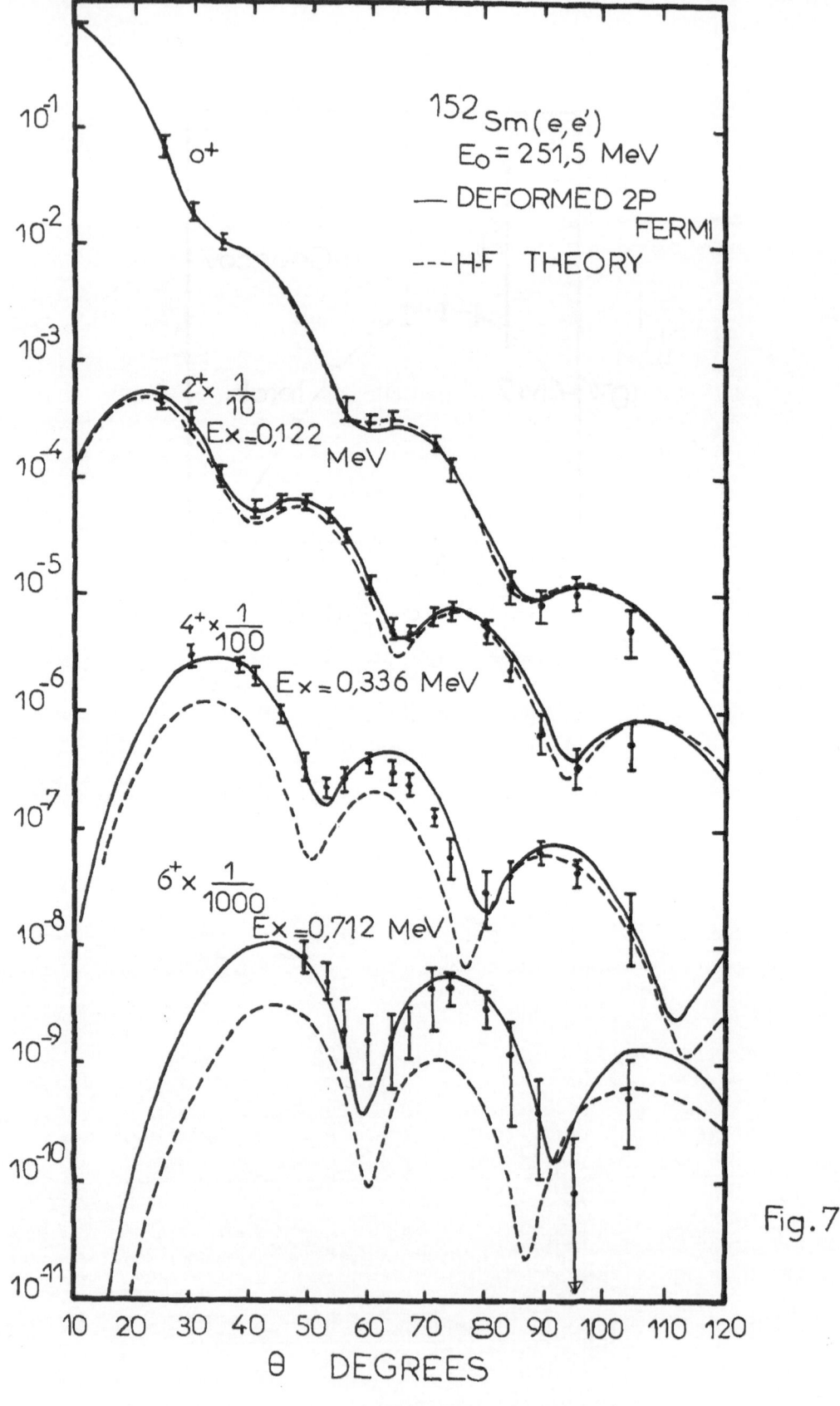

$^{152}Sm(e,e')$

$E_0 = 251,5$ MeV

— DEFORMED 2P
       FERMI

--- H·F THEORY

$0^+$

$2^+ \times \frac{1}{10}$

$Ex = 0,122$ MeV

$4^+ \times \frac{1}{100}$

$Ex = 0,336$ MeV

$6^+ \times \frac{1}{1000}$

$Ex = 0,712$ MeV

Fig. 7

$\theta$   DEGREES

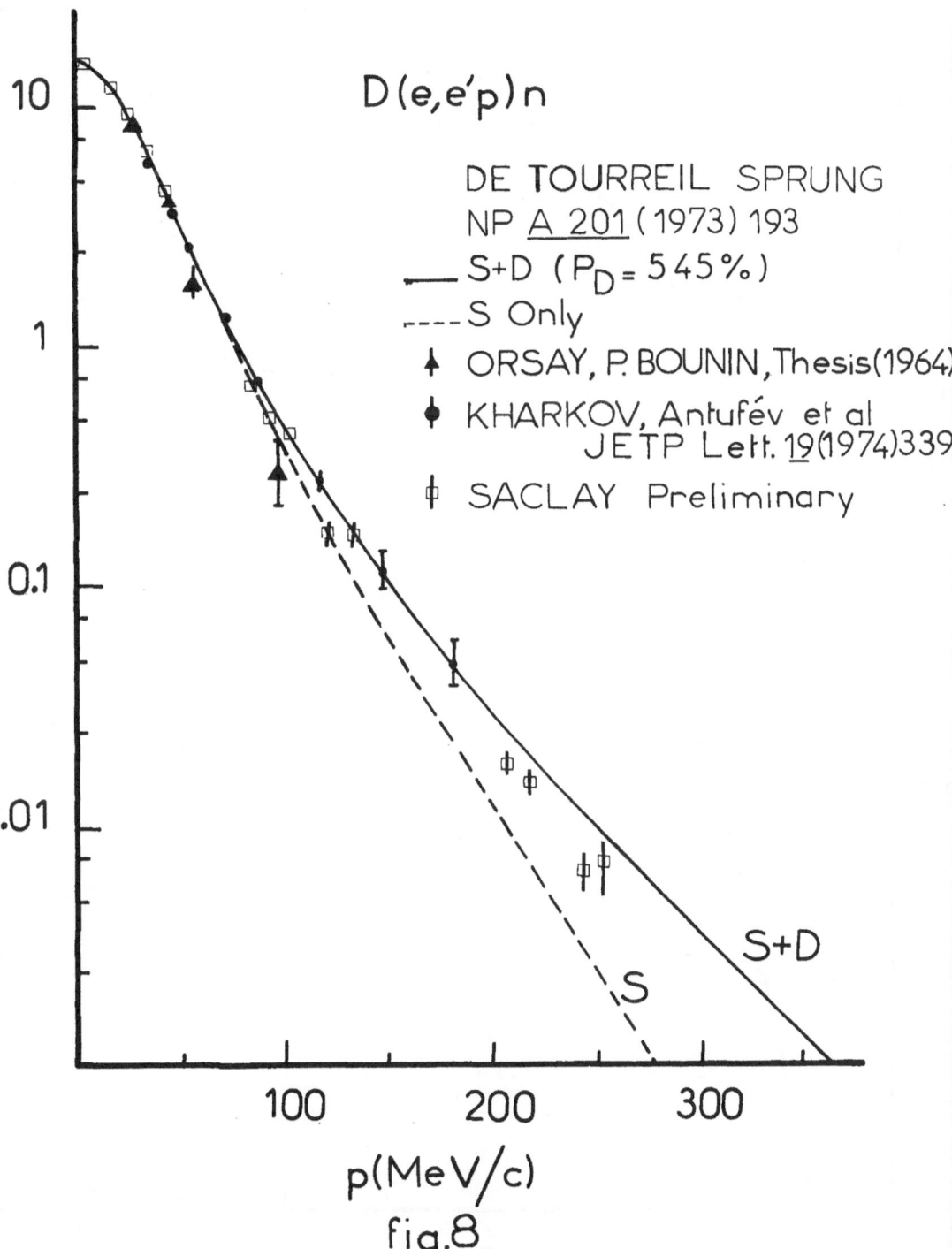

fig.8

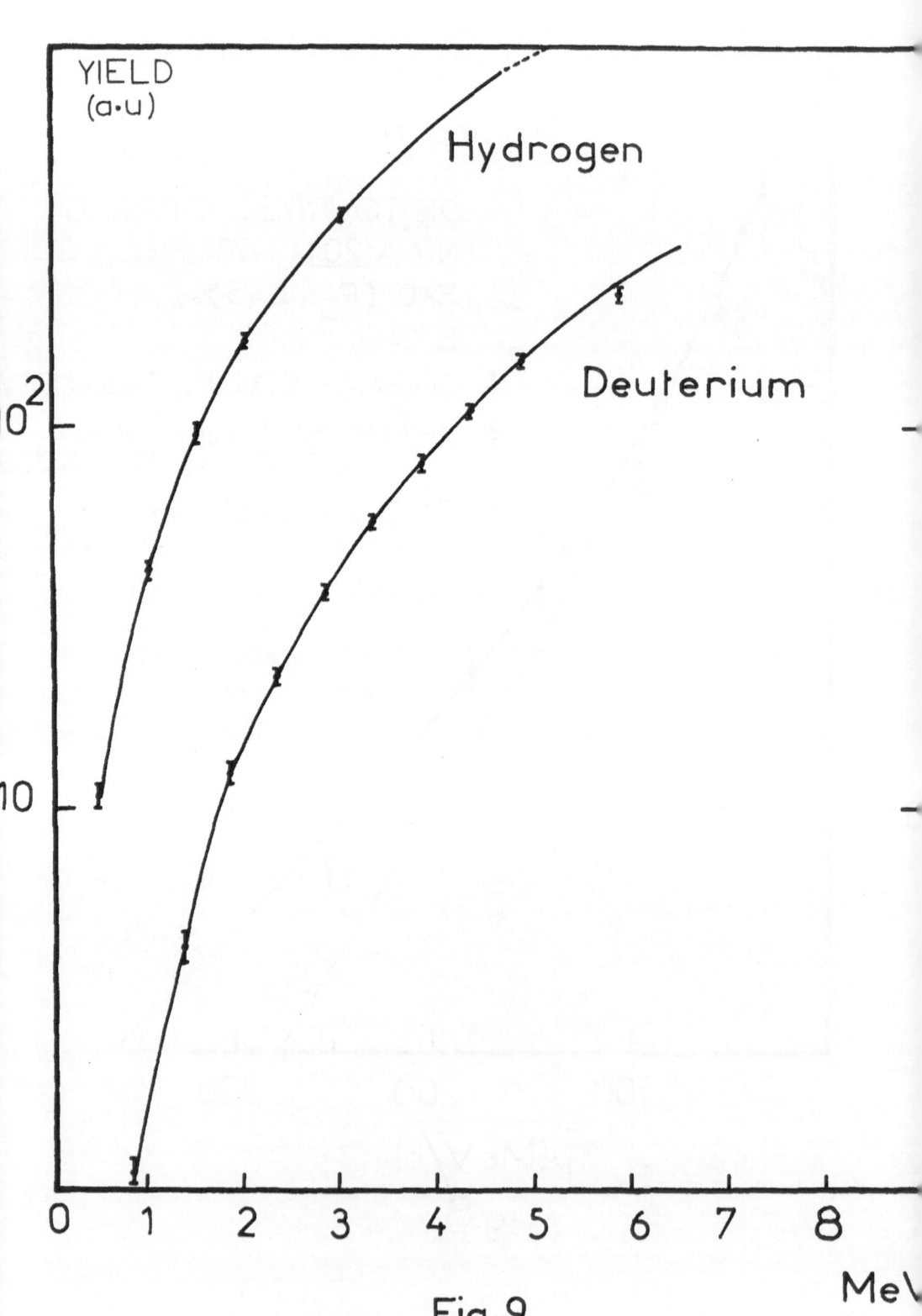

Fig.9

"ELECTRON SCATTERING WORK AT AMSTERDAM;
past, present and future activities".

C. de Vries
Institute for Nuclear Research

Introduction.

In this talk I will present the activities at Amsterdam in the
electron scattering field. These activities are typical for an
Institute which is in the process of switching from low energy
(<100 MeV) physics to intermediate energy ($\leq$500 MeV) physics.
The low energy physics has been entertained since 1967 with a
85 MeV linear electron accelerator[1] until the end of 1975. Then
this machine has been closed for further experiments mainly because of manpower problems. The Institute, namely, is engaged
with the construction of a large new facility for intermediate
energy experiments.
Although I will present my remarks according to the main lines:

    A)  recent experimental results
    B)  planned future experiments
    C)  design of instrumentation for intermediate
        energy physics

it is worthwhile to first present some technical details for the
existing low energy machine and for the intermediate energy
machine presently under construction.

Table I.

|                      | 85 MeV linac | 500 MeV linac |
|----------------------|--------------|------------------------------|
| energy (nominal)     | 85 MeV       | 500 MeV                      |
| duty factor          | 0.1%         | 2½% at 500 MeV               |
|                      |              | 10% at 250 MeV               |
| repetition rate      | 200 pps      | 2000 pps                     |
| pulse width          | 5 μsec       | 50 μs                        |
| intensity (average)  | 100 μA       | 500 μA                       |
| beam power           | 10 kW        | 250 kW                       |
| secondary beams      | photons      | photons                      |
|                      | neutrons     | neutrons                     |
|                      |              | pions                        |
|                      |              | muons                        |
| completion date      | 1967         | 1979                         |
| shut down            | 1975         | –                            |

In Table I. are indicated the most important features for both
machines and it is appropriate to defend some of those parameters in more detail. After all, shutting off a reliable good
machine in order to allow future work with another machine needs
explanation.

1. As one can see the energy of the new machine is 5x the older
   one, meaning that one can probe the nucleus for details in
   the order of 0.2 fermi and research on the structure of the
   nucleon and even the pion becomes possible.

2. The duty factor for the new machine is a factor of 25-100
   higher than was available with the old accelerator. Although
   most of you are familiar with the importance of a large duty
   factor (rep.rate x pulse width) let me at this point make
   the statement that the duty factor chosen for the new machine
   allows not only much more refined single channel experiments
   but also coincidence experiments which were hitherto very
   difficult not to say impossible at lower duty factors.
   We will come back on this point later.
   The different numbers quoted for the duty factor are related
   to accelerator technology and I will refrain from going into
   this.

3. The increase of electron beam intensity is mainly due thanks
   to the larger duty factor which is achieved essentually by
   dropping the peak intensity relative to the average intensity. However, although the 250 kW (intensity x energy) indicated can be extracted from the machine, handling of such an
   enormous beam power by collimators, slits and targets is another matter. The expectation is that for electron scattering
   experiments one never will use higher than 100 kW. For the
   generation of pions on the other hand the highest attainable
   beam power is requested in order to have useful incoming fluxes
   for pion (or muon) induced experiments.

4. Secundary beams of pions and muons will be used for the future
   experimental program. I will refrain here from going into
   more details because it does not fit into the motivation of
   this school.

5. The 85 MeV machine is already exit before the new machine is
   working. The reason for this was mentioned already. However,
   in order to make the beamless period at the Institute as short
   as possible, all the 85 MeV experimental equipment[2] is being
   transferred to the substation at the 140 MeV point along the
   new machine ( see fig. 1. ).
   The expectation is that reinstalling the equipment there will
   be completed early 1977 so that in the course of 1977 the
   experimental program with the high duty factor beam will start.
   Some changes in the 85 MeV instrumentation will be made to
   accommodate the 10x longer pulse length. For instance, a proto-
   type multiwire proportional counter will be tested in the focal
   plane of the existing double focussing spectrometer[3] to re-
   place the present overlapping scintillator counter[4].

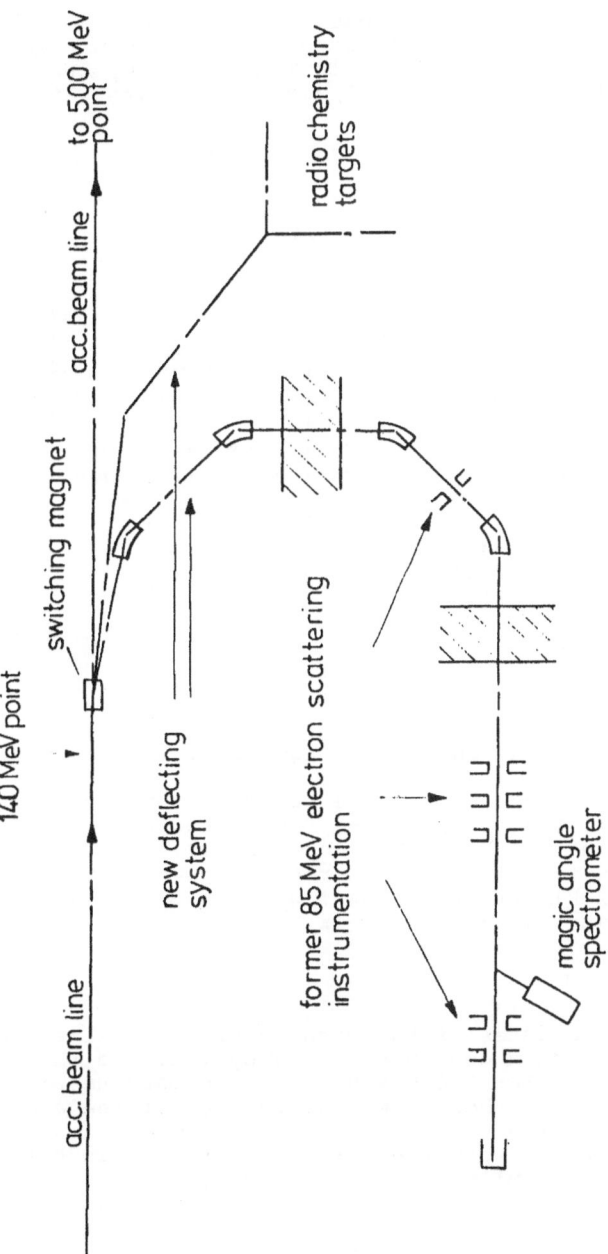

Substation at the 140 MeV point of the 500 MeV accelerator

fig. 1

Before going on to the next subject of my talk I will shortly
discuss the lay-out of the new accelerator. The parameters chosen
have already been discussed but in Table II they are compared
with those of the Saclay and MIT high duty factor linear electron accelerators.

Table II.

|  | Saclay | MIT | Amsterdam |
|---|---|---|---|
| energy | 640 MeV | 400 MeV | 500 MeV |
| duty factor | 2% at 400 MeV<br>1% at 640 MeV | 6% at 200 MeV<br>1,5% at 400 MeV | 10% at 250 MeV<br>2.5% at 500 MeV<br>(planned) |
| secondary beams | positrons<br>photons<br>pions<br>muons | photons | photons (for<br>radio chemistry)<br>pions<br>muons |

Fig. 2. shows the lay-out of the facility. The different halls
are indicated.
Fig. 3. shows the scheme for the central beam line equipment
of the machine. The sections are fabricated by Varian, according
to the design made for the MIT accelerator. The injection is a
6 MeV linac in itself, manufactured by Haimson Research Corpo-
ration. The 12 klystrons (Varian) are rated at 100 kW average
rf-power and at 1-4 MW peak power.

Other special features of the machine are the all-solid-state
modulators (designed and constructed by IKO) and the complete
computer control. The 140 MeV part of the machine will deliver
a beam next year and beam tests at the end of the accelerator
are expected in 1979.
Fig. 4. and fig. 5. show the already installed injector and
some parts of the buildings, respectively.

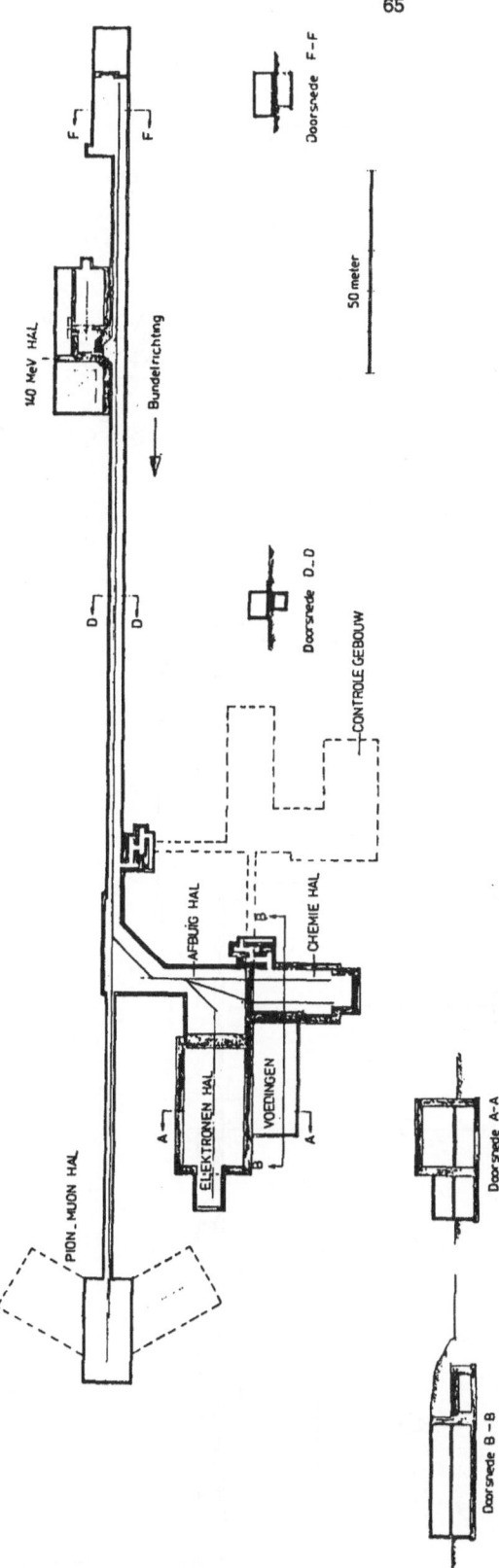

Bundelrichting

140 MeV HAL

50 meter

Doorsnede F–F

Doorsnede D–D

CONTROLE GEBOUW

AFBUIG HAL

CHEMIE HAL

PION – MUON HAL

ELEKTRONEN HAL

VOEDINGEN

Doorsnede A–A

Doorsnede B–B

fig. 2

66

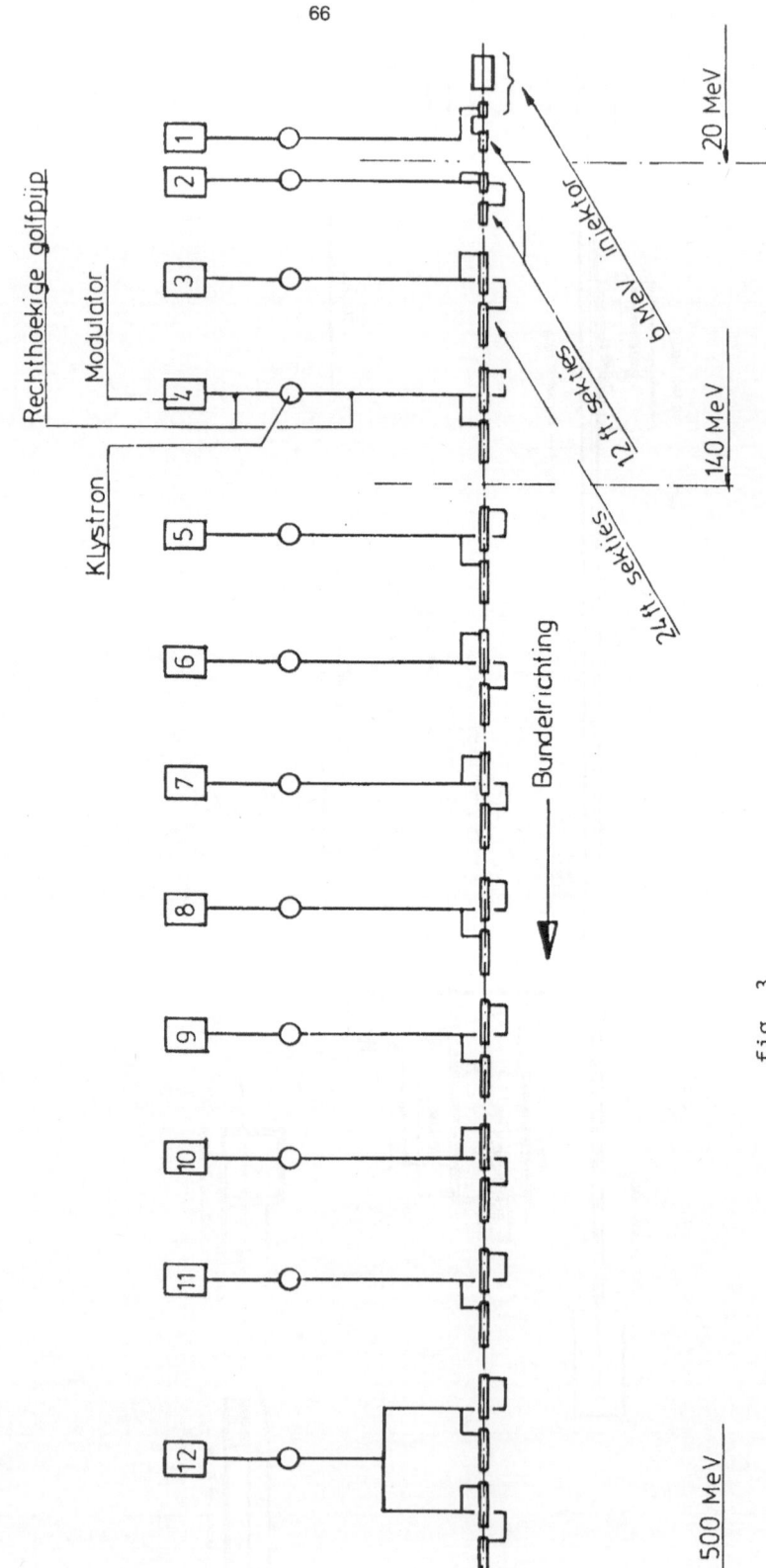

fig. 3

## A.  RECENT EXPERIMENTAL RESULTS.

Although experimental results are available from elastic charge
scattering and inelastic discrete level scattering I will focus
on magnetic moment distribution experiments for two reasons.
First of all the latter type of experiments have been dealt with
mainly during the last period of the experimental program with
the 85 MeV machine for reasons of exclusivity, and secondly,
the ongoing collaboration with Saclay involves high-q experiments which are to be considered as an extrapolation of the
Amsterdam work on magnetic scattering.
Before describing these experiments and their results let us
briefly go into some formulae which illustrate the type of
information to be obtained from electron scattering experiments
at backward (including 180°) angles.
Neglecting distortion of the electron waves by the Coulomb field
of the nucleus - in other words: in plane wave Born approximation (PWBA) - elastic scattering of relativistic electrons of
incident energy $E_0$ through an angle $\theta$ from a nucleus with mass M,
charge Z and ground-state spin J is given by:

$$\frac{d\sigma}{d\Omega} = \sigma_M \{F_L^2 (q) + (\tfrac{1}{2} + tg^2(\theta/2)F_T^2 (q)\} \tag{1}$$

where

$$\sigma_M = \text{Mott cross section} = (Z\alpha/2E_0)^2 \cos^2(\theta/2)/\sin^4(\theta/2)\frac{1}{\eta}$$

$$\eta = 1 + \frac{2E_0}{M} \sin^2(\theta/2)$$

In PWBA the longitudinal (L) and transverse (T) formfactors
depend on the momentum transfer q ($=2E_0\sin(\theta/2)$ only.

In terms of the different multipoles contributing to the magnetic
scattering:

$$F_T^2 = \frac{4\pi}{2J+1} \sum_{\lambda=1,odd}^{2J} |<J|\hat{T}_\lambda (q)|J>|^2 \tag{2}$$

where $\hat{T}$ is the magnetic multipole operator defined as follows:

$$\hat{T}(q) = \int d\vec{x}\, j_\lambda (qx)\vec{Y}^\mu_{\lambda\lambda 1} \cdot \vec{J}(\vec{x}) \tag{3}$$

$\vec{J}(\vec{x})$ is the nuclear current distribution. Hence from the experimental determination of $F_T^2$ one can derive the magnetization distribution of the nucleus.
When comparing these experimental data with theory, model assumptions have to be made to evaluate the magnetic multipole operator (formula 3). In the single particle (sp) model one can
rewrite formula 2 as follows:

$$F_T^2 (q) = \sum_{\lambda=1,odd}^{2J} N_\lambda^{1j} (\frac{qM\lambda}{m_p})^2 \{<j_{\lambda-1}>^{n1} + A_\lambda^{1j} <j_{\lambda+1}>^{n1}\}^2$$

where N and A are given constants for the sp model, $j_\lambda$ are the
radial integrals to be obtained from the radial wave functions
and M$\lambda$ is the static multipole moment.

It is well known that the Coulomb interaction (rising with Z)
dominates in most cases the magnetic interaction which is due to
nucleons in the valence shell. As can be seen from formula 1,
only at backward angles one is able to determine experimentally
the magnetic part of the cross section. At low energies (<200MeV)
one is even obliged to measure at 180°. At higher energies also
"normal", say >150°, backward angles experiments will allow to
extract this information.
From combining low and high energy experiments one then yields
the formfactor $F_T^2$ for a sufficiently extended region of the
momentum transfer q to map out the contributions from all multipoles. An example is shown in fig. 6.

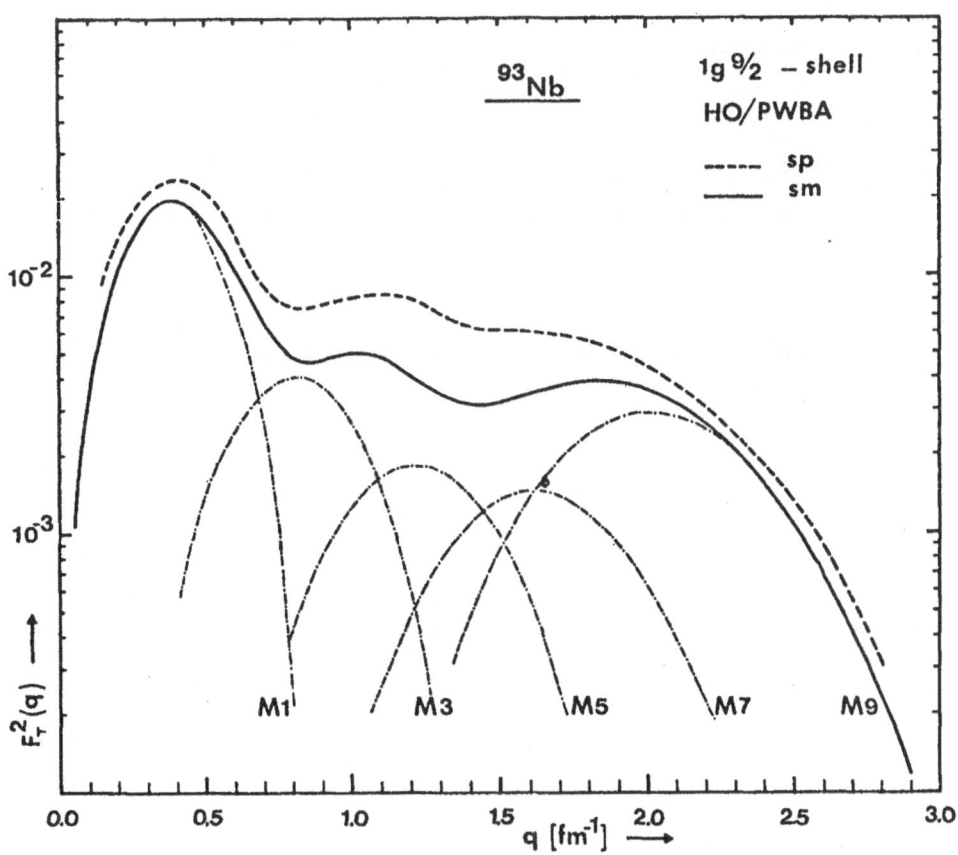

fig. 6

The formfactor $F_T$ is presented in PWBA for the nucleus $^{93}$Nb, the magnetization of which is caused by the odd 1g9/2 proton. Clearly one sees the contributions of the different multipoles (see formula 2), which contributions are calculated in the single particle model using HO wave functions. Most striking is that the contributions peak at different q-values and that the height of all maxima is of the same order of magnitude.

Hence accurate information for all magnetic multipoles can be obtained from magnetic moment scattering if one covers a sufficiently large q-region.

This is in contrast with the observed behaviour of Coulomb multipoles, where for instance the quadrupole contribution effects can be observed only, if at all, in the minima of the monopole contribution.

It has been said before that in the lower region of q, 180° scattering experiments are requested. Fig. 7. show the experimental equipment at Amsterdam, especially designed for this purpose[5].

Recent developments in this field are the coverage of an extended q-region by the data obtained at Saclay in a Saclay-IKO collaboration. At an angle of 155° and at energies ranging from 170 MeV to 270 MeV results were obtained for $^{51}$V, $^{59}$Co (ref.10.) $^{93}$Nb (ref. 11.) and $^{209}$Bi. These data are presently analyzed in conjunction with the low energy data obtained at Amsterdam for $^{51}$V and $^{59}$Co (ref.9.) and for $^{23}$Nb (ref. 12.). Fig. 8. shows the preliminary result of such an analysis for $^{93}$Nb.

Fig. 9. shows recent results obtained at Saclay.

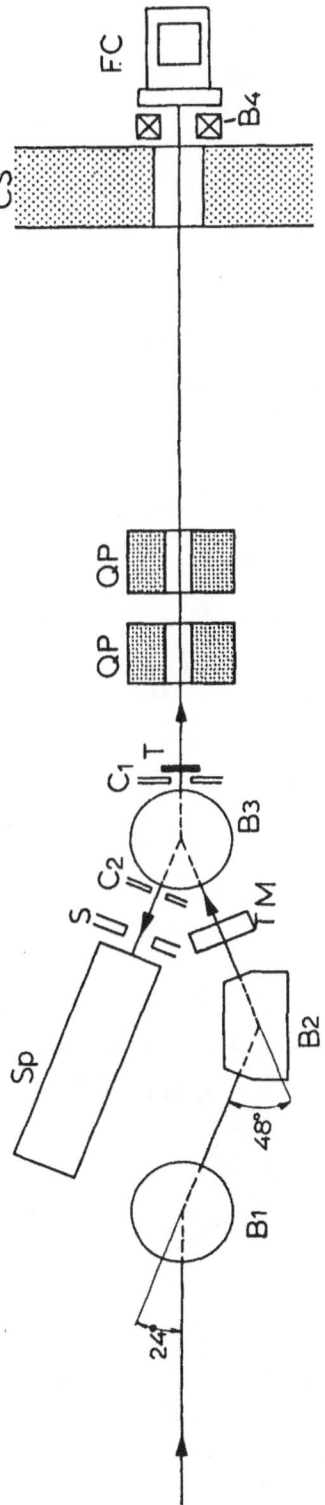

fig. 7

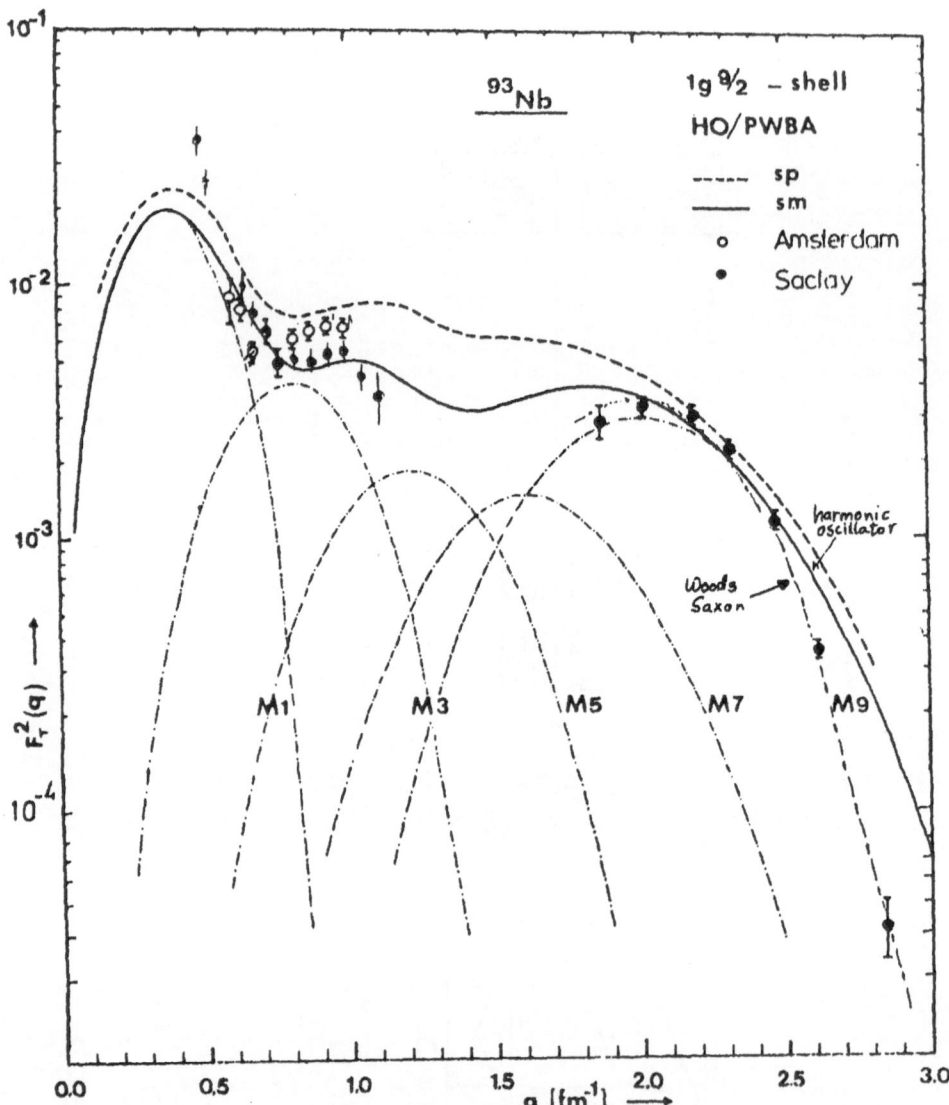

fig. 8

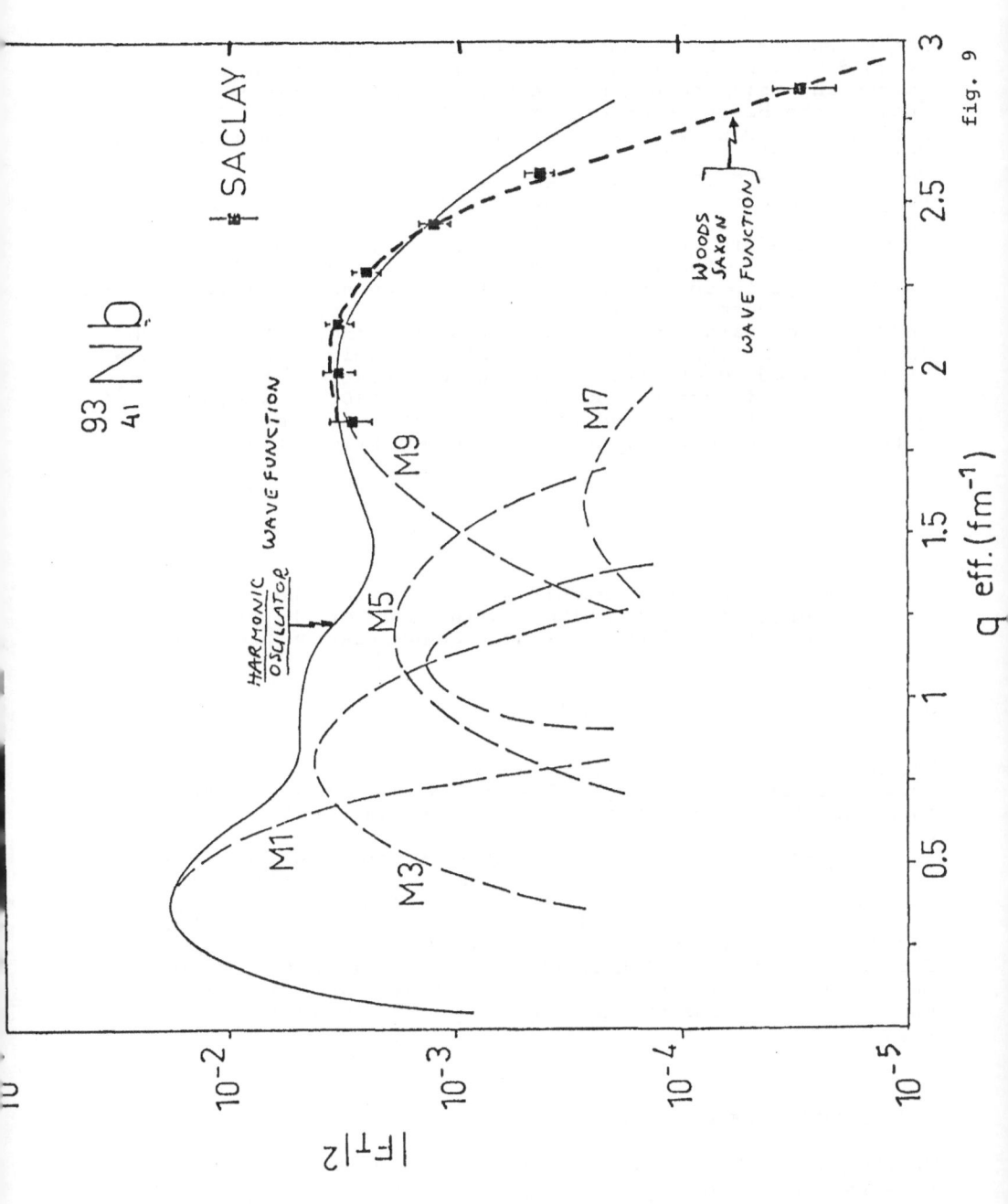

fig. 9

Due to special techniques applied to the Amsterdam 180° data such as to diminish background and to determine the remaining charge scattering using comparison measurements from neighbouring spin 0 nuclei, one has been able to measure the magnetization distribution for nuclei with low Z as well as for medium heavy nuclei (up to $^{115}$In).

Results in terms of nuclear ground state magnetization distribution parameters have been published for $^7$Li (ref.6.), $^9$Be and $^{13}$C (ref.7.), $^{27}$Al (ref.8.), $^{45}$Sc and $^{51}$V and $^{59}$Co (ref.9.). I will therefore refrain here from presenting these results and limit myself to the remark that from these low energy experiments one was able to obtain precise information not only on the magnetic dipole moment but also on the magnetic octupole moment. Furthermore, data have been obtained for $^{25}$Mg, $^{63,65}$Cu, $^{93}$Nb and $^{115}$In. Analysis is in progress. Preliminary results have been obtained. Most striking is the observation of the rising of the magnetic cross section for $^{39}$K at 180° as a function of q, which can be explained to be the result of a destructive interference of orbital and spin magnetization at q = o.†

†      Namely in the sp model one can ascribe the nuclear magnetization distribution in terms of the orbital current $\vec{j}$ and the intrinsic magnetization $\vec{\mu}$ of the odd nucleon.

The highest q-points (at 2.5. and 2.7 $f^{-1}$) can be seen to be very sensitive to the type of wave functions used and therefore illustrate how powerful a tool magnetic scattering is to determine single particle wave functions.

It is clear from the above remarks that magnetic elastic electron scattering is the only way so far to obtain experimental information on higher order magnetic moments and that information on single particle wave functions can be obtained in a rather straight forward manner.

Further experiments of this kind are to be expected at intermediate energy machines at backward angles including 180° at which angle the information is obtained in a relatively clean way.

B.  PLANNED FUTURE EXPERIMENTS.

In the following I have indicated in a brief survey what type of experiments can be undertaken with the electron- and photon beams from an intermediate energy machine like the Amsterdam accelerator. By no means do I claim that all of these experiments are planned for the intermediate future. However, such a survey is needed in order to design the very costly experimental equipment as versatile as possible.

## B1. Single channel electron scattering experiments.

### B1a. *Introduction.*

The new high duty-cycle, high-intensity medium-energy accelerators, such as MEA will allow a vast improvement in electron beam quality. Further important advances will stem from the incorporation of the energyloss  technique and other instrumental innovations (multiwire proportional chambers, forward-angle spectrometer). A significant breakthrough in the data analysis has been the development of the model-independent analysis (MIA), which has the advantage that a comparison between theoretical calculations and experimental results becomes more meaningful. On the other hand ther improved experimental accuracy requires more precise estimates of the contributions  from short-range correlations, exchange currents, virtual nuclear excitations and the relativistic motion of the nucleons.

### B1b. *Elastic scattering from the charge distribution.*

Experiments should be concentrated a) on lighter nuclei where theoretical calculations yield more trustworthy results, b) on accurate measurements at large q-values which yield information on the fine structure of the charge distribution, on short-range correlation effects and on dispersion effects (by comparing the behaviour of the tail of $\rho$ (r) determined from the experiment with theoretical expectations. Emphasis should be given to scattering experiments from even-even nuclei, where no disturbing contributions due to higher-order multipole moments are present.

### B1c. *Elastic scattering from the magnetization density distribution.*

Here, the higher beam intensities to be expected will yield a significant improvement. A fixed-angle spectrometer might be needed to yield a more accurate normalization. Up to $q \approx 2.5 \text{fm}^{-1}$ a $180^0$ scattering facility will be mandatory, at higher q-values Rosenbluth separation can be used. The separation of the contributions from different multipoles is easier but still causes some model dependence of the results.

### B1d. *Inelastic scattering.*

This very extensive field will benefit significantly from the higher beam intensities and higher resolution. This will make it possible to study the formfactor behaviour accurately over a large q-range. A model-independent analysis of such data will not only yield more reliable values for spectroscopic properties, such as transition probabilities, but also allow sensitive tests of nuclear models. Some interesting subjects are:

> 1) Deformed nuclei. Analysis of data from the ground-state rotational band (including elastic scattering) yields information on the intrinsic deformation.

2) The giant resonance region. Important features are
   the giant E2 and magnetic resonances, separation
   of spin-isospin components, and isospin splitting.
   A study of very high spin states will yield information which can be compared directly with theory
   since these states are expected to be nearly pure
   particle-hole configurations.
3) Sum rules. These will yield a direct estimate of the
   effect of exchange currents.


## B2. Coincidence electron scattering experiments.

B2a: *Introduction.*

The technique of (e,e'x) coincidence experiments allows the study
of the single-particle aspects of nuclear structure. The determination of properties of nucleons bound in the nucleus, directly
tests the predictions of the shell model. Such experiments have
yet been carried out only for a few, mostly light, nuclei and
with limited experimental resolution and statistics.
The coincidence cross section is measured as a function of the
angle and momentum of the emitted particle x and of the scattered
electron e'. Using the Plane Wave Impulse Approximation one can
extract from the cross sections the complete spectrum of hole states
in the target nucleus as a function of missing momentum s and
missing energy $\varepsilon$. The dependence on s directly yields the momentum distribution of the particle x in a shell model orbital of
the target nucleus, whereas the single particle energies of
these orbitals are determined from the positions of the peaks
in the energy spectrum. Since the corresponding hole states in
the residual nucleus are in general not eigen states, these peaks
can have a non-zero width, related to the strength of the residual interaction.
The competing reaction (p,p'x) yields larger cross sections, but
here all particles in the in- and outgoing channels are distorted
or partially absorbed by the optical potential of the nucleus.
Our incomplete knowledge of the strong interaction and especially
of its offshell behaviour makes the analysis much more complicated
than that of (e,e'x) reactions, where only one hadron takes part.
An accurate determination of the small coincidence cross sections
requires a high duty factor f, since the ratio real/random coin-
cidences is proportional to f. In the following we only list
some experiments which are particularly relevant from a theoretical point of view.
Obviously a large number of experiments is already possible by
simply choosing different target nuclei and outcoming particles x.
This kind of systematic research should certainly be undertaken
and it will yield interesting results. However, we do not mention
these experiments explicitly in the research proposed below.

B2b. *(e,e'p) experiments.*

1) On very light nuclei.

Experiments on the very lightest ($A \lesssim 4$) nuclei are theoretically
the easiest to interpret. The tri-nucleon problem and three-body

forces can be studied by a comparison of (e,e'p) reactions on $^3$He and $^3$H . The reaction $^2$He(e,e'p)n would yield detailed information on the electro-magnetic formfactors of the neutron. Obviously these experiments require the design and construction of gas targets.

2) On light nuclei.

Earlier experiments on light nuclei ($6 \leq A \leq 58$) have been performed with a resolution in missing energy down to 1 MeV. Improved resolution may allow to distinguish decay to different states in the residual nucleus.

3) On heavier nuclei.

Extension of the (e,e'p) experiments to heavier nuclei could yield results comparable with those of pick-up reactions like (d,$^3$He). The variation of the binding energy of 1s states with nuclear mass is sofar unknown for heavy nuclei. The importance of deformed target nuclei or polarised targets should be studied in more detail.

4) Special effects.

a) The validity of radiative corrections and corrections for the distortion of the outgoing proton can be tested by using the freedom given by the kinematics. b) A measurement of the energy spectrum at high values of the missing energy can yield information on short-range correlations and exchange currents in the nucleus. c) The results of (e,e'p) experiments yield the distribution of the individual protons in momentum space. By summing over all protons and Fourier transforming to coordinate space one can compare these results with those of "ordinary" elastic electron scattering. d) Other (energy weighted) sum rules may be tested directly by comparing integrated coincidence cross sections with theoretical estimates. e) The determination of occupation probabilities from coincidence spectra which has already yielded surprising results for 1p-shell nuclei, should also be undertaken for heavier nuclei.


B2c. *(e,e'n)experiments.*

Comparison of the results of (e,e'n) and (e,e'p) experiments gives direct information on the difference between proton and neutron distributions and on Coulomb energy differences in nuclei. An experimental problem is the detection of neutrons in the presence of many other particles.

B2d. *(e,e'd) and (e,e'α) experiments.*

The clustering of nucleons inside the nucleus has been studied with other methods. Although the theoretical interpretation of (e,e'd) and (e,e'α) reactions is not yet clear, these experiments might yield valuable new information on correlations. The same experimental apparatus can be used as for (e,e'p) experiments, but the background may be a problem.

B2e. *(e,e'np) and (e,e'2p) experiments.*

These or other triple coincidence experiments are rather difficult to perform, but they are, e.g. useful to study correlations
in the initial or final state. A careful comparison with (γ,np)
and (γ,2p) reactions should be made before initiating this type
of experiments.

B2f. *(e,e'π) experiments.*

Electroproduction of pions on the proton yields in principle information on the electromagnetic structure of the proton, the
neutron, the pion and even the (3-3) resonance, if one is able
to determine partial cross sections for each of the different
reaction mechanisms. The neutron structure is not easy to
determine in this way, but the information is obtained indepen-
dently of the deuteron structure which complicates the analysis
of electron-deuteron scattering.
The determination of the total cross section for charged pion production  on nuclei can also yield information on nuclear structure. From $\pi^0$ production the mass distribution in the nucleus
can be determined.


B3. <u>Photon induced reactions.</u>

In photon induced reactions the interaction takes place through
the absorption of a real photon, whereas electron scattering is
governed by the transfer of a virtual photon. This has among
others two important implications. In the first place, the energy
balance determines the momentum transferred to the nucleus with
the consequence that one particle less has to be observed in a
study of break-up andproduction processes than in electron induced
reactions. Secondly, irrelevant processes such as the radiative
tail in electron scattering play no rule. Thus, a photon beam
with a reasonably small energy spread is a desirably facility.
There are several possible solutions for the production of a
quasimonochromatic photon beam in the 500 MeV end station.
1). A conventional bremsstrahlung facility yields an effective
energy spread of a few MeV, from subtracting the data obtained
with a medium-Z conversion target from those obtained at a
slightly larger end-point energy. 2) With a positron converter
in the end station, followed by a magnetic energy selection
channel, one can obtain a photon beam with an energy spread
of 2 MeV or less through positron annihilation in flight. This
solution has the disadvantage that the maximum usable photon
energy is only about  one third of the primary electron energy,
that the intensity of the photon beam is quite low and that the
photon beam is contaminated by bremsstrahlung induced photons
which necessitate a subtraction run. 3) In a tagging facility
one detects the electron which has produced the photon in a
bremsstrahlung converter in coincidence with the photon induced event(s). Although in principle an extremely good resolution (of the order $10^{-3}$ or even better) can be obtained in this
way, one has abandonned one of the advantages of photons, i.e.
that one can suffice with the detection of one particle less.

4) Finally, the optimal solution is to use a positron-conversion target at the beginning of the accelerator, then accelerate the positron beam and finally create a photon beam through annihilation in flight. However, this solution had to be dropped from the present accelerator proposal because of budgetary reasons. At present the most acceptable solution is either a tagging or a bremsstrahlung facility in the end station. Nevertheless, the possibility to install a positron converter at some place along the accelerator should be considered as a future solution. Possible experiments with photon beams at energies above 100 MeV involve the detection of knock-out nucleons or of pions. Because a photon beam produces much less background than an electron beam, solid state or scintillator detectors can be used close to the target covering a large solid angle. Photon induced knock-out reactions yield information on short-range ground-state correlations in nuclei due to the fact that conservation laws prevent a free nucleon from absorbing a photon. Generally speaking a study of photon induced reactions yields the same type of information as the coincidence experiments described in section 2.2. However, the expected intensities of photon beams are much smaller than those of the electron beams. Nevertheless, this disadvantage will be more than compensated by the special characteristics of the photo-absorption mechanism in certain types of experiments.


C. DESIGN OF INSTRUMENTATION FOR INTERMEDIATE ENERGY PHYSICS.

From the above given survey of experiments which can be executed with an intermediate energy electron accelerator one can derive that the instrumentation should contain the following equipment:

1. A high resolution electron spectrometer.
   Considering the maximum energy of the machine and the request for resolving spectrum peaks at a level density of, say, 50keV or less the resolution should be better than one part in $10^4$.

2. A "hadron" spectrometer.
   This instrument should be used also in conjunction with the electron spectrometer for coincidence experiments at a resolution of one part in $10^3$

3. A highly accurate beam deflecting system.
   In addition to the usual requirements of beam deflecting system - accurate definition of the primary beam energy, control over the energy spread of the beam at the target by means of slits, purifying the beam at the target by means of deflecting the beam away from the accelerator center line and the use of shielding walls - one is presently obliged to design the beam handling system with

4. A dispersion matching system.
   Due to the very low cross sections one deals with at higher energies, conventional techniques of obtaining a small $\Delta E/E$ by means of slits - at the expense of a vast amount of

primary current - are no longer adequate. One should introduce a dispersion matching system. Such a system allows to
use large currents spread out over a line-type focus at the
target if a very accurate energy-position correlation can
be achieved. In other words, even for an extended $\Delta E/E$ of
the primary beam at the target (say 0,5%) - as long as there
is a very accurate energy-position correlation within the line
focus - an overall resolution in the order of, say, $1 : 10^4$
can still be achieved.

5. A multi-channel detection system in both spectrometers.
   Due to the very low cross sections involved and due to the
   expenses related with intermediate energy physics a multi-
   channel counter array has to be used in the focal plane of
   the spectrometers.

6. Elaborate shielding arrangements.
   Even more than for low energy electron scattering equipment
   the demands on shielding are severe for intermediate energy
   physics due to the combination of extremely low cross sections
   and intense background radiation related with the high-
   energy and high intensity of the beam.

Obviously many more components - like different type of monitors,
target chamber and beam stopper - have to be incorporated but
I will refrain here to discuss them. Rather I will elaborate
on the above mentioned pieces of equipment as they are presently under construction at Amsterdam.
Before doing so, let me briefly describe what kind of equipment
is presently in use at Saclay[13] and MIT[14].
At SACLAY the most sophisticated equipment in use for electron
scattering contains a 600 MeV/c large solid angle (6.5 mster)
moderate resolution ($\sim 1 : 10^3$) Elbeck-type spectrometer and a
900 MeV/c high resolution ($< 2 : 10^4$) magic angle ($169^0.7$) spectrometer. The field indices for the spectrometers are $n = \frac{1}{2}$,
$\beta = 3/8$ and $n = \frac{1}{2}$, $\beta = 1/6$, respectively. Interference of the
two elaborate detection systems and their shielding construc-
tions is avoided by bending the scattered particles upwards
and downwards (into a pit) respectively.
High resolution ($1-2 : 10^4$) spectra have been measured under extremely beautiful condition of background reduction: cross
sections as low as $10^{-38}$ $cm^2$/ster could be detected. This is to
a large extend due to the heavy iron shielding around the focal
planes.
No dispersion matching system is available mainly due to the
fact that such systems have only become operational at other
electron scattering facilities (Darmstadt, MIT, Mainz) at a
much later stage than the Saclay equipment. Hence the high
resolution spectra achieved at Saclay are obtained in the normal
point to point focussing operation of the spectrometer (at the
expense of primary current)
At MIT one has installed a very large homogenuous field $90^0$
dipole magnet with entrance and exit curvatures. The focal
plane is curved in both directions. The spectrometer has achieved an extremely high resolution ($0.5 \times 10^{-4}$) in the so-called
line-to-point focussing mode which requests a dispersion matching
system as part of the beam handling system. The requested $90^0$

rotation of the disperion plane at the target has been achieved
with a set of 5 quadrupoles. No second spectrometer is available
so that coincidence experiments at MIT are not possible so far.

After careful studies of the above mentioned and other spectrometers as well as the physics requirements set forward under B
the Amsterdam group decided to design a combination of a
Los Alamos energy-loss type[15] QDD spectrometer for the "electron"
arm and a QDQ spectrometer for the "hadron" arm. Both spectrometers will bend the particles upward. This has the drawback of
forsaking the advantage of a pit for background rejection.
Considering, however, the costs involved for such a pit under
Amsterdam soil conditions it was felt mandatory to follow this policy.
The most important parameters of both spectrometers are presented in Table III.

Table III.

| | Electron spectrometer | Hadron spectrometer |
|---|---|---|
| type | QDD | QDQ |
| max. momentum | 550 MeV/c | 750 MeV/c |
| max. field | 13.1 kG | 13.9 kG |
| mean radius | 1.40 m | 1.80 m |
| deflection angle | $2 \times 75^0$ | $90^0$ |
| gap | 7 cm | 14.4 cm |
| momentum acceptance | 10% | 10% |
| intrinsic acceptance | $<1 \times 10^{-4}$ for $\frac{\Delta p}{p} = 2\%$ <br> $<1 \times 10^{-3}$ for $\frac{\Delta p}{p} = 10\%$ | $10^{-3}$ |
| transverse-plane focussing | point-point | point-point |
| median-plane focussing | parallel-point | point-point |
| dispersion | 6.82 cm/% | 8.27 cm/% |
| focal-plane angle | $42^0$ | $50^0$ |
| solid angle | 5.6 msr | 17.6 msr |

82

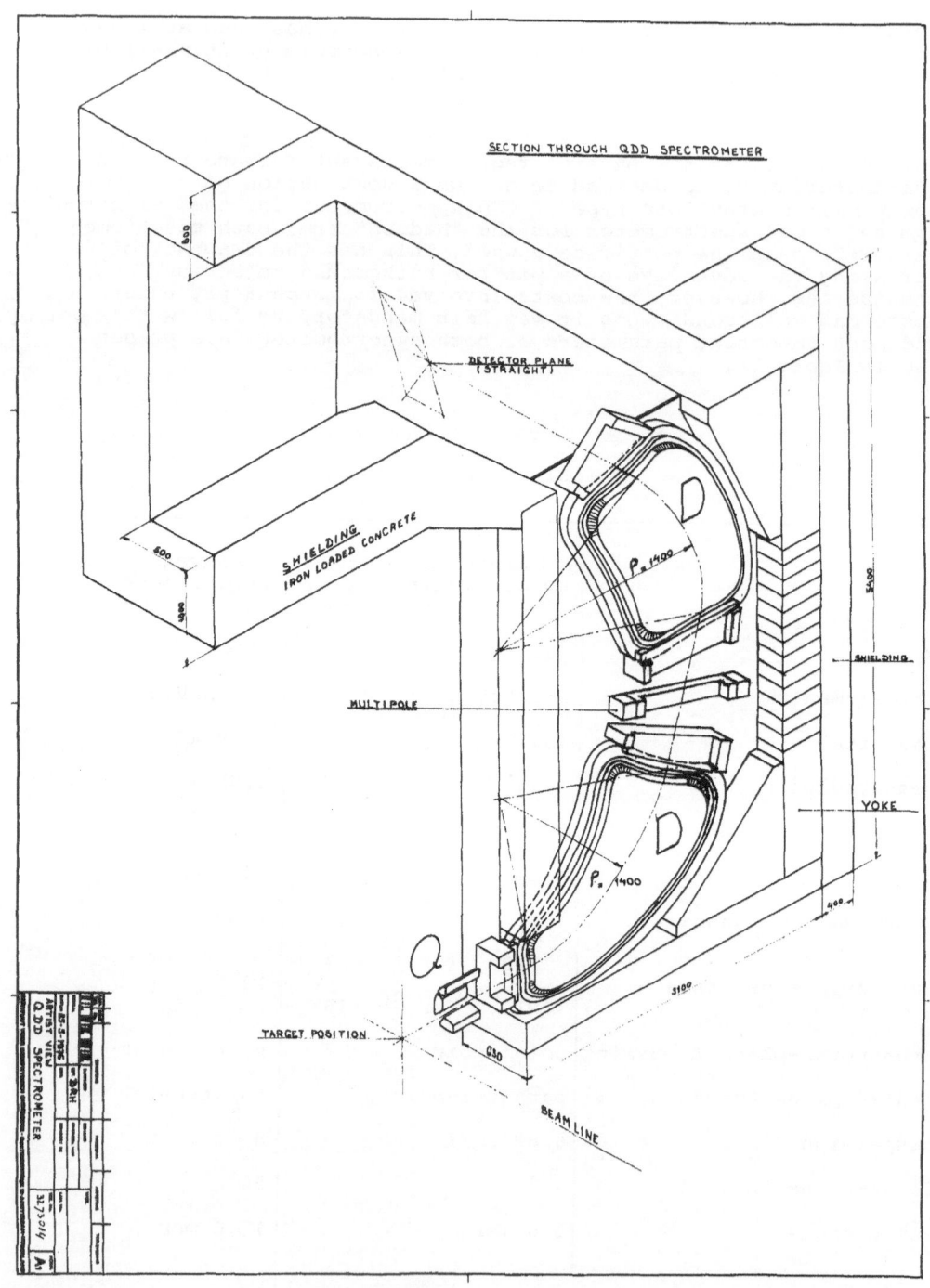

fig. 10

83

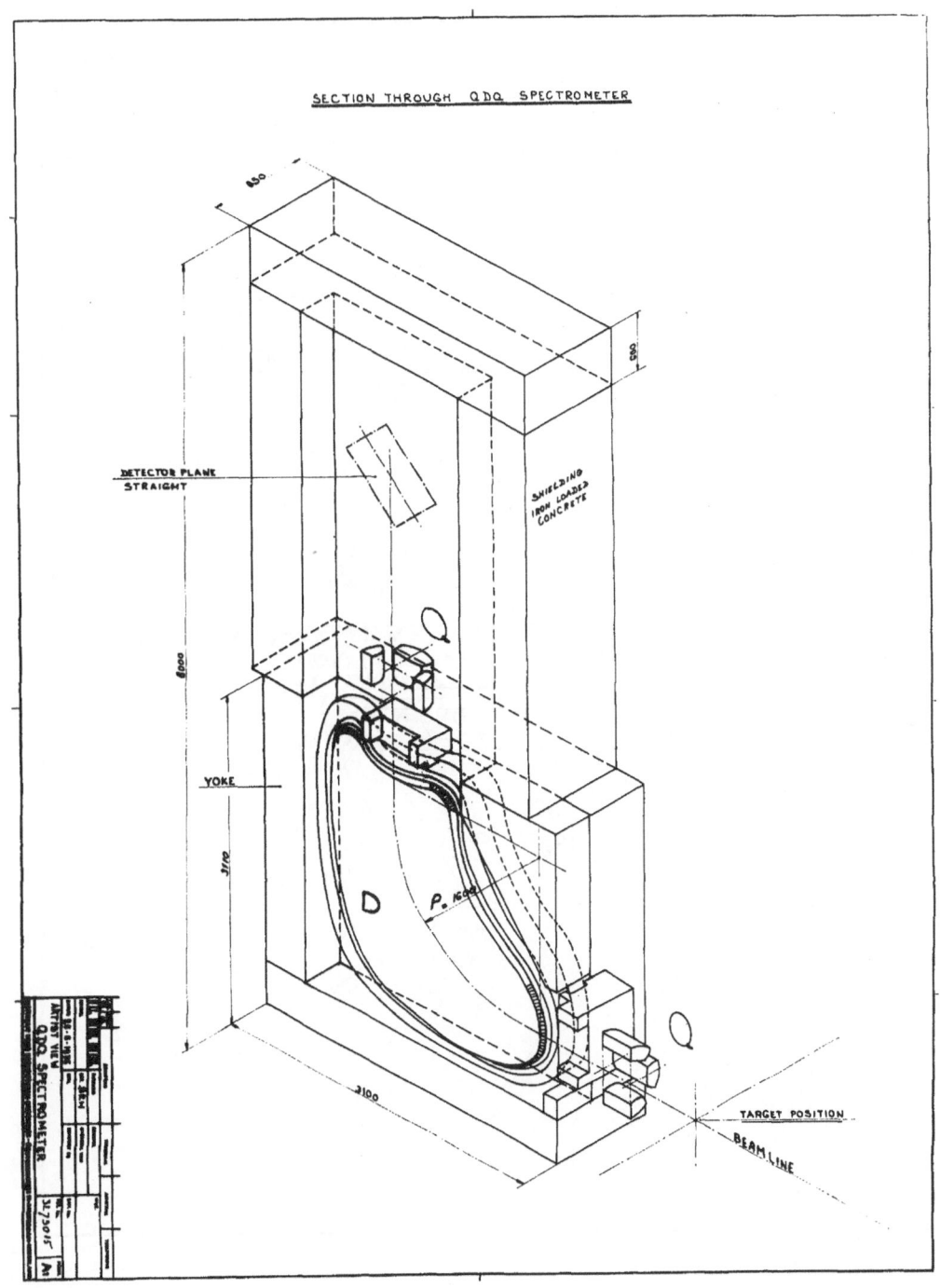

fig. 11

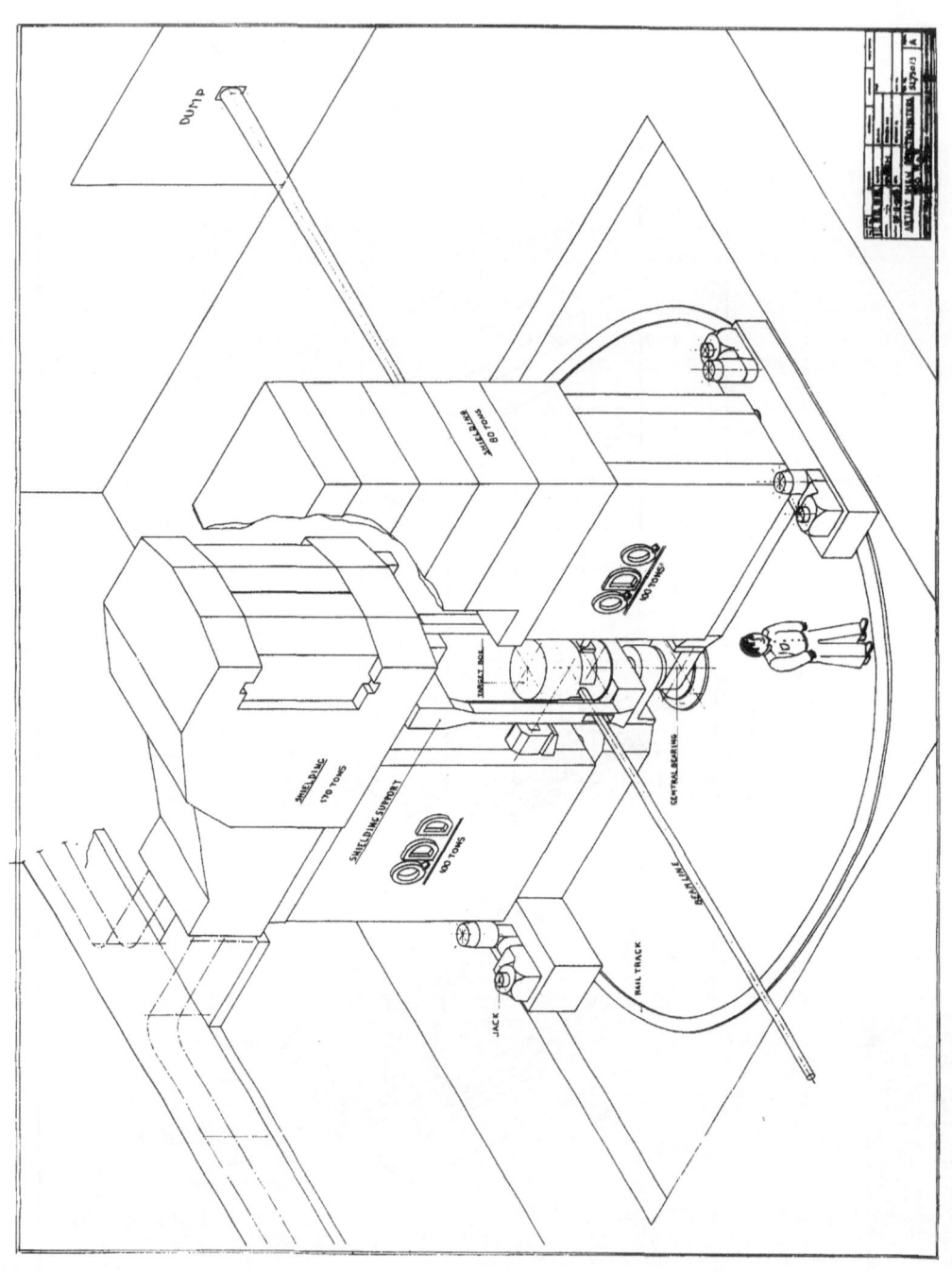

fig. 12

It is gratefully acknowledged that Prof. Enge (MIT) has attributed
to a large extend to the optical study involving among others
the proper shape of entrance and exit faces for higher order
corrections. One important achievement has been that for both
spectrometers a straight focal plane showed to be feasible which
is important for large detection system arrays.
The numerical data indicated in Table III are self explaining.
The technical lay-out on the basis of these specifications is
under study now. Fig. 10, 11, 12 show artist's view on the
QDD, the QDQ and the combined set, respectively. At this stage
it is too early to quote any constructional details.

However, it is interesting to point out the IKO design for
the support of the very heavy shielding (iron-loaded concrete)
around the detection plane of the QDD. It was considered un-
acceptable to have this (150 ton) shielding hanging on the spectrometer construction itself. Therefore it was decided to have
a separate bearing - mounted well above the beam line - for
the shielding construction which obviously has to rotate syn-
chronously with the spectrometer. Fig. 12. clearly shows how
this bearing is supported and how it is correlated with the
central bearing for the two support constructions of the spectrometers themselves.

The optical design of the beam handling system has been completed.
We would like to acknowledge the expertise from Dr. K. Brown,
who has helped considerably in the final stage of the design to
improve its quality especially with a view on the ease of tuning
it properly according to the stringent conditions requested for
line-type focussing at the target. The system indicated in fig.13
involves:
a) a 90° deflecting system which delivers a monochromatic beam
   either into the radio-chemistry hall or into the radio
b) a second 90° deflecting system which has the property of de-
   livering a beam either in the normal mode (point focussing)
   or in the energy-loss mode (line focussing) into the electron
   scattering hall
c) a rotator system (5 quadrupoles) for rotating the dispersion

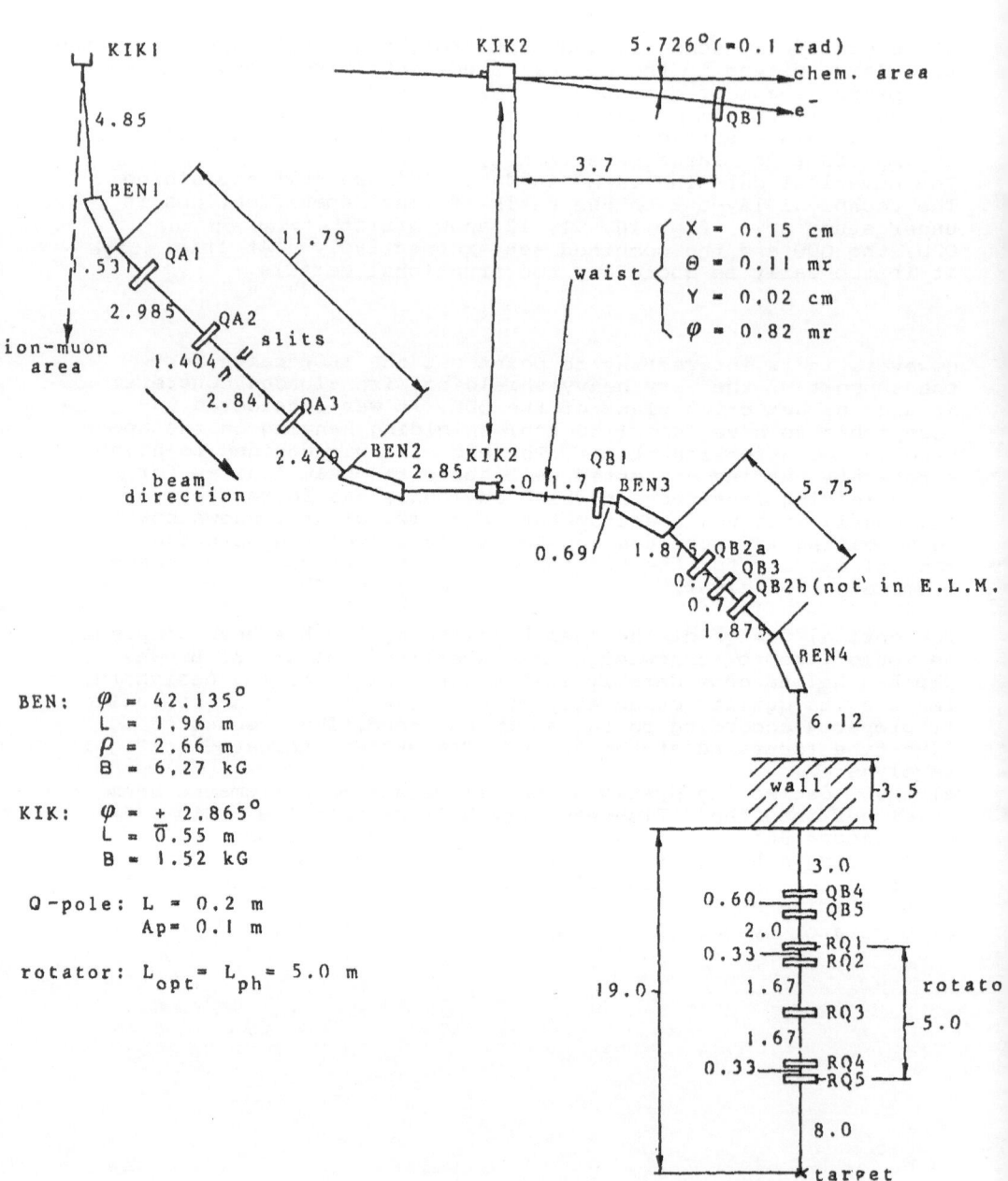

fig. 13

plane by 90° as is requested for the vertical plane bending
spectrometer.
d. Two switching magnets for easy of accommodating in an alter-
nating way different experimental halls on a short term
basis.

The magnetic components and their powersupplies for the switchyard and the rotator have recently been ordered. Installation
will take place before summer 1977. Considering all the other
components required the expectation is that the beam handling
system will be completed early 1978. Although the contract for
the spectrometers has not been placed yet the expectation is
that construction can start fall this year. With a delivery
time of about 2 years this means that early 1979 one can expect
first trial runs with the complete system. This means of course
that also the detection systems should be then available.
The present status of activities in this direction is that the
construction is underway of a multiwire proportional counter
(64 wires) to be installed in the focal plane of the 140 MeV
station spectrometer. Initial tests will take place in the course
of 1977. The present technology has shown that with 2 mm spacing
of the wires and the application of drift techniques a position
accuracy of one quarter of the wire distance can be achieved.
Under these conditions the dead time of the counters is
$\sim$ 20 nano sec. This means in relation to the specific beam conditions of the IKO accelerator that counting rates of 25 per
burst can be obtained under a 1% dead time correction condition.
These specifications are sufficient to match both the resolution requirement ($\sim 5 \times 10^{-5}$) in the focal plane of the QDD as
well as the fast counting condition.

The author wishes to thank all members of the electron scattering
group in Amsterdam for their invaluable contribution to the
ongoing physics as well as for their enthousiastic and able
support requested to build a new facility for intermediate
energy physics. These members are: J.C. Bergström (on sabbatical leave from Saskatoon, Canada), J.E.P. de Bie, G. Box,
P.J.T. Bruinsma, J.H.F. Distelbrink, R.S. Hicks, C.W. de Jager,
L.Lapikás, R.Maas, J.G. Noomen, H. de Vries, H.v.d.Watering and
P.K.A. de Witt Huberts.

Thanks are due to Prof. A.H. Wapstra for his continuous interest.
This work is part of the research program of the Institute for
Nuclear Research (IKO), made possible by financial support from
the Foundation for Fundamental Research on Matter (FOM) and
the Netherlands Organization for the Advancement of Pure
Research (ZWO).

References.

1. P.J.T.Bruinsma, J.G. Noomen, C. de Vries, "EVA, the 85 MeV
   linear electron accelerator at Amsterdam", Nucl. Instr.
   74(1969)1.

2. C. de Vries and P.J.T. Bruinsma, "The 100 MeV electron scattering
   facility at Amsterdam". Nucl. Instr. 74(1969)5.

3. C.W. de Jager, F.Th. Douma, P.J.T. Bruinsma, C. de Vries, "The
   magic-angle electron spectrometer at Amsterdam", Nucl. Instr.
   74(1969)13.

4. P.K.A. de Witt Huberts, H. de Vries, G.J. van Niftrik,
   G.A. Peterson, "An overlapping scintillator detection system
   for electron scattering", Nucl. Instr. 74(1969)27.

5. G.J.C. van Niftrik, H. de Vries, L. Lapikás, C. de Vries,
   "A three-magnet system for measurement of electron scattering
   through $180^0$", Nucl.Instr. 93(1971)301.

6. G.J.C. van Niftrik, La.Lapikás, H. de Vries, G. Box,
   "Magnetization distribution of the $^7$Li nucleus as obtained from
   electron scattering through $180^0$. The electric quadrupole
   moment of $^7$Li", Nucl.Phys. A174(1971)173.

7. L. Lapikás, G. Box, H. de Vries, "Magnetic elastic electron
   scattering from $^9$Be and $^{13}$C", Nucl.Phys. A253(1975)324.

8. L.Lapikás, A.E.L. Dieperink, G, Box, "Elastic electron scattering
   from the magnetization distribution of $^{27}$Al",Nucl. Phys.
   A203(1973)609.

9. H. de Vries, L. Lapikás, G. Box, G.J.C. van Niftrik, "Results
   on elastic $180^0$ scattering from $^{45}$Sc, $^{51}$V, $^{55}$Mn, $^{59}$Co",
   Phys.Letters 33B(1970)403 and Proc. Int. Conf. on Photonuclear Reactions and Applications, Asilomar(1973)179.

10. P.K.A. de Witt Huberts, J.B. Bellicard, B. Frois, M.Huet,
    Ph.LeConte, A.Nakada, Phan Xuan Ho, S.Turck, L.Lapikás,
    H. de Vries, I. Sick, "Magnetic elastic electron scattering
    of high multipolarity from $^{59}$Co", Proceedings of the June
    1976 Cracau conference on "Radial shapes of Nuclei", to be
    published.

11. P.K.A. de Witt Huberts, J.B. Bellicard, B.Frois, M.Huet,
    Ph.LeConte, A.Nakada, Phan Xuan Ho, S.Turck, L. Lapikás,
    H. de Vries, I. Sick, "Elastic electron scattering from the
    magnetization distribution of $^{93}$Nb, Phys. Lett. 60B(1976)157.

12. G. Box, "Elastic electron scattering from the magnetization
    distribution of $^{93}$Nb, $^{115}$In" , PhD thesis (1976) University
    of Amsterdam, to be published.

13. P. Barreau, J.B. Bellicard, P. Bounin, C. Grunberg. P.Leconte,
    J. Millaud, J. Mougey, A. Tomasso, "Two high performance
    spectrometers for electron scattering at the 600 MeV linear
    accelerator of Saclay (France)", International report
    C.E.N., Saclay, 1972.

14. S. Kowalski, W. Bertozzi, C.P.Sargent, "High resolution
    energy loss spectrometer design study", Proc. MIT Summer-
    study, MIT-2098(1967)39

    S. Kowalski, H,Enge, Proc. Int. Conf. on Magnet Technology,
    Brookhaven(1972)182.

    S. Kowalski, "High resolution spectrometers", Int. Conf. on
    Photonucl. reactions and applications" Asilomar(1973)227.

15. H.A. Enge, S.B. Kowalski, Proc. 3rd Int. Conf. on Magnet
    Technology, Hamburg(1970)366.

Figure captions.

fig. 1    Beam lines into the 140 MeV substation for radio chemis-
          try and electron scattering. The latter beam line
          is instrumented with equipment (including the spec-
          trometer) which has been utilized for ten years in
          conjunction with the 85 MeV accelerator (shut down as
          of Jan 76).

fig. 2    Lay-out of the intermediate energy physics facility
          under construction at Amsterdam.

fig. 3    Central beam line equipment of the 500 MeV linear elec-
          tron accelerator.

fig. 4    Picture of the 6 MeV injector of the new accelerator.
          In conjunction with the first 3.6 m accelerator section,
          20 MeV beam tests will be performed shortly.

fig. 5    The injector hall (foreground), first part of the modu-
          lator hall (on top of the underground accelerator
          vault) and the 140 MeV substation.

fig. 6    PWBA form factor for scattering from the ground-state
          magnetization distribution of $^{93}$Nb. The q-dependence
          of the different multipole form factors (dot-dash
          curves) is calculated from HO wave functions. The
          dashed curve gives the single particle (SP) form factor,
          the solid curve is obtained from the shell model (SM)
          calculation  D.H. Gloeckner, private communiaction.
          For clarity secondary maxima of the individual multi-
          pole contributions (dashdot curves) have not been drawn.

fig. 7    180$^0$ scattering equipment which has been used with the
          85 MeV linear electron accelerator (shut down as of
          Jan 76).

fig. 8    Preliminary analysis of the combined results obtained
          at low q-values (Amsterdam) and at high q-values
          (Saclay).

fig. 9    Preliminary analysis of only the high q-data obtained
          at Saclay. Notice the sensivity for the particular wave
          function used in the analysis.

fig.10    Artist's view of the QDD high resolution energy-loss
          type "electron" spectrometer (technical lay-out
          presently under study).

fig.11    Artist's view of the QDQ large solid angle "hadron"
          spectrometer (technical lay-out presently under study)

fig. 12  Artist's view of the combined QDD and QDQ spectrometers.
Clearly shown is the central pivot common for both
spectrometers and the static frame for the QDD focal
plane shielding. The bearing allowing rotation of the
shielding synchronuously with the spectrometer is mounted
on top of this static frame. For special purposes the
static frame can be rotated away from the vertical plane
through the beam line.

fig. 13  Lay-out of the beam deflecting system for the 500 MeV
accelerator. Indicated are a) switching magnets (KIK 1
and KIK 2), dipole bending magnets (BEN1,2,3,4),
quadrupole magnets (QA1-3, QB1-6), rotator quadrupoles
(RQ1-5). Distances are in m.

ELECTRONUCLEAR SUM RULES

D. Drechsel

Institut für Kernphysik

Universität Mainz

D-6500 Mainz

## 1. Introduction

Sum rules for electroexcitation are a generalization of the more familiar
sum rules of photonuclear physics. While sum rules have their origin in
prehistoric quantum mechanics, there has been a revived interest in this
field recently in connection with internal degrees of freedom of nucleons
and mesonic degrees of freedom, meson exchange, nucleon isobars and relativistic effects in nuclear structure.

By summing the experimental or theoretical transition probabilities over
the final states of the nucleus, the dependence on the final states may
be removed by closure relations. The remaining relations do not require
knowledge of the complicated and detailed properties of the excitation
spectrum, but concentrate on more fundamental aspects of nuclear physics,
such as internucleon correlation functions and the nuclear forces, particularly its exchange behaviour in the nuclear ground state.

The classical dipole sum rule of Thomas, Reiche and Kuhn[1] (TRK) was extended by Heisenberg[2] to include retardation effects and expressed in
terms of two-body correlation functions. Feenberg and Siegert[3] demonstrated first the effects of exchange forces on sum rules. Exchange forces have been reinvestigated in many theoretical contributions since
then.

The calculations had converged to a "canonical" additional 40% contribution of exchange forces, until the results of the Ziegler group[4] showed that the experimental sum rule, integrated to meson threshold, is of the order of twice the "classical" sum rule. Though some recent calculations indicate that exchange and tensor forces[5,6] lead to a further increase of the sum rule as required by the experiments, the situation is still under discussion[7]

Using causality and dispersion relations in connection with experimental data as well as assumptions on the high-energy scattering amplitude, Gell-Mann, Goldberger and Thirring (GGT)[8] obtained a value of about 140% of the classical dipole sum rules, again the "canonical" result. The puzzle that the experiments of ref.4 lie above this seemingly model-independent value was resolved by Weise[9]. He suggested that even at high energies a nucleus may not be considered to be an ensemble of free nucleons, as had been assumed by GGT, but that hadronic components in the photon propagator lead to considerable screening effects.

While the GGT-sum rule includes absorption of all multipoles and may be directly related to the experimental absorption cross section, the TRK-sum rule considers the non-retarded dipole absorption only. It had been shown by Gerasimov[10] that higher multipole contributions and dipole retardation cancel, if the scattering amplitude is analytic in the energy plane except for the usual cut along the real axis and, furthermore, the integrated total cross section converges. As shown by ref.7, there is considerable experimental evidence against the latter assumption. As far as the analyticity of the scattering amplitude is concerned, recent explicit calculations using non-relativistic models[11,12] have demonstrated that the cancellation of the Gerasimov theorem does not hold. In particular, non-relativistic models give rise to anomalous branch cuts at relativistic energies.

The different analytic structure of scattering amplitudes or form factors
in non-relativistic vs. relativistic theories has also been pointed out
by Friar and Fallieros[13] and by Woloshyn[14]. Explicit comparison of the
relativistic nucleon-nucleon bremsstrahlung amplitude in OBE-approximation
and the non-relativistic approximations indicate, that the precise treatment of the nucleon propagator or the wave function in the intermediate
states is essential and of much more importance than the usual relativistic corrections, such as vertex corrections and pair current[15]. The mentioned ambiguities and open problems should be kept in mind in any discussion of sum rules.

In the case of electron scattering, the preceding photonuclear sum rules
may be extended into the region of space-like momentum transfer to the
nucleus. For a review of previous work see the conference reports of
O'Connell[16] and Czyz[17].                        The kinematical variables for electroexcitation are explained in fig.1.

The connection of photo- and electronuclear sum rules is shown in fig.2a
for the transverse sum rule. The amplitude for electroexcitation is a
function of both excitation energy $\omega$ and momentum transfer q, restricted
only to a region of space-like momentum transfer ($|\underline{q}| > \omega$). Obviously,
the photonuclear results are recovered in the limit $|q| \rightarrow \omega$. Electroex-
citation sum rules are usually discussed at $\underline{q}$ = const., summing over all
excitation energies $\omega$. More general curves $\omega = \omega(\underline{q})$ may be of similar
physical interest, but are more difficult to calculate theoretically,
e.g., a summation along a straight line $\omega = \alpha|\underline{q}|$ (photonuclear sum ru-
le $\alpha$ = 1!) or along a curve $\omega^2 - q^2 = -\mu^2$ (virtual photons with constant
mass).

Different experimental techniques have been proposed, in particular:
1. "low-resolution" sum rules, keeping $\epsilon_i$ and $\theta$ constant and summing
   over all values $\epsilon_f$;

2. "high-resolution" sum rules, which determine the scattering amplitudes
   in the $(q, \omega)$-plane.

In the latter case, the experiments may be performed either

a) at constant $\epsilon_i$, varying $\theta$ and $\epsilon_f$ or

b) at constant $\theta$, varying $\epsilon_i$ and $\epsilon_f$.


There are two obvious reasons for the lack of systematic experimental investigations in electronuclear sum rules. For one reason the experiments are tedious, time-consuming and therefore not immediately as gratifying as an experiment on a discrete nuclear state. Secondly, the large bremsstrahlung background is known only in Born approximation, which limits the experiments to light nuclei and introduces, even in these cases, problems for hard photons. In both cases, however, the situation is comparable to photonuclear physics: The few measured integrated cross sections are the result of quite a few man-years of the Ziegler group and are limited in accuracy by the background of pair production.


Let me express at this point the hope that the enormous interest, which these few photonuclear numbers have met over the past years, may eventually trigger a corresponding large-scale investigation in electroexcitation. Such sum rules in the $(q, \omega)$-plane will ultimately give basic information on momentum and spatial distribution of exchange forces and internucleon correlations.

## 2. Correlation functions

For simplicity let us restrict the discussion for the moment to electro-excitation via the static Coulomb potential of the nucleus. In this case the differential cross section is proportional to the response function [18]

$$(2.1) \qquad R(q,\omega) = \sum_n |\rho_{no}(q)|^2 \, \delta(E_n - \omega) .$$

Neglecting relativistic and finite size effects in the charge operator, we have

$$(2.2) \qquad \hat{\rho}(q) = \sum_k e_k \, e^{-i q \cdot r_k} \quad \text{or} \quad \hat{\rho}(r) = \sum_k e_k \, \delta(r - r_k) ,$$

where $e_k$ is the charge of nucleon $k$. A sum rule is constructed by integrating eq. (2.1),

$$(2.3) \qquad S_0(q) = \int d\omega \, R(q,\omega) = \sum_n |\rho_{no}(q)|^2 .$$

With eq. (2.1) and the closure relation, we obtain

$$(2.4) \qquad S_0(q) = \sum_{k,l} \langle 0 | e^{i q \cdot (r_l - r_k)} | 0 \rangle .$$

Breaking the sum into direct ($k = l$) and interference ($k \neq l$) parts, we find

$$(2.5) \qquad S_0(q) = Z + Z(Z-1) \, f_2(q) ,$$

with

$$(2.6) \quad Z(Z-1)f_2(q) = \iint e^{i q \cdot (r-r')} \rho_2(r,r') \, d\tau \, d\tau'$$
$$= \sum_{k \neq l} \langle 0| \, e_k e_l \, e^{i q \cdot (r_k - r_l)} |0\rangle$$

and

$$(2.7) \quad \rho_2(r,r') = \sum_{k \neq l} e_k e_l \langle 0| \, \delta(r-r_k) \, \delta(r'-r_l)|0\rangle .$$

The two-proton density function $\rho_2$ gives the probability of finding one nucleon at $r$ and a second nucleon at $r'$. Similarly, spin and isospin correlation densities may be defined by introducing spin and isospin variables into eq. (2.7).

The sum rule (2.3) contains elastic (n = 0) and inelastic (n $\neq$ 0) contributions. Separating out the elastic contribution $|Z F(q)|^2$ we have

$$(2.8) \quad S_0(q) = Z^2 |F(q)|^2 + S_0^{inel.}(q)$$

with

$$(2.9) \quad S_0^{inel.}(q) = Z(1 - |F(q)|^2) + Z(Z-1)\left(f_2(q) - |F(q)|^2\right).$$

The first term in eq. (2.9) gives the uncorrelated contribution while the second term is the correlation function. If there is no correlation, $f_2(q) \rightarrow |F(q)|^2$ and the probability of finding two protons is simply the square of the one particle probability (in q-space).

The correlation function, $f_2(\underset{\sim}{q}) - F(\underset{\sim}{q})^2$, is determined by statistical ("Pauli") and dynamical ("hard core") correlations. In comparison with experimental data, also many-body currents ("meson exchange") will show up as deviation from the uncorrelated sum rule.

Since both $F(\underset{\sim}{q})$ and $f_2(\underset{\sim}{q})$ approach unity for vanishing momentum transfer, we have $S_0(0) = Z^2$, i.e., coherent elastic scattering in this limit. For momentum transfer going to infinity, the usual assumption has been that only the terms $k = l$ survive in eq. (2.4) and hence $S_0(q \to \infty) \to Z$, i.e., that incoherent scattering dominates. Such an assumption means that the nucleus behaves as an ensemble of Z independent protons and neglects shadowing effects, which have been found to be necessary in reconciling the GGT sum rule with the experimental data[9].

For the Fermi gas model (Fermi momentum $k_F$) the sum rule may be evaluated analytically:

$$(2.10) \quad S_0(q) = \begin{cases} Z^2 & \text{for } q = 0 \\ Z\left(\frac{3}{4}\frac{q}{k_F} - \frac{1}{16}\left(\frac{q}{k_F}\right)^3\right) & \text{for } 0 < q \leqslant 2k_F \\ Z & \text{for } q > 2k_F. \end{cases}$$

Note that in this case elastic scattering occurs only at $q = 0$ and the statistical correlations reduce the inelastic contribution in an intermediate region to less than the no-correlation value. For $q > 2\,k_F$, no correlations occur and the "asymptotic" incoherent value Z is obtained. Fig.3 shows the results for a Fermi gas as compared to a typical result for a finite nucleus.

Relations similar to eq. (2.5) may be obtained for the transverse contributions:

$$(2.5) \quad S_o^{Coul}(q) = Z + Z(Z-1)\, f_2(q)$$

$$(2.5\,a) \quad S_o^{conv}(q) = \frac{2}{m^2}\left\{ \tfrac{1}{3} Z\langle p^2\rangle + Z(Z-1)\, f_2^{conv}(q) \right\}$$

$$(2.5\,b) \quad S_o^{magn}(q) = \frac{q^2}{2m}\left\{ Z\mu_P^2 + N\mu_n^2 + A(A-1)\, f_2^{magn}(q) \right\},$$

where

$$(2.11) \quad Z(Z-1)\, f_2(q) = \sum_{k\neq\ell} e_k e_\ell \,\langle o|\, e^{i\vec{q}\cdot(\vec{r_k}-\vec{r_\ell})}\,|o\rangle$$

$$(2.11\,a) \quad Z(Z-1)\, f_2^{conv}(q) = \sum_{k\neq\ell} e_k e_\ell \,\langle o|\, p_{k,x}\, p_{\ell,x}\, e^{i\vec{q}\cdot(\vec{r_k}-\vec{r_\ell})}\,|o\rangle$$

$$(2.11\,b) \quad A(A-1)\, f_2^{magn}(q) = \sum_{k\neq\ell} \mu_k \mu_\ell \,\langle o|\, \sigma_{k,x}\, \sigma_{\ell,x}\, e^{i\vec{q}\cdot(\vec{r_k}-\vec{r_\ell})}\,|o\rangle.$$

In eqs. (2.5) and (2.11), m is the nucleon mass; $e_i$, $\mu_i$, $p_{i,x}$ and $\sigma_{i,x}$ the charge, magnetic moment and x-component of momentum or spin of the $i^{th}$ nucleon, respectively.

The momentum dependence of the three sum rules (2.5) is schematically shown in fig. 4. Assuming no correlations for $q\to\infty$, the functions $f_2$ vanish at infinity and the asymptotic behaviour of the sum rule follows immediately from eq. (2.5). At $q\to 0$, the Coulomb sum is purely elastic (value $Z^2$), the convection current gives a contribution proportional to the TRK- sum rule (see ref. 16 for further discussion).

Finally, we note that the response function (2.1) may be expressed by a
ground state expectation value, even without an integration over the excitation energy[19,20], by the relation

$$(2.12) \quad R(q,\omega) = \sum_n |\rho_{no}(q)|^2 \, \delta(\omega_n - \omega)$$

$$= \frac{i}{\pi} \, \text{Im} \int_{-\infty}^{\infty} dt \, e^{i\omega t} \langle o| \, \mathcal{T} \, \{\rho^+(t)\rho(o)\} |o\rangle$$

where

$$\rho^+(t) = e^{iHt} \, \rho^+(o) \, e^{-iHt}$$

is the charge operator in Heisenberg representation and $\mathcal{T}$ the time order-
ing operator. Recently, Quarati[21] has performed calculations of response
functions in a statistical treatment.

## 3. Sum rules for electroexcitation

Electron scattering sum rules as proposed by Drell and Schwartz[22] and by McVoy and Van Hove[23] require a summation over the nuclear excitation energies in order to use closure and eliminate the dependence on the excited spectrum. Energy-weighted sum rules at constant momentum transfer q are given by

$$(3.1) \quad \int_0^{\omega_{max}} \frac{d^2\sigma(q,\omega,\theta)}{d\Omega_2 d\epsilon_2} \frac{\omega^N}{\sigma_0} d\omega = \int \left[ A_N(q,\omega) + \tan^2(\tfrac{\theta}{2}) B_N(q\omega) \right] d\omega$$

$$= A_N(q) + B_N(q) \tan^2(\tfrac{\theta}{2}) \; .$$

where

$$A_N(q,\omega) = \sum_n \omega_n^N \left\{ \frac{q_\mu^4}{q^4} |\rho_{no}(q)|^2 - \frac{q_\mu^2}{2q^2} |\underline{j}_{no}^T(q)|^2 \right\} ,$$

$$(3.2)$$

$$B_N(q,\omega) = \sum_n \omega_n^N |\underline{j}_{no}^T(q)|^2 .$$

In eq. (3.1), the differential cross section has been divided by $\sigma_0$, which contains the Mott cross section, usually a recoil correction from phase space and sometimes an overall nucleon form factor.

The nuclear charge and (transverse) current operators, $\rho$ and $\underline{j}^T$, have to be evaluated between the nuclear ground state and all excited states n of the nucleus. If one wants to include relativistic and finite size vertex corrections, it is useful to represent the operators im momentum space,

$$(3.3) \quad \hat{\rho}(q) = \sum_j \left\{ \left(1 - \frac{q^2}{8m^2}\right) e^{i\underline{q}\cdot\underline{r}_j} G_j^E - i\underline{\sigma}_j \cdot (\underline{q} \times \underline{p}_j) e^{i\underline{q}\cdot\underline{r}_j} \frac{2G_j^M - G_j^E}{4m^2} \right.$$

$$\left. + [m^{-4}] \right\} ,$$

$$(3.4) \quad \hat{\underline{j}}(q) = \sum_j \frac{1}{2m} \left\{ (\underline{p}_j e^{i\underline{q}\cdot\underline{r}_j} + e^{i\underline{q}\cdot\underline{r}_j} \underline{p}_j) G_j^E + i(\underline{\sigma}_j \times \underline{q}) e^{i\underline{q}\cdot\underline{r}_j} G_j^M \right\}$$

$$+ [m^{-3}] \; .$$

where $G_j^{E,M}(q_\mu^2)$ are the electric and magnetic Sachs form factors of nucleon j. Since $q_\mu^2 = \omega^2 - q^2$, the charge and current operators depend on the excitation energy via the form factors.

In addition to the one-body charge and current operators, there are dynamical contributions from two-body currents as well as kinematical relativistic corrections.

While the theoretician prefers to extend the upper integration limit in eq.(1) to infinity, the experimental upper limit is given by $\omega_{max} = q$. If one wants to use closure over the final states, the momentum transfer should be chosen such that the main excitation spectrum (i.e., the quasi-elastic peak) is well included.

Eqs. (1) and (2) clearly show a longitudinal and transverse sum rule $A_N$ and $B_N$, respectively, which may be resolved by varying $\Theta$ at constant q. By integrating certain linear combinations of $A(q,\omega)$ and $B(q,\omega)$ with appropriate weight factors over the range of excitation energies $\omega$, various other sum rules may be obtained,

$$(3.5) \quad S_N^C(q) = \sum_n \omega_n^N \, \rho_{on}^* \, \rho_{no}$$

$$(3.6) \quad S_N^{C/L}(q) = \sum_n \omega_n^N \, \tfrac{1}{2} \left( \rho_{on}^* \, j_{no}^L + j_{on}^{L*} \, \rho_{no} \right)$$

$$(3.7) \quad S_N^L(q) = \sum_n \omega_n^N \, j_{on}^{L*} \, j_{no}$$

$$(3.8) \quad S_N^T(q) = \sum_n \omega_n^N \, j_{on}^{T*} \, j_{no}^T \quad .$$

In eqs. (3.5)-(3.8) the superscripts refer to Coulomb (C), longitudinal (L) and transverse (T) current. Note that the longitudinal current is related to the charge by the continuity equation. Further, it has been assumed that only the leading non-relativistic terms in eqs. (3.3) and (3.4) have to be considered and that the behaviour of the nucleon form factors as function of $q_{\mu}^2$ is known. Such sum rules have been obtained by ref.(24) and expressed in terms of ground state expectation values of nucleon-nucleon density correlation functions, spin-orbit correlations, kinetic energy correlations etc. Particularly for very light nuclei, an experimental investigation of the problem could shed some light on those interesting quantities and furthermore clarify the situation with regard to non-single particle currents.

As far as the latter effects are concerned, energy-weighted sum rules, in particular the familiar N = 1 sum rule, are a much more sensitive test of the role of exchange currents[22]. For the N > 0 sum rules, the energy dependence may be removed by introducing commutators and subsequently using closure:

$$(3.9) \quad \sum_n \omega_n \, \rho_{on}^* (q') \, \rho_{no} (q) = \tfrac{1}{2} \langle \sigma | \, [ \hat{\rho}^+(q'), [\hat{H}, \hat{\rho}(q)]] \, | \sigma \rangle .$$

In sect.4., the density operator $\rho(q)$ shall be further decomposed in a multipole series and eq. (10) is actually needed for $q' \neq q$. In the previous equations, of course, all charge and current operators have to be evaluated at a fixed momentum transfer $q$. Also sum rules for N < 0 have been considered, in particular the inversely energy weighted sum rule (N = -1), which expresses the nuclear polarizability.

A delicate problem in replacing the excitation energies by the commutator with the Hamiltonian is our lack of knowledge about true wave functions and true Hamiltonians except, possibly, in the case of the few nucleon problem. As has been discussed in ref.(25) the RPA-Hamiltonian may be used in evaluating the N = -1, +1 and +3 sum rules.

Following the original calculations[22,23] non-energy weighted sum rules for Coulomb ($S_0{}^C$) and transverse current ($S_0{}^T$) contributions have been calculated[26,27] with refined wave functions and compared to the existing experimental data[28]. Within the experimental (and theoretical) error bars, a satisfactory agreement is obtained.

The Coulomb sum rule, $S_0{}^C$, is mostly influenced by statistical correlations, particularly for small momentum transfer, and the effect of dynamical internucleon correlations is substantially suppressed (see figs. 5a-c). Note that the elastic contribution has been removed from the sum, hence $S^{inel.}(q = 0) = 0$. For high q values, the single particle model predicts incoherent scattering from Z protons, i.e., $S^{inel.}(q\rightarrow\infty)=ZG_p{}^2$ + relativistic corrections. Also within the single particle model, the transverse sum rule is strongly dominated by incoherent scattering on the magnetic moments of neutrons and protons (see fig.6a),

$$(3.10) \qquad S_0^T(q) \approx \frac{q^2}{2m^2}\left(Z\mu_p^2 + N\mu_n^2\right).$$

Statistical and dynamical correlations give only a negligible effect in this case and any big discrepancy between eq.(3.10) and the experimental data would indicate contributions of two-body currents ("quenching") to the magnetic moment of nucleons in nuclei. As may be seen in fig.6b the agreement is good, indicating that "quenching" is less than about 5%.

Energy weighted sum rules $S_1{}^C(q)$ have been calculated by Inopin and Roshchupkin[29] and Mekjian[30] in a Fermi gas model. For a Hamiltonian with
exchange mixture,

$$(3.11) \quad H = \sum_i t(i) + \sum_{i<j} (W + B P_\sigma^{ij} - H P_\tau^{ij} + M P_x^{ij}) V(\underline{r}_i - \underline{r}_j),$$

where $P^{ij}$ are the familiar spin, isospin or space exchange operators, the
commutator (3.9) may be evaluated. Neglecting relativistic corrections,

$$(3.12) \quad S_1^C(q) = \frac{Z q^2}{2m} (1 + \Delta(q^2)),$$

where

$$(3.13) \quad \Delta(q^2) = -\frac{8m(H+2M)}{Z q^2} \int K(\underline{r}) \sin^2\left(\frac{q\underline{r}}{2}\right) V(\underline{r}) d\tau,$$

with the correlation function ($\underline{r} = \underline{r}_1 - \underline{r}_2$, $\underline{R} = (\underline{r}_1 + \underline{r}_2)/2$)

$$K(\underline{r}) = \int |\rho(\underline{r}_1, \underline{r}_2)|^2 d\tau_R .$$

As known from photonuclear physics, the exchange effect is positive for
small q (using the usual exchange mixtures!), but decreases and eventually becomes negative for $q \gtrsim 5$ fm$^{-1}$ as may be seen from fig.7.

## 4. Angular momentum projected sum rules

More detailed information on nuclear structure may be obtained by a decomposition of the previously discussed sum rules into partial waves as well as isoscalar and isovector components.

In the following we shall restrict ourselves to the energy-weighted sum rule for the isoscalar charge density operator. The latter commutes with the nucleon-nucleon potential to a good approximation and the right-hand side of eq.(3.9 ) may be evaluated[31]

$$(4.1) \quad \sum_n \omega_n \, \text{Re} \left( \rho_{no}^*(q') \, \rho_{no}(q) \right) = \frac{\hbar^2 q \cdot q'}{2m} \left\langle \sigma | \hat{\rho}(q'-q) | \sigma \right\rangle .$$

From this progenitor sum rule, the relations of Fallieros et al.[32, 33] may be derived, which connect the transition densities of strong collective states with the ground state density.

Recently, this technique has been generalized and applied to transition densities of low-lying collective states of anharmonic vibrator nuclei[34]. Such generalized sum rules may be obtained by evaluating eq. ( 1) for the multipole components $F^\ell$ of the charge operator,

$$(4.2) \quad \hat{F}^\ell(q) = \int d\tau \, j_\ell(qr) \, Y^\ell(\Omega) \, \hat{\rho}(r) .$$

Coupling the double commutator of two tensor operators (of rank $\ell_1$ and $\ell_2$) with the Hamiltonian to overall angular momentum L, evaluating it between two nuclear states of spins $J_i$ and $J_f$ and inserting a complete set of states we obtain

$$(4.3) \quad \langle J_f \| [\hat{F}^{\ell_1}(q'), [\hat{A}, \hat{F}^{\ell_2}(q)]]^L \| J_i \rangle$$

$$= \sqrt{2L+1} \sum_n \left[ \omega_{ni} \left\{ \begin{matrix} J_f & \ell_1 & J_n \\ \ell_2 & J_i & L \end{matrix} \right\} F_{fn}^{\ell_1}(q') F_{ni}^{\ell_2}(q) + i \leftrightarrow f \right].$$

Note $F_{fn}^{\ell_1}(q') \equiv \langle J_f \| \hat{F}^{\ell_1}(q') \| J_n \rangle$ and $\omega_{ni} = \epsilon_n - \epsilon_i$.

Assuming that the isoscalar density operator commutes with the potential energy, the commutator may be evaluated explicity and a series of relations may be derived, in particular between

(I ) multipole moments $M^\ell$ in the limit q and q'$\rightarrow$ 0;

(II) Transition densities $\rho^\ell$ and multipole moments for q'$\rightarrow$0 but q finite.

Specializing to $\ell_1 = \ell_2 = \ell$ and introducing multipole operators

$$(4.4) \quad M^\ell = \lim_{q \to 0} \begin{cases} (2\ell+1)!! \; F^\ell / q^\ell & \text{for } \ell \neq 0 \\ -3 F^0 / q^2 & \text{for } \ell = 0 \end{cases}$$

we obtain

$$\sum_n \sqrt{2L+1} \, (\omega_{ni} + \omega_{nf}) \left\{ \begin{matrix} J_f & \ell & J_n \\ \ell & J_i & L \end{matrix} \right\} M_{fn}^\ell \rho_{ni}^\ell(r)$$

$$(4.5) \quad = (-)^{J_i + J_f + 1} \frac{e \hbar^2 (2\ell+1)}{2m \sqrt{4\pi}} \begin{pmatrix} \ell & \ell & L \\ 0 & 0 & 0 \end{pmatrix} r^{\ell-2} \left[ L(L+1) + 2\ell r \frac{\partial}{\partial r} \right] \rho_{fi}^L(r).$$

The sum rules obtained by Fallieros et al.[32,33] follow immediately for $J_i = J_f = L = 0$. As an example for the generalized relations, we evaluate eq. (4.5) for quadrupole transitions

(I ) as ground state expectation value ($J_i = J_f = L = 0$),

(II ) between ground state and first excited state ($J_i = 0$,

   $J_f = L = 2$),

(III) as expectation value in the first excited state ($J_i = J_f = 2$;

   $L = 0, 2$ or $4$).

The resulting equations are

$$(4.6) \quad \sum_n \omega_n M_{on}^{(2)} \rho_{no}^{(2)}(r) = -\frac{5e\hbar^2}{m\sqrt{4\pi}} r \frac{\partial}{\partial r} \rho_{oo}^{(o)}(r),$$

$$(4.7) \quad \sum_n (\omega_{no}+\omega_{n1}) M_{1n}^{(2)} \rho_{no}^{(2)}(r) = \sqrt{\frac{2}{7\pi}} \frac{5e\hbar^2}{m} \left(r\frac{\partial}{\partial r}+\frac{5}{2}\right) \rho_{10}^{(2)}(r).$$

The exhaustion hypothesis of refs.(32, 33 ) truncates the sum in eq.(4.6) to the "one-boson-state" $|2_1^+\rangle$. In the following we shall assume that two states contribute, a one-boson state and a two-boson state $|2_2^+\rangle$, the electromagnetic decay being dominated by the cascade $2_2^+ \to 2_1^+ \to 0^+$ with

$$(4.8) \quad M_{21}^{(2)} \approx \sqrt{2}\, M_{10}^{(2)}.$$

This further relation may be derived from the sum rule expectation value in the excited state $|2_1^+\rangle$ , assuming that the mean square radius remains the same for the excited state as for the ground state. Alternatively, the sum rule relation gives the radius of the excited state, whenever the ratio $M_{21}/M_{10}$ is known from experiment.

Eqs. (4.6)-(4.8) express the transition densities of excited states in terms of the experimental energies and multipole moments. Since the phase between $M_{10}$ and $M_{20}$ is not known from experiment, two solutions occur corresponding to a more prolate or oblate deformation. In agreement with previous collective models[35,36], electron scattering is very sensitive to the sign of the quadrupole moment. The transition densities are shifted to smaller radii for a prolate solution and to larger radii for an oblate solution.

The recent data of Neuhausen[37] for a series of Zn isotopes demonstrate that the transition densities of the "weak" transitions are indeed totally different from the "Tassie"-type transition densities predicted by Fallieros et al[35,36] (see figs. 8 and 9 ). In view of the obvious weakness of the exhaustion hypothesis (only a few percent of the transition strength is concentrated in surface vibrations, the dominant part is in the isoscalar giant resonances!) the agreement between experiment and the three level model is surprisingly good (figs. 10, 11). Note that all relevant quantities are integrals of the transition density weighted with at least $r^4$, which suppress the obvious deviations near the origin.

The ratio of the transition radii, $(R_{no})^2 = \int \varrho_{no} \, r^6 \, dr / \int \varrho_{no} \, r^4 \, dr$, in the two excited states is predicted in good agreement with the experimental data, e.g. for $^{64}$Zn we obtain $(R_{20}/R_{10})^2 = .76$ (experiment .75 .04).

Collectivity of a state in the sense of this approach has two aspects:
   (I ) Strong transition matrix elements ("exhaustion"),
   (II) Collective shape of the transition densities.

From the fact that the low-lying surface vibrations fulfil (II) in spite of the fact that aspect (I) is certainly not fulfilled, we have to conclude that the giant resonances share their strength in a rather even way with the surface vibrations.

The remaining shell model structure in transition densities has been demonstrated in RPA calculations (see, e.g., refs.38,39) and it is not surprising that the effect is yet stronger for the weaker transition to the "two-phonon" states. Obviously, these shell model fluctuations, which are already present in ground state densities, are out of range of a model-independent sum rule approach. However, the sum over all excited states including the giant resonances should follow closely the sum rule prediction - unless if there should exist strong exchange contributions for the isoscalar sum rule, too. Fig.12 is an outline of the history of RPA calculations for the strongly collective $3^-$ level in $^{208}Pb$, which shows only little remaining shell model structure.

In conclusion, electroexcitation combined with sum rule techniques makes it possible to determine detailed aspects of the nuclear shape in excited states, e.g. the static quadrupole moment including its sign (oblate vs. prolate) or changes in the radius as compared to the ground state distribution.

## 5. Spin-isospin correlations and electropion production

Electroproduction of pions from nuclei has been the subject of many contributions[43,44]. Experiments in this field are of interest from the standpoints of both pion physics and nuclear structure studies. In this contribution we shall report on some preliminary experimental results and compare them with preliminary calculations[45]. While the usual interest of photo- or electro-pion production concentrates in the region of the (33)-resonance, the experiments are performed close to threshold and analyze energy and angular distributions of charged pions with time-of-flight[46] or other detection[47] techniques.

At these energies S wave production dominates, which is fairly well understood. Other contributions, e.g., from the (33)-resonance, give a typical 10% correction[44], which one should be able to estimate to sufficient accuracy.

Final state interactions, however, play a very important role. Since the optical pion-nucleus potential is not sufficiently well known at these energies[48], the measurement of both $\pi^+$ and $\pi^-$ cross sections is a valuable tool to determine the optical potential from the strong interference of Coulomb and nuclear interactions.

Once the optical potential is fixed, the cross section for pion production may be related to spin-isospin correlation densities.

The double differential cross section is[44]

$$(5.1) \quad \frac{d^4\sigma}{d\varepsilon_2 d\Omega_2 d\omega_\pi d\Omega_\pi} = \sum_n \frac{\alpha^2}{4\pi^2} \frac{\varepsilon_2}{\varepsilon_1} \frac{q\omega_\pi}{q^4} \delta(\omega-\omega_\pi-E_n)$$

$$* \left[ \underset{\sim}{B} \cdot \underset{\sim}{H} \ \underset{\sim}{B} \cdot \underset{\sim}{H}^* + (1 \leftrightarrow 2) - \frac{q^2}{2} \underset{\sim}{H} \cdot \underset{\sim}{H}^* \right].$$

where $q_\pi$, $\omega_\pi$ and $\Omega_\pi$ are momentum, energy and angle of the produced pion. The nuclear matrix elements are

$$(5.2) \quad \underset{\sim}{H} = \int \frac{d\tau}{\sqrt{2\omega_\pi}} \, e^{i\underset{\sim}{q}\cdot\underset{\sim}{r}} \, \varphi_\pi^*(q_\pi, r) \, \langle n | \underset{\sim}{\hat{M}} | i \rangle ,$$

where $\varphi_\pi \rightarrow e^{i\underset{\sim}{q}_\pi\cdot\underset{\sim}{r}}$ if the final state interaction may be neglected. In this case, eq. (5.2) gives the Fourier transform of the operator $\underset{\sim}{M}$ with regard to the momentum $\underset{\sim}{t} = \underset{\sim}{q} - \underset{\sim}{q}_\pi$, which has been transferred to the nucleus. In impulse approximation the operator $M$ has the structure

$$(5.3) \quad \underset{\sim}{\hat{M}} = \frac{q\sqrt{2}}{2m} \sum_{j=1}^{A} (A_j + i\, D_j \, \underset{\sim}{\sigma}_j + i\, B_j \, \underset{\sim}{\sigma}_j\cdot\underset{\sim}{q}_\pi + i\, C_j \, \underset{\sim}{\sigma}_j\cdot\underset{\sim}{q}) \\ * \, \delta(\underset{\sim}{r} - \underset{\sim}{r}_j) \, \tau_\pm(j) .$$

where $A_j$ etc. are connected with the production amplitudes on the individual nucleon [44].

Upon integration over the electron coordinates, the cross section has a structure of the type

$$(5.4) \quad \frac{d^2\sigma}{d\Omega_\pi d\omega_\pi} \propto \sum_n \sum_{k\ell} f(E_n) \langle 0 | \sigma_k \tau_{k,\pm}^\dagger \, e^{-i\underset{\sim}{t}\cdot\underset{\sim}{r}_k} | n \rangle \\ * \, \langle n | \sigma_\ell \tau_{\ell,\pm} \, e^{i\underset{\sim}{t}\cdot\underset{\sim}{r}_\ell} | 0 \rangle .$$

The explicit summation over the excited states has been performed previously for the Fermi gas model [44]. For a more realistic shell model and close to threshold an expansion about a mean excitation energy is quite useful,

$$f(E_n) |n\rangle = \left[ f(\bar{E}) + \frac{\partial f}{\partial E}\bigg|_{\bar{E}} \, (\hat{H} - \bar{E}) + \dots \right] |n\rangle .$$

Determining the mean excitation energy such that the second term of the expansion vanishes, the summation over the excited spectrum may be performed. In this approximation the pion production cross section is proportional to the spin-isospin two-particle density in the ground state,

$$(5.5) \quad \frac{d^2\sigma}{d\Omega_\pi d\omega_\pi} \sim \langle o | \sum_{k,\ell} \vec{\sigma}_k \vec{\sigma}_\ell \tau_{k,\mp} \tau_{\ell,\pm} e^{i\vec{k}\cdot(\vec{r}_k - \vec{r}_\ell)} | o \rangle .$$

This is the equivalent of the photonuclear sum rule ( eq. 2.4    ), which is connected with the two-proton density in the ground state. For the experiments under discussion, scattering of 300 MeV electrons and subsequent observation of the produced pions only, we obtain an average excitation energy which is considerably above the dipole giant resonance region. This result is contrary to our previous results[44] which closely followed   photo-pion production calculations, taking into account the "$1\hbar\omega$-states" only.

Fig. 13 shows the dependence of the measured cross section on the atomic number A. The linear increase with A ("volume production") indicates that near threshold the absorptive potential is relatively weak. This is contrary to the resonance region where "surface production" occurs.

The preliminary calculations of the $\pi^+/\pi^-$ ratio (fig. 14 ), which, however, do not yet include a full partial wave calculation, lie considerably above the experimental data. This may indicate that the "standard" optical potentials have to be modified near threshold.

In view of the shortcoming of the optical potentials the agreement with experimental angular distributions (fig. 15 ) may be termed satisfactory. The cross section peaks at backward angles contrary to previous calculations, which included only a few low-lying states (dashed curve).[44]

Note that the inclusion of an effective mass, $m^* = m/1.4$, which produces the "canonical" enhancement of the photonuclear sum rule, does not seem to improve the agreement.

Summarizing we conclude that electropion production is very sensitive to the optical pion-nucleus potential near threshold. Furthermore, it is closely connected to the spin-isospin correlations in the nuclear ground state and gives information complementary to the photonuclear sum rules. Before drawing definite conclusions, however, we have to investigate more closely the finite state interactions in a complete phase shift analysis.

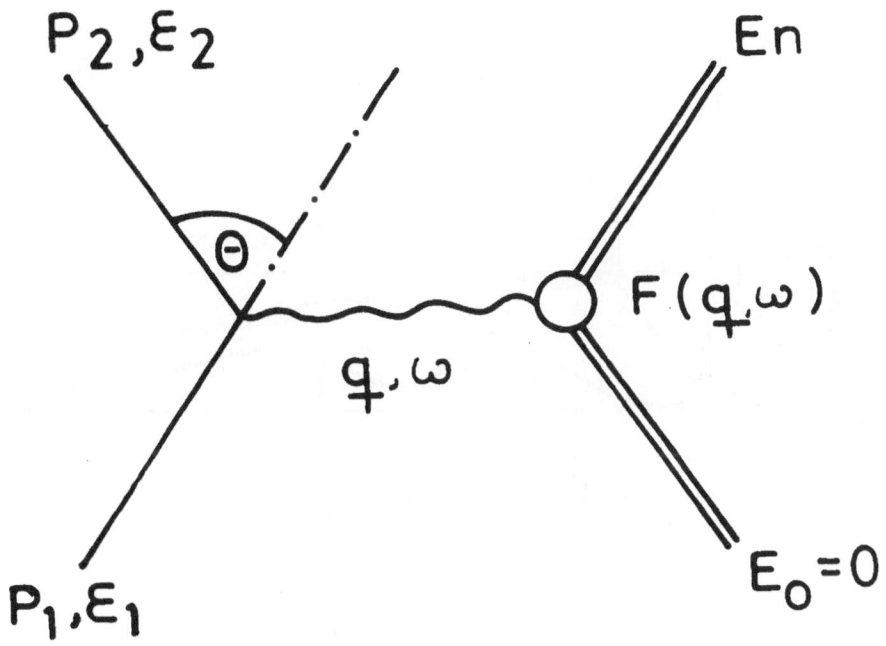

FIG. 1

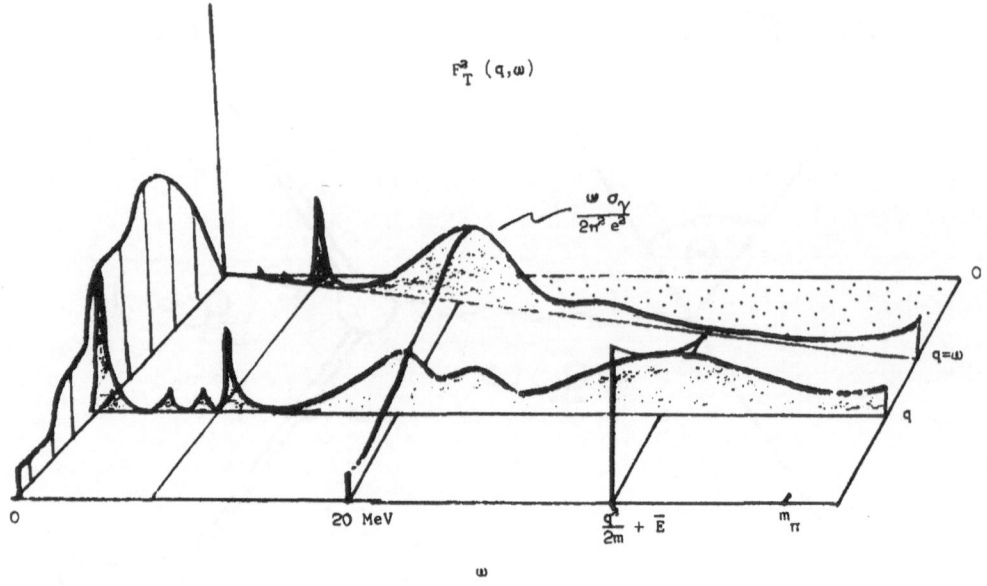

$$F_T^2 (q,\omega)$$

$$\frac{\omega}{2\pi^2} \frac{\sigma_\gamma}{e^2}$$

$q = \omega$

$q$

0      20 MeV      $\frac{q^2}{2m} + \overline{E}$      $m_\pi$

$\omega$

FIG 2a

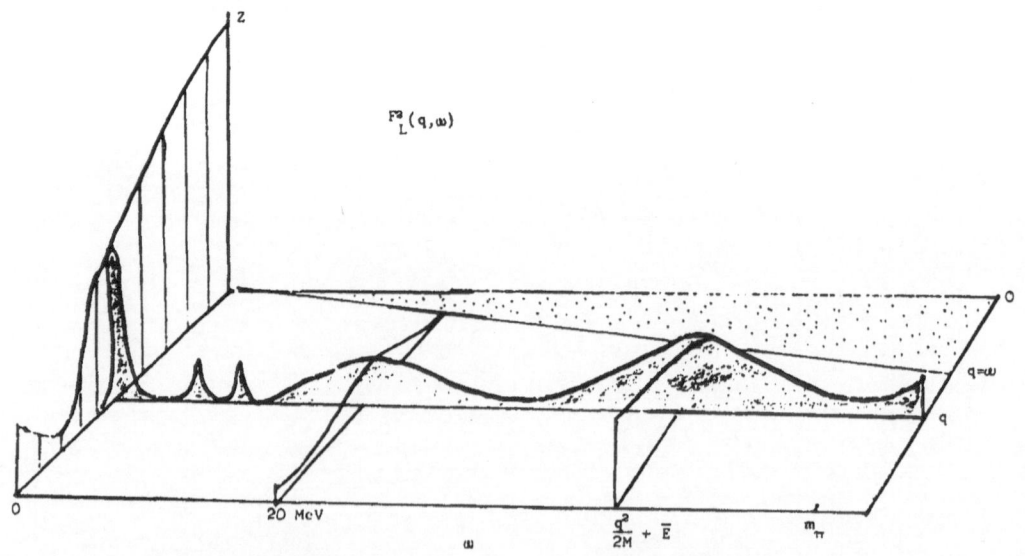

$z$

$$F_L^2 (q,\omega)$$

$q = \omega$

$q$

0      20 MeV      $\frac{q^2}{2M} + \overline{E}$      $m_\pi$

$\omega$

FIG. 2b

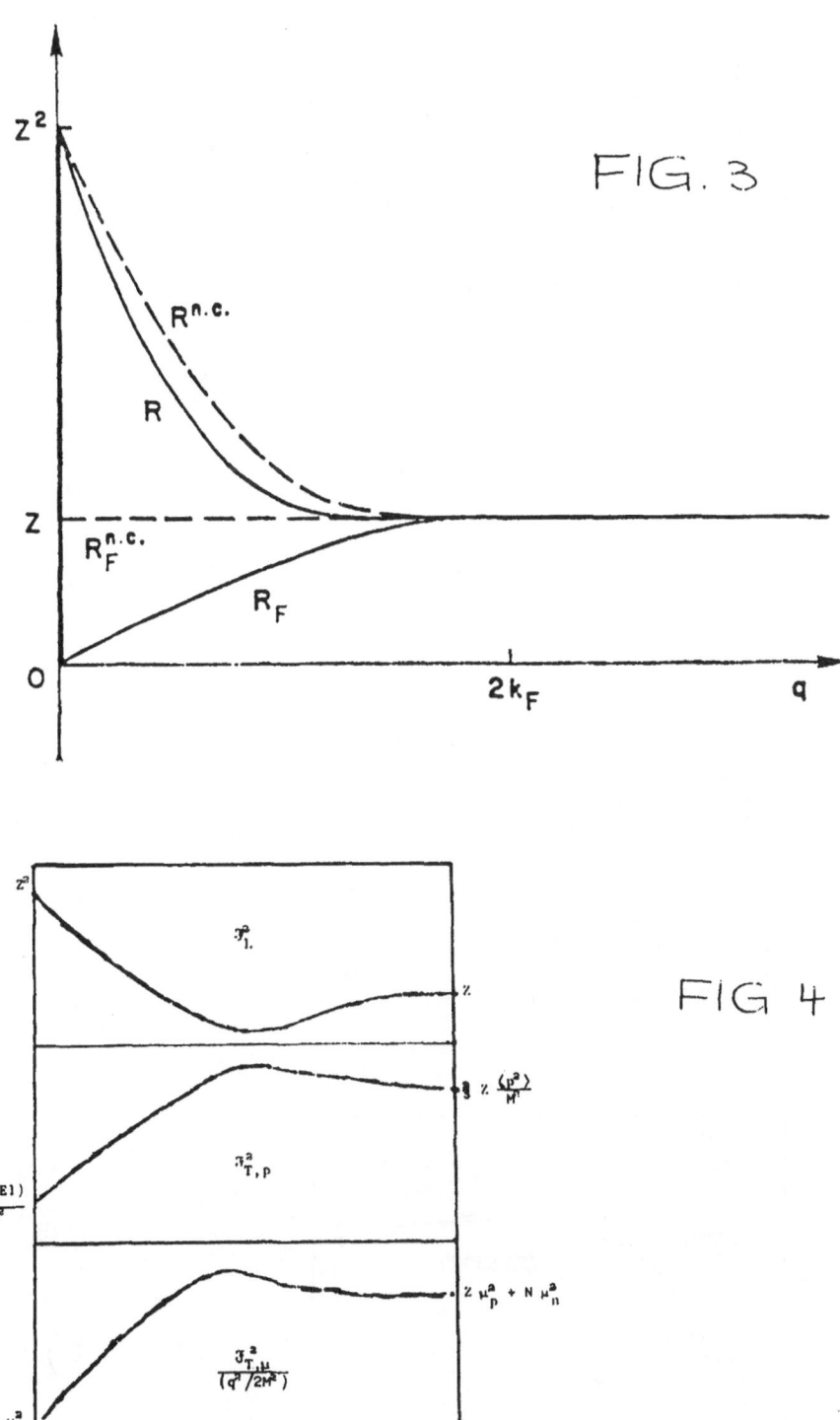

FIG. 3

FIG 4

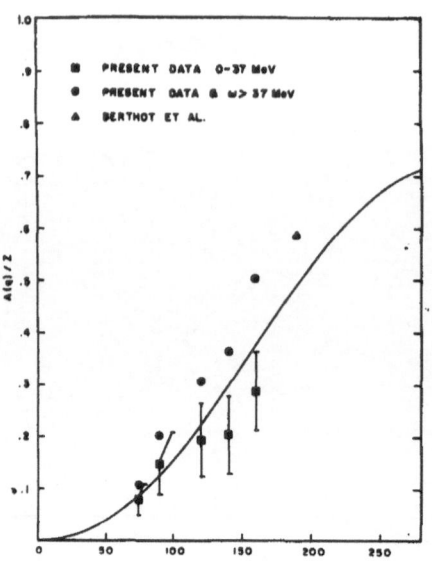

FIG. 5

a)

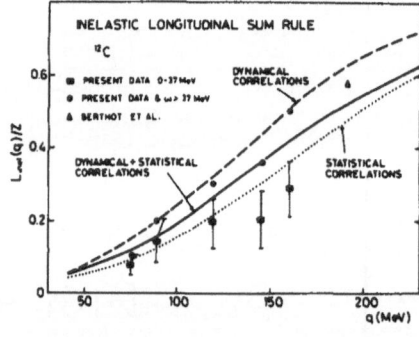

b)

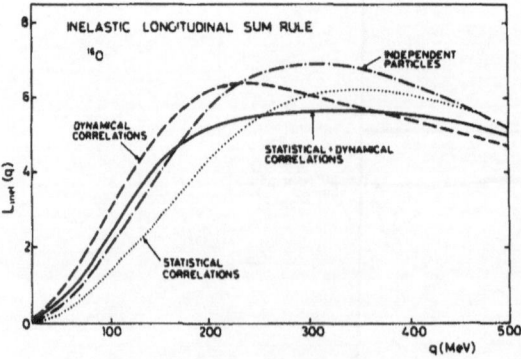

c)

FIG 6

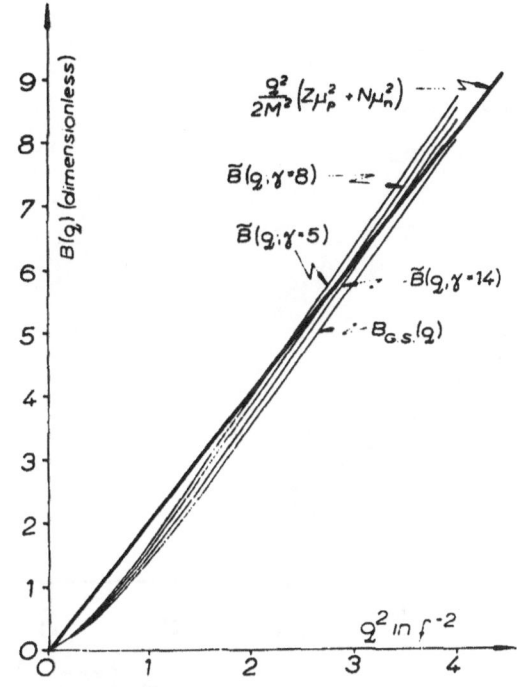

a)

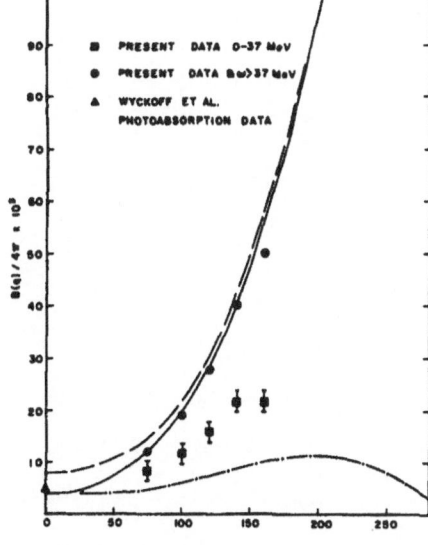

b)

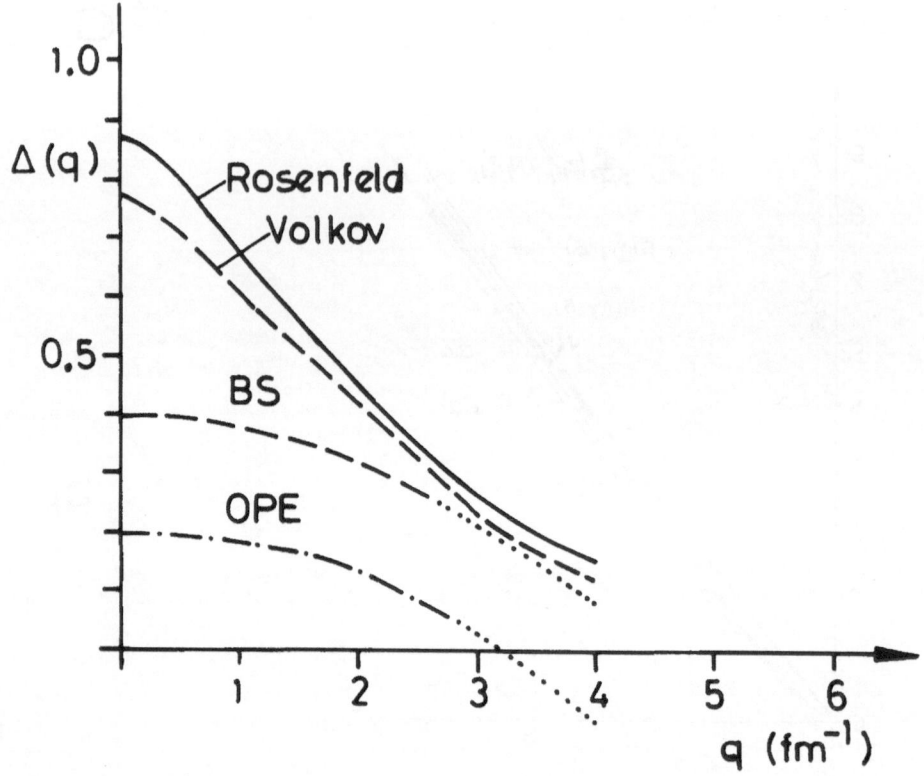

FIG. 7

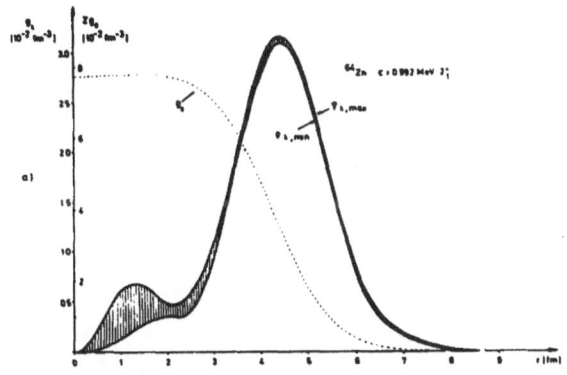

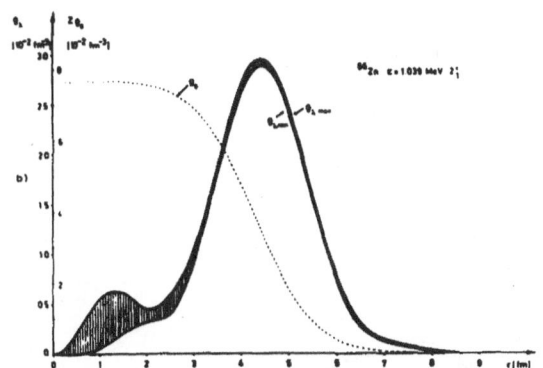

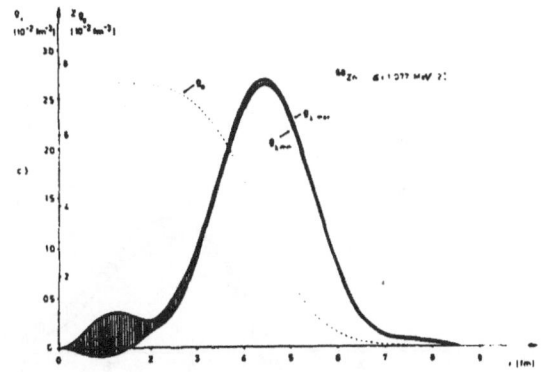

FIG. 8

122

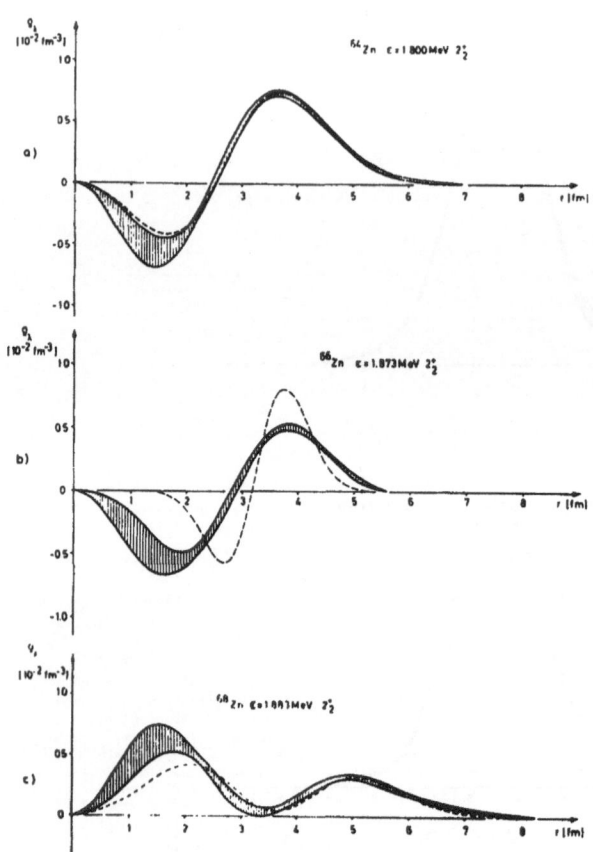

FIG. 9

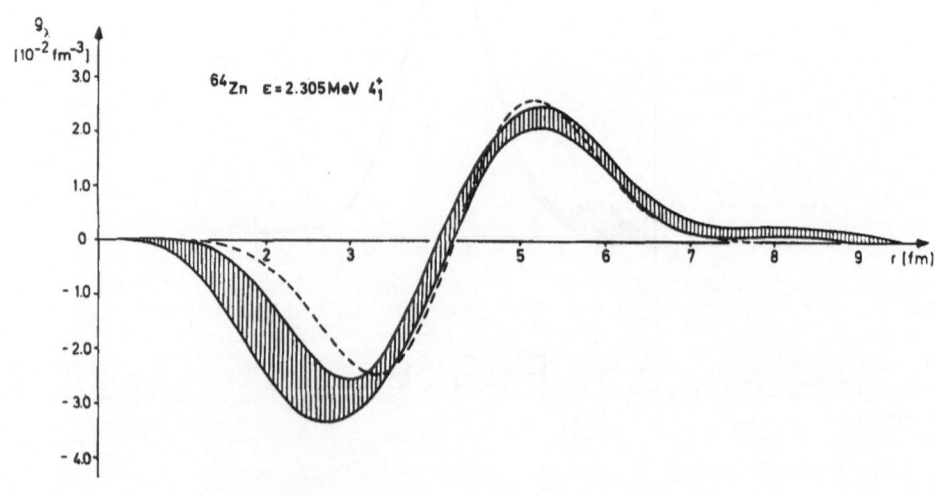

123

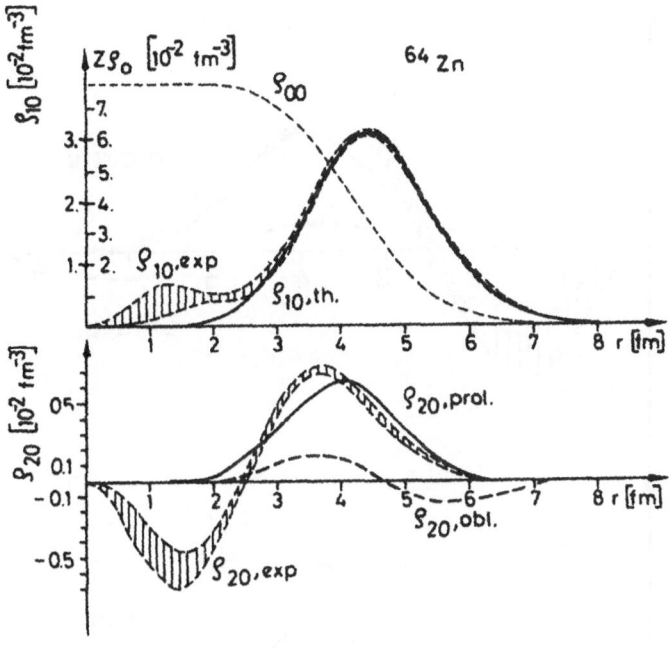

FIG. 10

124

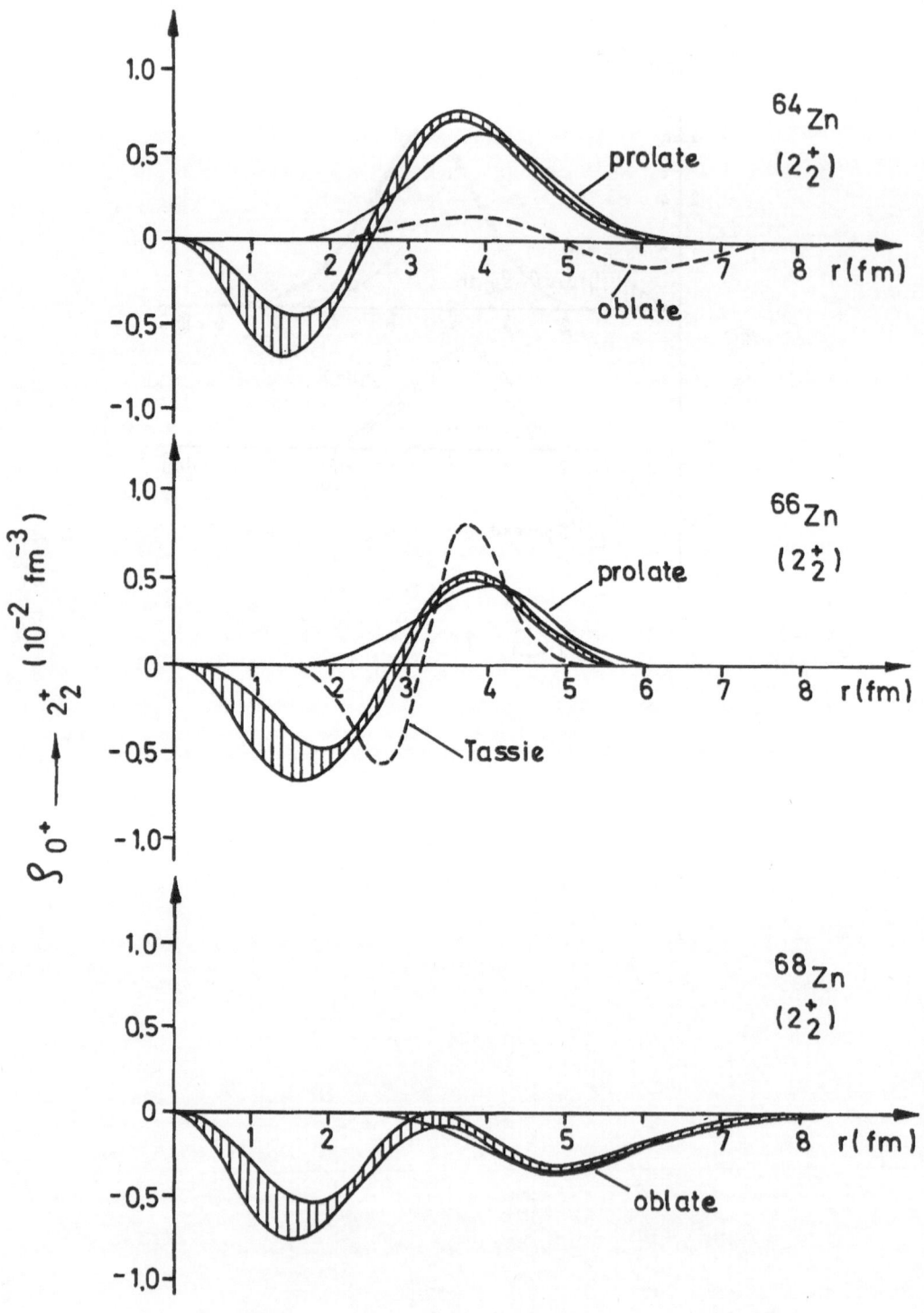

FIG. 11

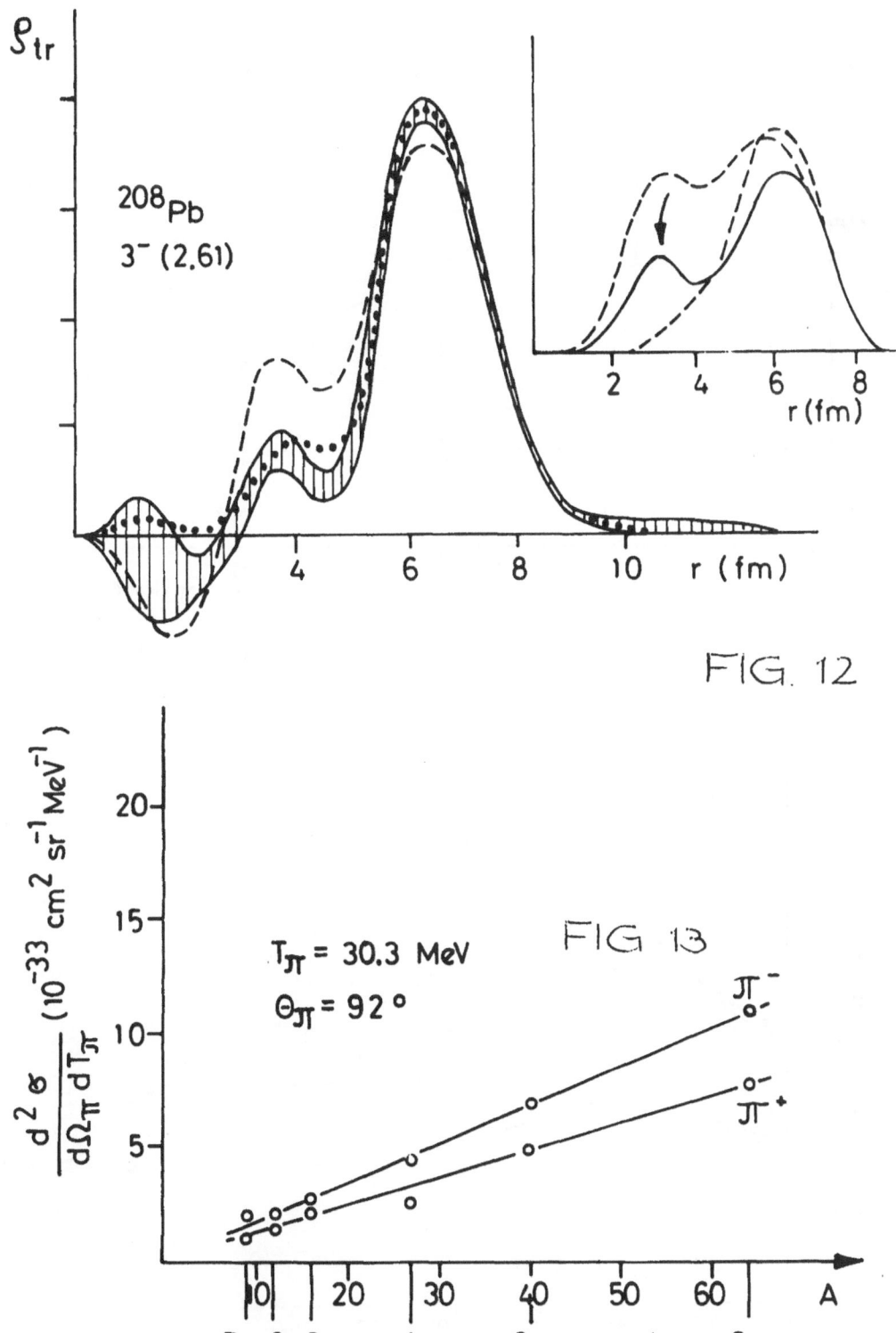

FIG. 12

FIG 13

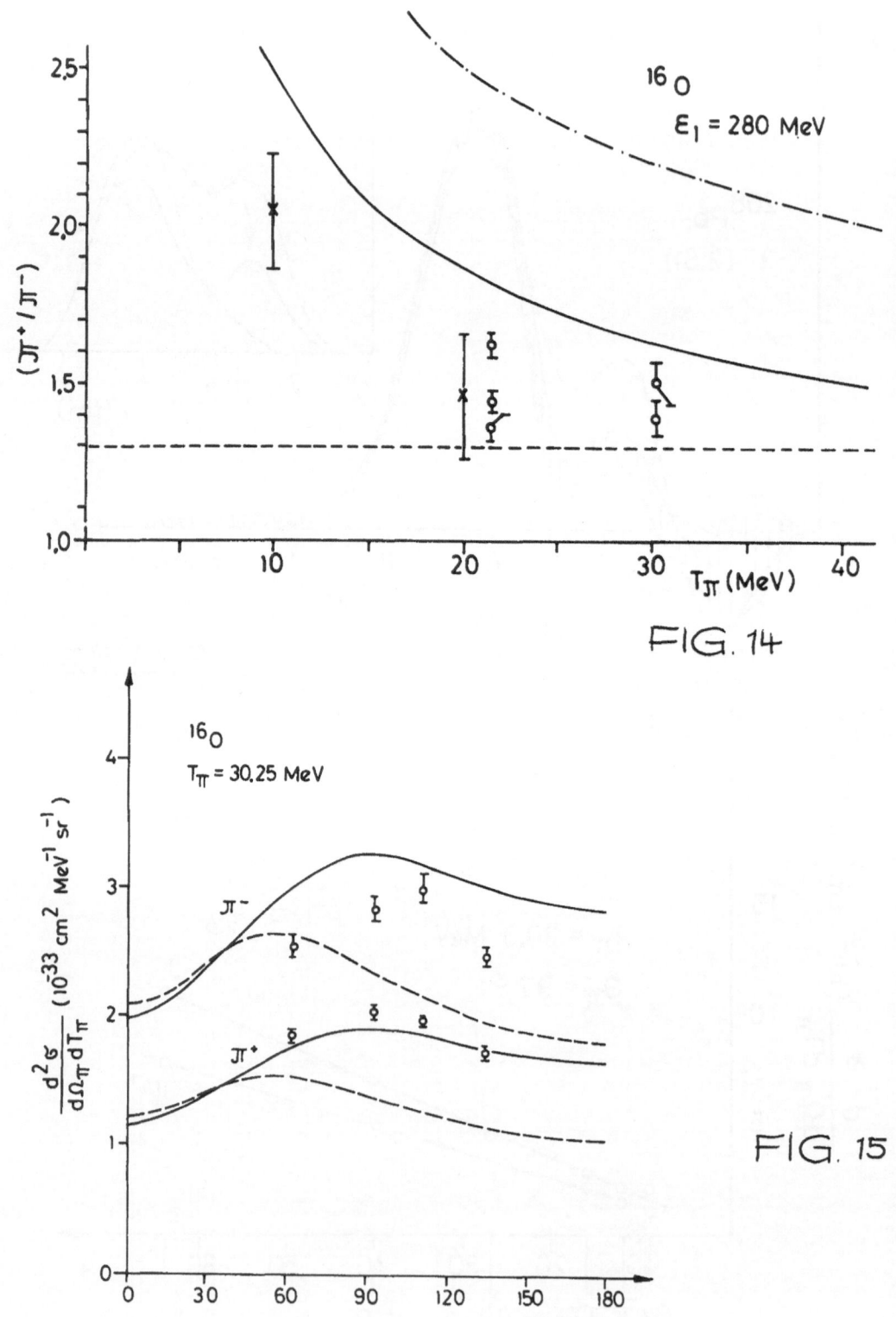

FIG. 14

FIG. 15

## Figure Captions

Fig.1.: Kinematical variables for electron scattering.

Fig.2.: Transverse (fig.2a) and Coulomb (fig.2b) form factors as function of momentum transfer q and energy transfer $\omega$. The excitation spectrum shows elastic scattering ($\omega = 0$), scattering to discrete levels and to giant resonances ($\omega \approx 20$ MeV), followed by the quasielastic peak and the meson threshold ($\omega = m_\pi$). The transverse form factors at $q = \omega$ are related to the photoabsorption cross section. The figure has been taken from ref.16.

Fig.3.: The nuclear response as function of momentum transfer q for the Fermi gas model (lower two curves) and a more realistic model (upper two curves). The dashed curves are obtained without correlations, the full curves including correlations. The figure has been taken from ref.18.

Fig.4.: Longitudinal and transverse (convection and magnetization) sum rules as function of momentum transfer q. The asymptotic limits are described in the text and in ref.16, from which the figure has been taken.

Fig.5.: Inelastic longitudinal sum rules as function of momentum transfer q. Fig.5a: Experimental data for $^{12}$C (ref.28), fig.5b and 5c calculations of ref.27 for $^{12}$C and $^{16}$C respectively.

Fig.6.: Transverse sum rule as function of momentum transfer q. Fig.6a various shell model calculations for $^{16}$O (ref.27) which differ little from incoherent scattering from the magnetic moments (straight line). Fig. 6b shows the experimental data for $^{12}$C (ref.28).

Fig. 7.: The exchange contribution Δ as function of momentum transfer q for
a Fermi gas model. The upper two curves are the predictions of ref.
29 for Rosenfeld and Volkov forces, respectively. The lower two
curves have been obtained by Mekjian[30] with Bryan-Scott and One-
Pion-Exchange potentials, respectively.

Fig. 8.: Transition densities for allowed ("one-phonon") transitions in Zn-
isotopes as function of radius. The experimental data (hatched
area) are from Neuhausen[37].

Fig. 9.: Transition densities for forbidden ("two-phonon") transitions in
Zn-isotopes as function of radius. The experimental data (hatched
area) are from Neuhausen[37], the dashed lines are various model
fits.

Fig.10.: Static density $\rho_{00}$, and transition densities to the first ($\rho_{10}$)
and second ($\rho_{20}$) excited state for $^{64}Zn$ as function of radius.
Hatched area: experimental data of ref.37, full curve: sum rule
prediction for the transition densities to the first and second
excited state (prolate shape). The dashed curve, obtained for
an oblate shape, does not agree with experiment.

Fig.11.: Transition densities for two-phonon transitions in Zn-isotopes as
function of radius. Hatched area: Data of ref.37, compared to sum
rule predictions on the Tassie model as indicated. Note the transition from prolate to oblate solutions.

Fig.12.: History of transition densities of the collective 3⁻state at 2.61
MeV in $^{208}Pb$. Right upper corner: The two dashed curves show qualitatively the collective (sum rule: one peak) and shell model
density (ref.41, two peaks), Nagao[42] pointed out that density-
dependent forces may decrease the inner shell model peak (full

curve). Main figure: The experimental data analyzed by Rothaas et al.[40] give the hatched area. The two curves show more recent calculations including density dependent forces and large configuration space (dashed curve: ref.38, dotted curve: ref. 39).

Fig.13.: Differential cross section for electroproduction of pions as function of the nuclear mass. The data[47] are very preliminary, subject to change - particular with regard to normalization - and should not be quoted.

Fig.14.: Ratio of positive to negative pions produced by scattering of 280 MeV electrons on $^{16}O$ as function of kinetic energy of the pions. The calculations are for a pure Coulomb field (dot - dashed curve), an optical potential (full curve, preliminary calculation!) and the case of no final state interaction (dashed curve). The preliminary data are from refs.46 (crosses) and 47 (circles) - see previous figure caption!

Fig.15.: Angular distribution of positive and negative pions by electron scattering on $^{16}O$. The calculations are for a Fermi gas model (full curves: normal nucleon mass m, dashed curves: effective mass m/1.4). The data of ref.47 are preliminary, see figure caption 13.

References:

1. W. Kuhn, Z. Phys. 33 (1925) 405; F. Reiche and W. Thomas, Z. Phys. 34 (1925) 510

2. W. Heisenberg, Physik. Z. 32 (1931) 737

3. E. Feenberg, Phys. Rev. 49 (1936) 328; A.J.F. Siegert, Phys. Rev. 52 (1937) 787

4. J. Ahrens et al., Nucl. Phys. A251 (1975) 479

5. A. Arima, G.E. Brown, H. Hyuga and M. Ichimura, Nucl. Phys. A205 (1973) 27; W.T. Weng, T.T.S. Kuo and G.E. Brown, Phys. Lett. 46B (1973) 329

6. M. Fink, M. Gari and H. Hebach, Phys. Lett. 49B (1974) 20

7. H. Hebach, contribution to this conference; W. Weise, ibid.

8. M. Gell-Mann, M.L. Goldberger and W. Thirring, Phys. Rev. 95 (1954) 1612

9. W. Weise, Phys. Reports 13C (1974) 53

10. S.B. Gerasimov, Phys. Lett. 13 (1964) 240

11. T. Matsuura and K. Yazaki, Phys. Lett. 46B (1973) 17

12. H.P. Schröder, diploma thesis, Mainz 1976

13. J.F. Friar and S. Fallieros, Phys. Rev. C11 (1975) 274 and 277

14. R.M. Woloshyn, Phys. Rev. C12 (1975) 901

15. L. Tiator, diploma thesis, Mainz 1976

16. J.S. O'Connell, Proc. Int. Conf. on Photonuclear Physics and Applications, Asilomar/Calif. 1973, ed. by B.L. Berman

17. W. Czyz, MIT 1967 Summer Study, ed. by W. Bertozzi and S. Kowalski

18. T. de Forest and J.D. Walecka, Adv. Phys. 15 (1966) 1

19. H. Oberall, "Electron Scattering from Complex Nuclei", Academic Press, 1971

20. W. Czyz and K. Gottfried, Ann. Phys. 21 (1963) 47

21. P. Quarati, preprint

22. S.D. Drell and C.L. Schwartz, Phys. Rev. 112 (1958) 568

23. K.W. Mc Voy and L. Van Hove, Phys. Rev. 125 (1962) 1034

24. V.D. Efros, Sov. Journ., Nucl. Phys. 18 (1974) 607

25. J. Martorell, O. Bohigas, S. Fallieros and A.M. Lane, Phys. Lett. B (1976)

26. W. Czyz, L. Lesniak and A. Malecki, Annals of Phys. 42 (1967) 119

27. A. Malecki and P. Picchi, Phys. Lett. 43B (1973) 351

28. J.W. Lightbody, Phys. Lett. 33B (1970) 129

29. E.V. Inopin and S.N. Roshchupkin, Sov. Journ., Nucl. Phys. 17 (1973) 526

30. A.Z. Mekjian, Phys. Rev. C9 (1974) 2084

31. J.V. Noble, Ann. Phys. 67 (1971) 98

32. E.I. Kao and S. Fallieros, Phys. Rev. Lett. 25 (1970) 827

33. T.J. Deal and S. Fallieros, Phys. Lett. 44B (1973) 224

34. P.G. Reinhard and D. Drechsel, Phys. Lett. 56B (1975) 17

35. J.W. Lightbody, Phys. Lett. 38B (1972) 475

36. E. Borie, K. Lezuo and D. Drechsel, Nucl. Phys. A211 (1973) 393

37. R. Neuhausen, report KPH 22/74, (Inst. f. Kernphysik, Universität Mainz)

38. W. Theis and E. Werner, Phys. Lett 44B (1973) 481

39. G. Bertsch and S.F. Tsai, Phys. Lett. 50B (1974) 319 and priv. comm.

40. H. Rothaas, J. Friedrich, K. Merle and B. Dreher, Phys. Lett. 51B (1974) 23

41. V. Gillet, A.M. Green and E.A. Sanderson, Nucl. Phys. 88 (1966) 321

42. M. Nagao, Proc. Conf. on "Nuclear Structure Studies Using Electron Scatter-
    ing and Photoreaction", Sendai/Japan, (1972) ed. by K. Shoda and H. Ui

43. W. Czyz and J.D. Walecka, Nucl. Phys. 51 (1964) 312;

    V. Devanathan, Nucl. Phys. 87 (1966) 256 and 397

    E. Moniz, Phys. Rev. 184 (1969) 1154

    J. Eisenberg and H.J. Weber, Phys. Lett. 34B (1971) 107

44. E. Borie, H. Chandra and D. Drechsel, Phys. Lett. 47B (1973) 291

45. R. Rosenfelder and W. Haxton, priv. comm.

46. P. Jennewein and B. Schoch, priv. comm.

47. F. Borkowski et al., priv. comm.

48. M. Thies, preprint

A REVIEW OF PRESENT PHOTONUCLEAR RESEARCH AT LUND
AND FUTURE ACCELERATOR PLANS

K Lindgren
Department of Nuclear Physics
Lund Institute of Technology
Sölvegatan 14, S-22362 LUND, Sweden

## 1. INTRODUCTION

In this talk the activity at Lund in the field of photonuclear research both at low and intermediate energies will be presented. This
research is performed with the aid of a small microtron accelerator and
a 1 GeV electron synchrotron. During the last years studies of a new
high duty-factor accelerator system have been carried out. Parts of it
are presently under construction. This accelerator system will also be
described.

## 2. LOW-ENERGY PHOTOFISSION IN $^{236}$U

Photofission near threshold has received a new interest during the
last few years because of its importance in studying the properties of
the double-humped barrier. As pointed out by Bohr [1] photofission of
even-even nuclei offers a very simple case, since photon absorption takes
place mainly through electric dipole and quadrupole absorption. This
results in compound nuclei with the total angular momentum (J) and parity
( ¶ ) equal to $1^-$ and $2^+$. The $1^-$ compound nucleus can fission through
K = 0 or 1, where K is the projection of J on the nuclear symmetry axis.
The $2^+$ compound nucleus can fission through K = 0, 1 or 2.

In the present experiment [2] a thin $^{236}$U target was irradiated with
bremsstrahlung from a microtron and the end-point energy was varied between 5.2 and 6.4 MeV. The fission fragments were detected in thin plastic foils and after etching the foils were analysed by the spark scanning
method [3].

The angular distributions measured were fitted in the usual way [4]
by the expression

$$W(\theta) = a + b \sin^2\theta + c \sin^2 2\theta .$$

Some typical exemples of measured distributions and their decomposition

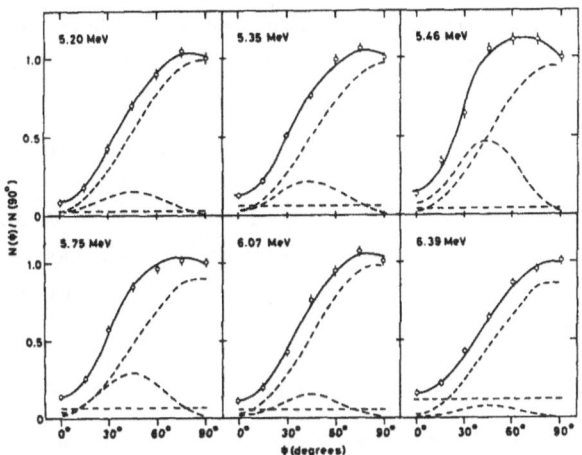

Fig. 1. Experimental angular distributions (o) normalized at 90°. The dashed curves are the components given by the best fit of a, b and c. The solid curves show the sum of these components.

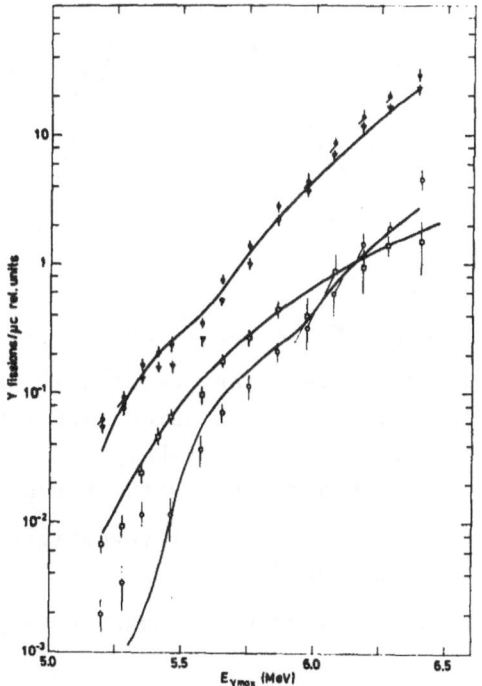

Fig. 2. The experimental yields $Y_{tot}$ (●), $Y_a$ (o), $Y_b$ (▽) and $Y_c$ (□). The solid curves are calculated to give the best fit to the experimental points.

into different components are shown in fig. 1. The partial yields
$Y_a \propto a$ , $Y_b \propto 2/3$ b , $Y_c \propto 8/15$ c and the total yield are shown in fig. 2
The b/a and c/b ratios were also determined as a function of bremsstrah-
lung end-point energy. Maxima were obtained at about 6 MeV and 5.6 MeV
in the b/a and c/b curves respectively.

The experimental results were compared with model calculations from
which it was possible to determine fission barriers for the different
channels $Y_a = 3/2\ Y^{1\ -1}$ , $Y_b = Y^{1\ 0}$ and $Y_c = Y^{2\ 0}$. A double-humped
barrier consisting of three smoothly jointed parabolas was used and the
height of the two humps and the second minimum was determined from a
fit to the experiment. The result is shown in fig. 2. For $Y_c$ and $Y_b$ the
calculated curves fit well to the experimental points. The calculated
b/a and c/b ratios reproduce the maxima obtained in the experimental data

The deduced $1^-1$ and $2^+0$ fission barriers are shown in fig. 3.

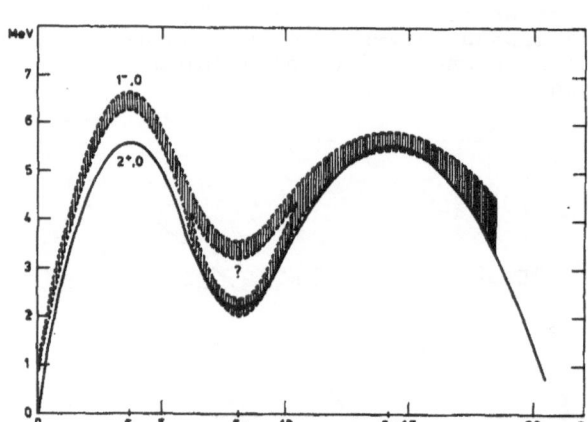

Fig. 3. The shape of the double-humped barriers.

The barrier for the $2^+0$ channel is in agreement with the ground-state
barrier determined in particle induced fission. The $1^-0$ barrier is almost equal to the $2^+0$ barrier at the outer hump and approximatively 1
MeV higher at the inner hump in agreement with theoretical predictions.

It should also be mentioned that similar investigations are in progress /5/ for the nuclei $^{234}$U and $^{238}$U.

3.  PHOTONUCLEAR REACTIONS AT INTERMEDIATE ENERGIES ($E_\gamma > 100$ MeV)

3.1 Introduction

In the experiments presented, cross sections for different reaction
channels have been studied mainly in the $\Delta$ (1236) resonance region to
get information about reaction mechanisms and nuclear structure. In this

energy region the photon absorption is assumed to take place through quasideuteron absorption and pion production on individual nucleons. The different types of reactions studied can grossly be divided into four groups, namely: (i) production of pions from complex nuclei, (ii) direct and semi-direct reactions, (iii) spallation and fragmentation and (iv) fission. The studies cover nuclear masses from deuteron to uranium. Some exemples of these types of experiments recently carried out will be presented. Besides at Lund, experiments have been carried out in collaboration with the DESY, Bonn and Kharkov laboratories.

## 3.2 Electro- and photoproduction of charged pions on $^{27}Al$ and $^{51}V$

Pion photoproduction on complex nuclei in the resonance region has been extensively studied at our laboratory. In a recent experiment carried out in collaboration with DESY the $\pi^+$ production on $^{27}Al$ and $^{51}V$ was investigated. A stack of thin foils was irradiated with electrons in the energy range 130 to 580 MeV and the yields of the product nuclei $^{27}Mg$ and $^{51}Ti$ were measured in each foil. With this technique it was possible to determine both the electro- and photoproduction cross sections. The photopion cross section for vanadium is shown in fig. 4.

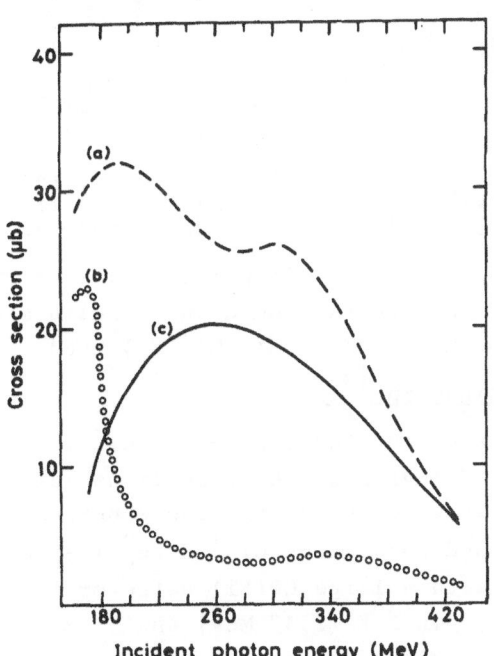

Fig. 4. Comparison of theory and experiment for the reaction $^{51}V(\gamma,\pi^+)^{51}Ti$. Curve (a) is calculated without f.s.i., curve (b) with f.s.i. and curve (c) is the experimental result.

The experimental results have been compared with calculations /7/ based on the impulse approximation. In this energy region pion absorption is supposed to be of great importance. In the calculations pion final state distorsion was taken into account through a realistic optical potential. Calculated cross sections for $^{51}$V with and without final state interactions are shown in fig. 4.

The $\sigma_q / \sigma_e$ ratios for different reactions and target nuclei are shown as a function of electron energy in fig. 5. Besides the Lund/DESY data also very recent data at higher energies from a Kharkov/Lund collaboration /8/ are shown. The experimental data agree both in form and magnitude

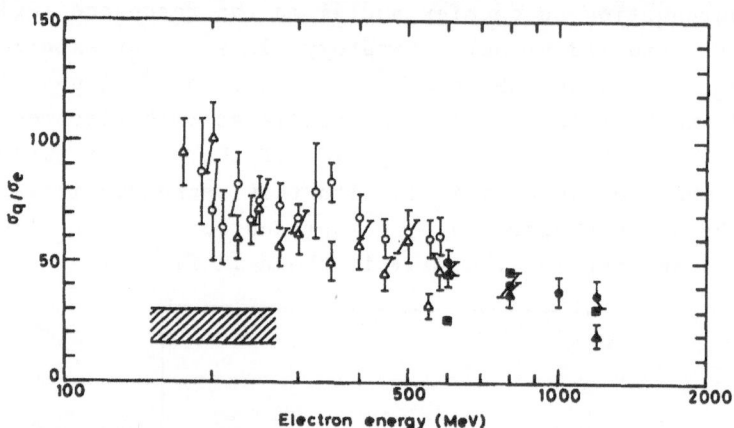

Fig. 5. $\sigma_q / \sigma_e$ ratios. (o) 27Al 27Mg, (Δ) 51V 51Ti Lund/DESY; (●) 27Al 27Mg, (▲) 51V 51Ti, (■) 11B 11C Kharkov/Lund; ///// 11B and 27Al data from ref. /9/.

with calculations with different multipoles in the virtual photon spectrum. Due to the high reaction threshold the results are almost insensitive to the different multipoles (E1, M1, E2) used.

## 3.3 Photonucleon reactions in $^{40}$Ca

A method to study gamma one-nucleon reactions is to detect $\gamma$-rays from short-lived excited levels in the product nuclei. In this experiment /10/ natural calcium was used as target. The experimental arrangement was the same as that used in a study of $^{16}$O reported earlier /11/. A $\gamma$-ray spectrum taken with a large Ge(Li) detector is shown in fig. 6. The first excited state in $^{39}$K (2.47 MeV) and $^{39}$Ca (2.52 MeV) are single hole states and may be assigned to be $(2s_{1/2})^{-1}$. In the present experiment the yield of these two states as a function of maximum bremsstrahlung energy was studied. From the yield due to pion production the number of nucleons available in the two reactions was determined.The upper limit

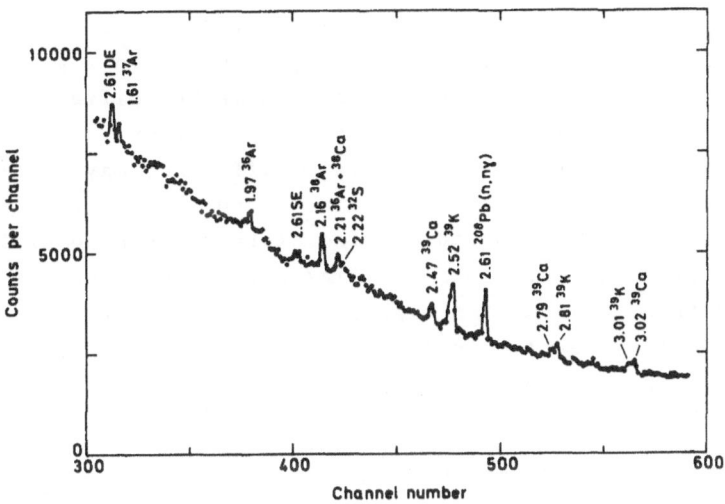

Fig. 6. A γ-ray spectrum from Ca irradiated with 360 MeV bremsstrahlung.

for this number was found to be 1.3 and 2.7 for the neutron and proton reaction respectively. These values can be compared with results from pick-up reactions.

## 3.4 The $(\gamma,2n)$ reaction in $^{12}$C and $^{16}$O

The $(\gamma,2n)$ reaction has been studied /12/ in the two nuclei $^{12}$C and $^{16}$O by the activation method. Due to the short half-lives of the product nuclei a pneumatic system was used to transport the irradiated targets to a Ge(Li) detector. In fig. 7 the yield curve for carbon is shown.

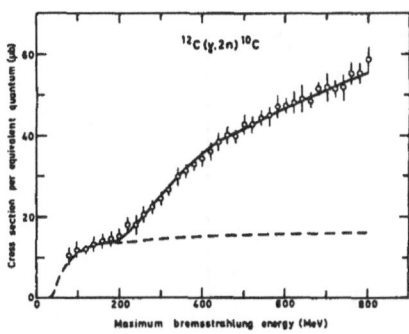

Fig. 7. Measured yield for the $(\gamma,2n)$ reaction in $^{12}$C. The dashed curve shows the low-energy contribution from ref./13/

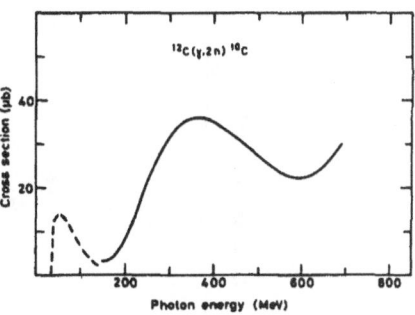

Fig. 8. Deduced cross section for the $(\gamma,2n)$ reaction in $^{12}$C. The dashed curve shows data from ref. /13/.

138

A marked increase in the yield at the pion threshold is observed. The
deduced cross section for the same nucleus is shown in fig. 8. Similar
results were obtained for $^{16}$O but there is a marked difference in the
cross section magnitude. For oxygen the cross section in the pion region
is less than 1/3 of that for carbon. This difference may be explained
by the fact that in $^{14}$O there is only one bound level (the ground state)
but in $^{10}$C there is also a bound excited state. This also explains why
cascade-evaporation calculations /12, 14/ give good agreement with experiment in the carbon case but too high a value in the oxygen case. Comparison with (p,p2n) reaction data for the same nuclei /15/ suggests that
the final state interactions are similar in photon and proton induced
reactions at these energies.

### 3.5 Deuteron photodisintegration and yields of photoprotons from Be

Results of a comprehensive series of measurements of the cross section
for deuteron photodisintegration from 139 to 832 MeV have been earlier
reported /16/. At energies below 300 MeV, the results were in marked
disagreement with the data from Bonn /17/ and Orsay /18/.

The measurements described here /19/ have been intended (i) to check
upon the reproducibility of the earlier data when the measurements are
repeated with major alterations in the experimental set-up and (ii) to
carry out a high resolution study in the energy region around 150 MeV
in an attemt to see if the cross section goes to zero somewhere at these
energies. This latter point is motivated by the Mainz total absorption
data /20/ for Li and Be, where the cross section is zero just below the
pion threshold.

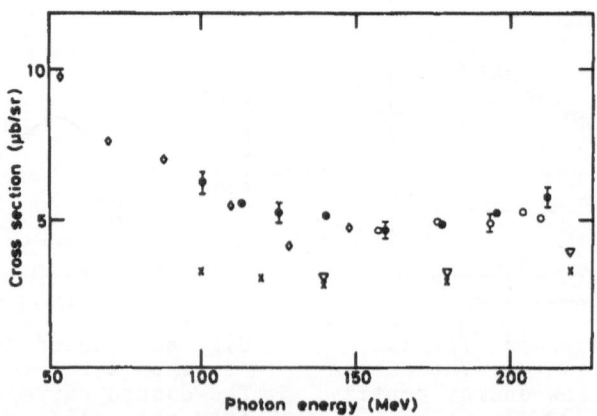

Fig. 9. Cross section for deuteron photodisintegration at a c.m.
angle of 113.6°. (●) new Lund data ref. /19/, (○) Lund data
ref. /16/, (◇) ref. /21/, (▽) ref. /17/, (×) ref. /18/.

The experimental method used was the same as employed earlier. The recoil protons from a liquid deuterium target were detected in a scintillator counter telescope. Some of the new results are shown in fig. 9. The present data reproduce the earlier ones and confirm the systematic discrepancies between different laboratories. For the 150 MeV region the data show that there is certainly no evidence for a 'Mainz hole' effect in the deuteron photodisintegration.

The possibility of a zero cross section in the 150 MeV region has been further checked upon by measurements of photoprotons from Be /22/ with high resolution ( $\Delta E_{max}$ = 5 MeV). The measured yields are shown as a function of the logaritm of the maximum bremsstrahlung energy in fig. 10. A zero cross section should show up as a plateau in the yield curves. As seen there is no evidence of a plateau in the present data.

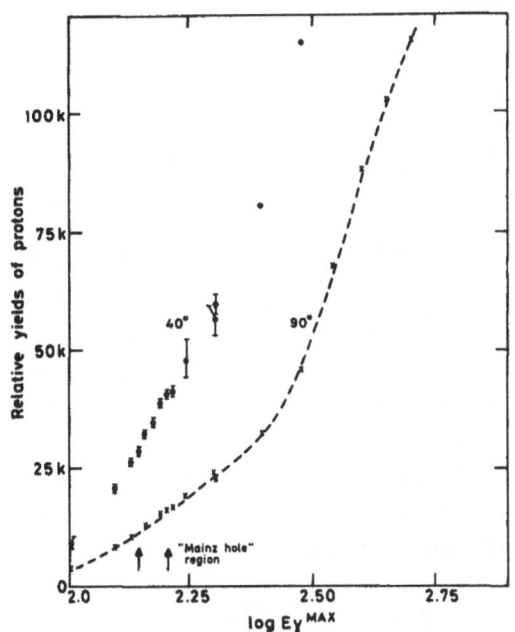

Fig. 10. Relative yields of photoprotons from Be for proton c.m. angles of 40° and 90°.

## 3.6 $\alpha$-clustering effects in the photodisintegration of Cu, Ag and Au

The emission of $\alpha$-particles from Cu, Ag and Au irradiated by 500 MeV bremsstrahlung has been studied /23/ with a telescope of surface barrier detectors. Energy- and angular distributions have been measured. A comparison with cascade-evaporation calculations /14/ shows that some $\alpha$-particles must be emitted during the cascade step. Energy- and angular distributions resulting from knock out of $\alpha$-particles by cascade nucleons were calculated. The properties of the cascade particles were taken from Monte Carlo calculations and the parameters for the potential and height of the Coulomb barrier were taken from $\alpha$-scattering experiments /24/. The number of $\alpha$-clusters on the nuclear surface, defined as the outermost 2 fm of the nucleus, was taken to be 3 for Cu and Ag and 4 for Au. Fig. 11 shows a comparison between the experimental angular distributions for Ag and Au and the calculated ones for evaporated and directly emitted $\alpha$-particles.

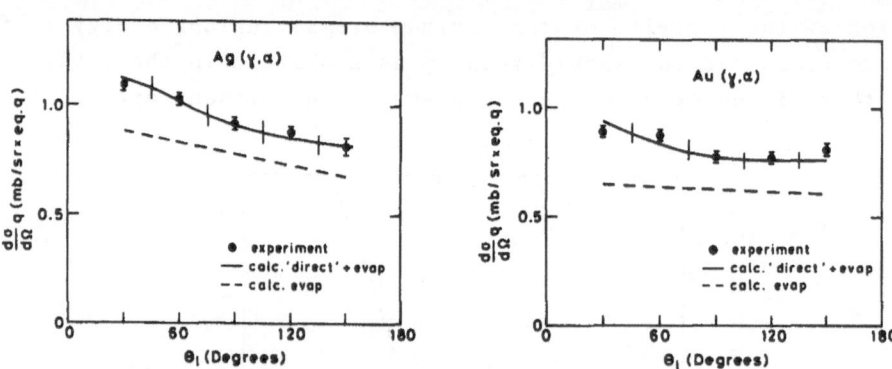

Fig. 11. Comparison between experimental and calculated angular distributions of $\alpha$-particles from Ag and Au.

## 3.7 Photospallation reactions in $^{45}$Sc and $^{nat}$Cu

Photon-induced spallation yields and cross sections have been measured for a large varity of target nuclei and incident energies. The yields have been analysed with a formula first applied to particle induced spallation reactions /25/. Taking into account the target mass ($A_t$) and energy ($E_{max}$) dependence of the parameters the formula applied to photon-induced spallation /26/ can be written:

$$\sigma_q(A,Z) = \frac{\hat{\sigma}_q P d^{-2/3} A_t^{-2/3} e^-}{1.79 \left[ e^{PA}t \left(1 - \frac{2e^-}{3PA_t}\right) - 1 + \frac{2e^-}{3} + \frac{2e^-}{3PA_t}\right]} \times$$

$$\exp\left[PA - R|Z - SA + TA^2|^{3/2}\right]$$

with  $P = 7.66 A_t^{-0.89}$  ($E_{max} > 600$ MeV)

$\hat{\sigma}_q = (0.81 + 0.184 \ln E_{max}) A_t^{1.13}$  ($E_{max}$ in MeV)

$R = d^- A^{-e^-}$

d´ = 11.8 ;   e´ = 0.45
S  = 0.486;   T  = 0.00038

Recently the yields of a large number of reaction products from $^{45}$Sc
and $^{nat}$Cu irradiated with 2 GeV bremsstrahlung have been measured /27/
by the activation method. The experimental yields have been compared
with those predicted by the above formula. The result for copper is shown
in fig. 12. Most of the measured yields are in very good agreement with

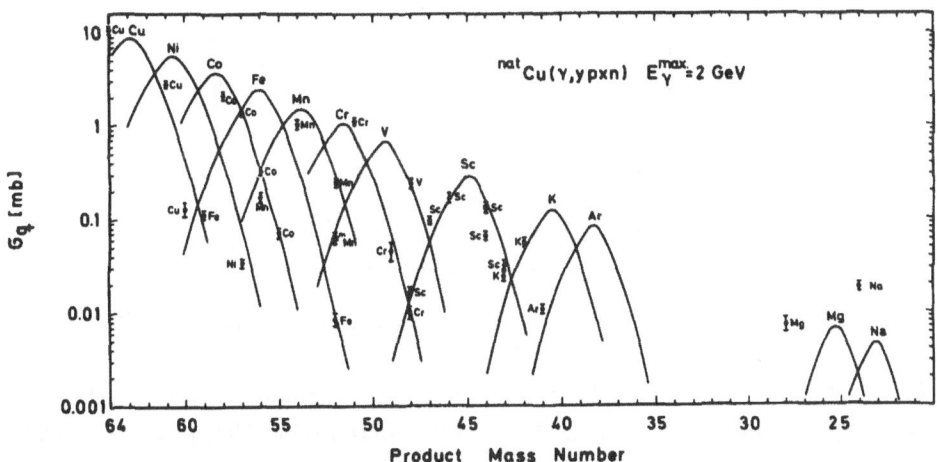

Fig. 12. Experimental yields of different products from Cu. The
curves show predictions by the above formula.

the predicted ones. It is, however, interesting to observe that the yields of the lightest elements (Mg and Na) exceed the expected values by
more than one order of magnitude indicating the onset of a new mechanism
(possibly a $\varrho$-meson effect).

3.8 <u>Photoproduction and recoil energies of $^{24}$Na from heavy nuclei</u>

The photoproduction of $^{24}$Na from targets with mass number $27 \leqslant A_t \leqslant 65$
has been studied earlier at this laboratory /28/ with the aid of the
activation method. These investigations have now been extended /29/ with
measurements on 9 target elements in the mass region $79 \leqslant A_t \leqslant 238$. The
mean cross sections ($\bar{\sigma}$) below 1 GeV were deduced and are plotted as a
function of $A_t$ in fig. 13 where also the earlier results are given.

The exponential decrease of $\bar{\sigma}$ with increasing $A_t$ for $A_t < 65$ is in
agreement with the assumption that $^{24}$Na is produced in a spallation
reaction. For $A_t > 65$ the decrease is weaker and for $A_t > 118$ the cross
section increases with increasing $A_t$. This indicates that a new process
is gradually taking over. The process could be e. g. binary fission,

ternary fission and/or fragmentation.

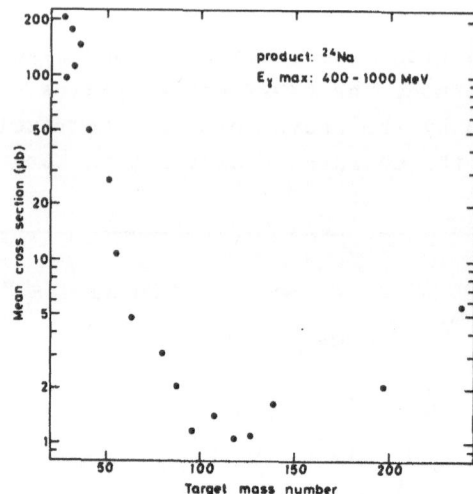

Fig. 13. Mean cross section for production of $^{24}$Na from different target nuclei.

For the target elements Cu, Ag and Au the mean range of the $^{24}$Na product was measured at $E_{max}$ = 800 MeV using the thick-target thick-catcher technique. From the measured ranges the kinetic energies were determined and 16.0, 34.8 and 66.5 MeV were obtained for the three target nuclei Cu, Ag and Au respectively. These results have been compared with recoil energies calculated for binary fission of a deformed nucleus where one of the fragments is $^{24}$Na. The calculated energies were found to be about 20% larger than the experimental ones.

3.9 Photofission of U, Th and Bi

Among the experimental methods used to study the fission process one of the most successful ones has been the application of surface barrier detectors. With two heavy-ion detectors the complementary fragments from a given fission event are registered simultaneously. The number of applications of this technique to photofission studies have been few so far.

Products from photofission of U, Th and Bi induced by 600 MeV bremsstrahlung have been studied /30/ with such a detector system. The kinetic energies of the two fragments were analysed event by event to yield primarily the single kinetic energies of the two fragments and the total kinetic energy associated with the particular fission event. As an example of information obtained the total kinetic energy distributions from the three target nuclei are shown in fig. 14. It was also possible to

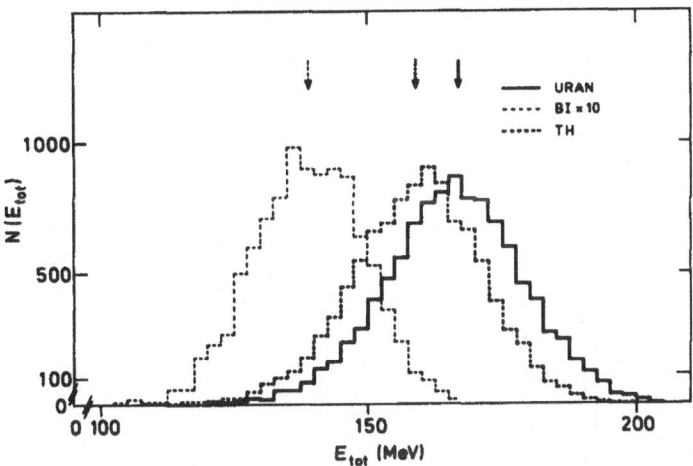

Fig. 14.Total kinetic energy distributions for U, Th and Bi.

obtain data about the variation of the total and single kinetic energies
with the product mass number and correlations between the different quantities studied.

## 3.  AN ELECTRON ACCELERATOR SYSTEM WITH HIGH DUTY-FACTOR

### 3.1 Introduction

During the last years plans for new electron accelerators with high
duty-factors have been presented at many laboratories. To obtain continous beams different techniques have been suggested, namely: (i) superconducting linacs, (ii) use of brute power and (iii) pulsed accelerators
in combination with pulse-stretcher storage rings.

At Lund studies of a high duty-factor accelerator system /31/ started
about 2 years ago. The system consists of a 100 MeV race-track microtron
and a pulse-stretcher ring.

### 3.2 The race-track microtron

As accelerator a race-track microtron was chosen due to its small
energy spread, compactness and low cost. In such a machine the accelera-
tion is achieved by a linac situated between two bending magnets, which
recycle the beam under resonant conditions. The microtron proposed is
shown in fig. 15. The electrons will be recycled up to 18 times with an
energy gain of 5.33 MeV per turn to reach the maximum energy of 100 MeV.

Due to the non-relativistic velocity of the electrons in the first
orbit the resonant condition is not fulfilled. To overcome this problem

two methods have been studied. One method is to inject the electrons into the first orbit via an injection linac. Another method is to let the electrons reenter into the linac from the opposite direction. The phase in the first orbit can then be adjusted by moving the linac along its axis. This latter method will be utilized.

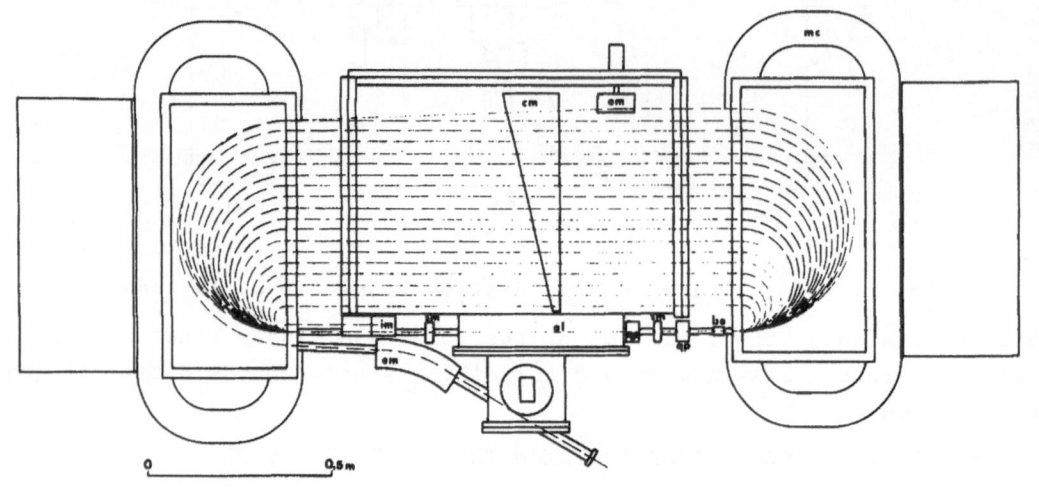

| im | Injection magnet | be | Bellow |
|----|------------------|----|--------|
| sm | Stearing magnets | cm | Correction magnet |
| qp | Quadrupoles | em | Extraction magnets |
| al | Acceleration linac | mc | Magnet coils |

Fig. 15. The race-track microtron.

The linac is a side-coupled standing wave $\pi/2$ mode structure with 8 cavities. The RF source will be a 5 MW klystron operating at 2998 MHz.

The 180° bending magnets are designed to give proper orbit lengths and suitable axial focusing. This is obtained by using a reversal field close to the magnet fronts.

Focusing is achieved both in radial and axial direction in the linac. This focusing is, however, not sufficient and therefore a quadrupole doublet is introduced just after the linac to get proper radial focusing. Axial focusing is also achieved by the reversed magnetic field mentioned above

The extraction system consists of two magnets one of which is move-able perpendicular to the electron orbits. This system allows extraction from any orbit except for the first two ones.

The characteristics of the extracted beam are:

| Energy | 6 - 100 MeV |
|---|---|
| Energy spread | 0.1% |
| Mean current | ~50 μA at 100 MeV |
| Duty-factor | 0.15% |

## 3.3 The pulse-stretcher

To increase the duty-factor it is planned to put the electron beam from the microtron into a pulse-stretcher ring. This is shown schematically in fig. 16. The aim of the pulse-stretcher is to store the electron pulse and then extract the electrons continously until the next pulse enters. The magnet structure consists of four weak-focusing bending magnets separated by field-free straight sections. The straight sections give place for injection and extraction devices, a tuning quadrupole and correction sextupoles.

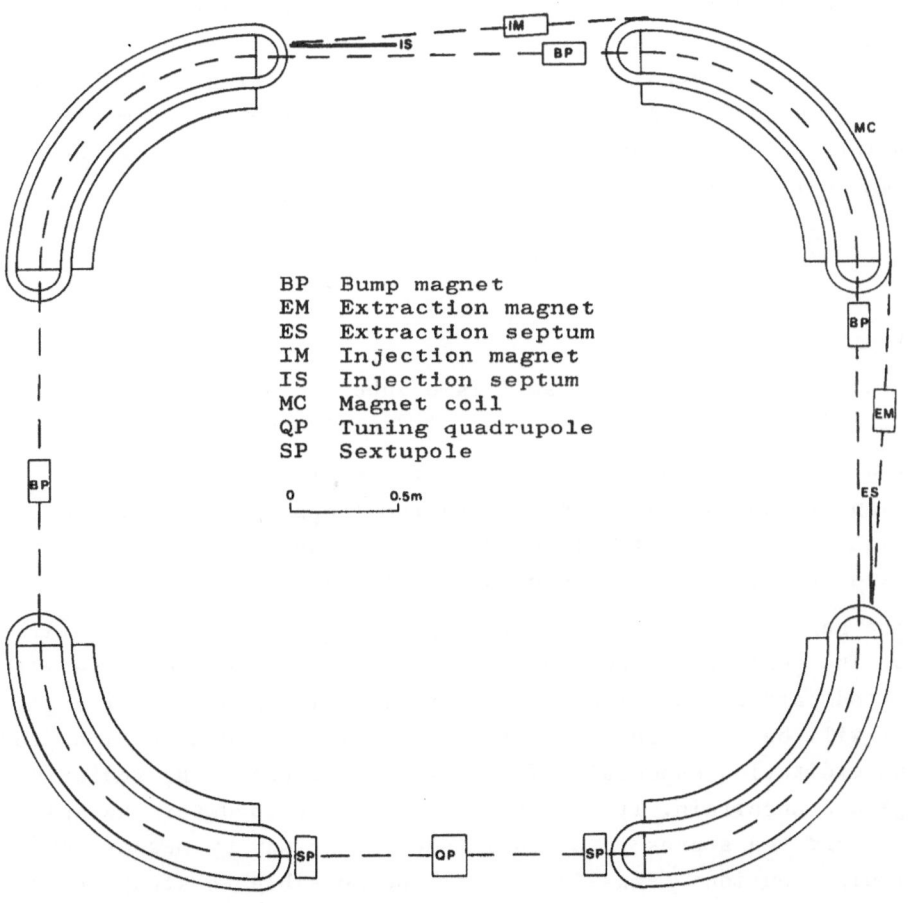

BP   Bump magnet
EM   Extraction magnet
ES   Extraction septum
IM   Injection magnet
IS   Injection septum
MC   Magnet coil
QP   Tuning quadrupole
SP   Sextupole

0        0.5m

Fig. 16. The pulse-stretcher

Since the circumference of the ring is 13.4 m the 300 m long pulse must be wound up in 23 turns at the injection. The inflector is a combination of a magnetic bend and an electrostatic septum. During the injection process, the closed orbit has to be continously moved away from the septum. This is accomplished by a local closed orbit distorsion (bump) which is made to shrink towards the central orbit as injection continous.

Due to the repetition rate (750 p/sec) the particles must be extracted within 1.3 msec. The extraction must be nearly constant during this time to maximize the duty-factor. The method to achieve this is commonly referred to as slow resonant extraction. Non-linear terms in the equations of motion are responsible for the extraction. The oscillation amplitude of that part of the beam which is at resonance will suddenly increase. When the amplitude has grown beyond a certain value, the particles are deflected into an extraction channel consisting of an electrostatic septum and a bending magnet. By tuning the ring with the aid of the quadrupole particles with initially smaller amplitudes will be brought into resonance and extracted. The extracted beam will have the following characteristics:

Energy and energy spread the same as above
Mean current   $> 10 \ \mu A$
Duty-factor    $> 50\%$

## 3.4 Experimental program

The race-track microtron itself is planned to be mainly used for applications such as activation analysis and isotope production.

In the first step the beam from the pulse-stretcher will be used for production of monoenergetic $\gamma$-rays by the 'tagging' method. The experiments planned with this facility are elastic and inelastic photon scattering and studies of the de-excitation of the giant dipole resonance.

In the next step electron scattering coincidence experiments are planned. The scattered electron will be detected in coincidence with the emitted particle or cluster of particles. In this kind of experiments it will be possible to accurately determine cross sections for electron excitation especially for heavy nuclei where the radiation tail is large and uncertain. It will also be possible to determine spin and parity of excited states in a model independent way by measurements of angular distributions. Finally, it will be possible to study the de-excitation of multipole vibrations of more generalized type compared with real photon experiments.

The author wishes to thank the members of the photonuclear research group, the deuteron group and the accelerator group for their contributions of unpublished results to this report.

REFERENCES

1.  A. Bohr, 'Proc. Int. Conf. Peaceful Uses At. Energy' 2 p. 151 (1960), U. N., New York

2.  A. Alm and L. J. Lindgren, Nuclear Physics Report LUNP 7506 (1975)

3.  W. G. Cross and L. Tommasino, Rad. Eff. 5(1970)85

4.  R. Vandenbosch and J. R. Huizenga, 'Nuclear Fission', Academic Press, New York (1973)

5.  L. J. Lindgren and A. Sandell, private communication

6.  I. Blomqvist, H. Dinter, N. Freed, P. Janeček, G. G. Jonsson, P. Ostrander and K. Tesch, to be published

7.  N. Freed and P. Ostrander, Phys. Lett. 61B(1976)449

8.  I. Blomqvist, G. G. Jonsson, V. S. Kuz´menko, A. V. Mitrofanova, V. I. Noga, Yu. N. Ranyuk, P. V. Sorokin and Yu. N. Telegin, Kharkov/ Lund collaboration 1976, to be published

9.  V. I. Noga, Yu. N. Ranyuk, P. V. Sorokin and V. A. Tkachenko, Sov. J. Nucl. Phys. 14(1972)506

10. J.-O. Adler, B. Bülow, G. G. Jonsson and K. Lindgren, to be published

11. J.-O. Adler, G. Andersson, B. Forkman, G. G. Jonsson and K. Lindgren, Nucl. Phys. A171(1971)560

12. B. Johnsson, M. Nilsson and K. Lindgren, to be published

13. K. Kayser, W. Collin, P. Filss, S. Guldbakke, G. Nolte, H. Reich, J. O. Tier and W. Witschel, Z. Phys. 239(1970)447

14. T. A. Gabriel and R. G. Alsmiller, Jr, Phys. Rev. 182(1969)1035

15. L. Valentin, G. Albouy, J. P. Cohen and M. Gusakow, Phys. Lett. 7(1963)163

16. P. Dougan, T. Kivikas, K. Lugner, V. Ramsay and W. Stiefler, Z. Phys. 276(1976)55

17. R. Kose, W. Paul, K. Stockhorst and K. H. Kissler, Z. Phys. 202 (1967)364

18. J. Buon, V. Gracco, J. Lefrancois, P. Lehman, B. Merkel and Ph. Roy, Phys. Lett. 26B(1968)595

19. P. Dougan, V. Ramsay and W. Stiefler, Report LUSY 7506 (1975)

20. J. Ahrens, H. Borchert, K. H. Czock, H. B. Eppler, H. Gimm, H. Gundrum, M. Kröning, P. Riehn, G. Sita Ram, A. Zieger and B. Ziegler, Nucl. Phys. A251(1975)479

21. Iu. A. Aleksandrov, N. B. Delone, L. I. Slovokhotov and G. A. Sokol, Sov. Phys. JETP 6(1958)472

22. P. Dougan, private communication

23. J.-O. Adler, G. Andersson, H.-Å. Gustafsson and K. Hansen, to be published

24. G. Hauser, R. Lökhen, H. Rebel, G. Schatz, G. W. Schwimmer and J. Specht, Nucl. Phys. A128(1969)81

25. G. Rudstam, Z. Naturf. 21a(1966)1027

26. G. G. Jonsson and K. Lindgren, Physica Scripta 7(1973)49

27. N. M. Bachschi, P. David, J. Debrus, F. Lübke, H. Mommsen, R. Schoenmackers, G. G. Jonsson and K. Lindgren, Nucl. Phys. A264(1976)493

28. A. Järund, B. Friberg and B. Forkman, Z. Phys. 262(1973)15

29. A. Järund and B. Forkman, Nuclear Physics Reports LUNP 7607, 7608 (1976)

30. G. Andersson, M. Areskoug, H.-Å. Gustafsson, G. Hyltén and B. Schrøder, to be published

31. R. Alvinsson and M. Eriksson, Report TRITA-EPP-76-07 and LUSY 7601 (1976)

# A MONOCHROMATIC AND POLARIZED PHOTON BEAM FOR PHOTONUCLEAR REACTIONS. THE LADON PROJECT AT FRASCATI.

G. Matone, P. Picozza, D. Prosperi, A. Tranquilli
INFN, Laboratori Nazionali di Frascati, Frascati, Italy

R. Caloi, C. Schaerf
Istituto di Fisica dell'Università, Roma, Italy

S. Frullani, C. Strangio
Istituto Superiore di Sanità, Roma, Italy

Presented by G. Matone

## 1. - INTRODUCTION.

In 1962, R. H. Melbourne[1] and F. R. Arutyunian[2] pointed out that backward Compton scattering of an intense polarized Laser light beam by high energy electrons would produce useful yields of nearly monoenergetic, polarized photons.

Because the usual Compton reaction is a two body process, for a determined incident photon energy $\omega_1$ and a fixed incident geometry, the energy $\omega_2$ of the scattered photons depends only on the emission angle $\theta$ as measured with respect to the direction of the incident electron beam.

For a head-on collision (see Fig. 1) and for $\theta \ll 1/\gamma$, the differential cross section for the process can be written as follows[3]:

$$(1) \qquad \frac{d\sigma}{d\Omega} = 4 r_o^2 \gamma^2 \frac{1 + n^4}{(1 + n^2)^4} \quad ,$$

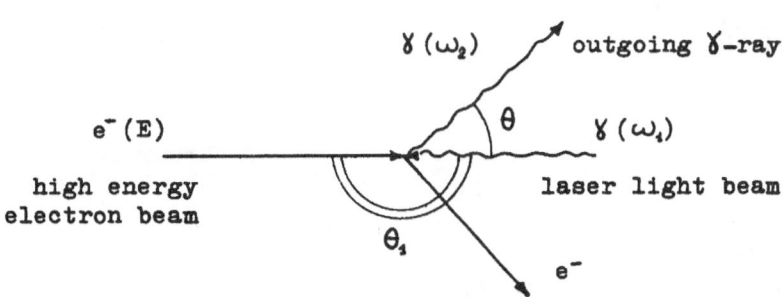

Fig. 1 - Schematic drawing of the head-on Compton scattering. E and $\omega_1$ are, respectively, the energies of the colliding eletrons and photons. $\omega_2$ is the energy of the outgoing photons.

where :

    $r_o$  is the classical electron radius,

    E   is the incoming electron energy,

    $\gamma$   $= E/m$,    $n = \gamma\theta$.

       Relation (1) descends directly from the original Arutyunian expression in the limit

$$\lambda = \frac{2\,\omega_1\,E}{m^2} \ll 1,$$

which is certainly fulfilled for all the high energy electron beams and lasers now operating (see Fig. 2).

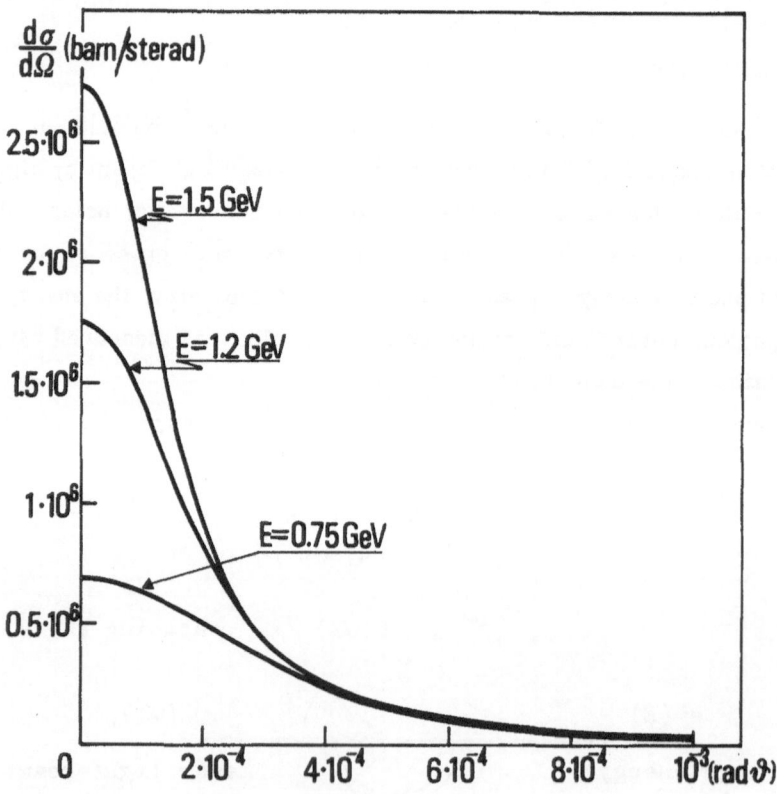

Fig. 2 – Differential cross section for Compton scattering on moving electrons when $\lambda \ll 1$ ($\omega_1 = 2.54$ eV).

       In the same approximation, the energy of the scattered photons is given by the expression:

(2)
$$\omega_2 = \frac{2E\lambda}{1 + n^2 + 2\lambda},$$

which reaches its maximum value:

(3)
$$\omega_2^{max} \simeq 4\,\omega_1\,\gamma^2$$

at $\theta = 0$.

Since the final photons come out in a very small cone along the inital electron direction suitable beam collimation selects narrow energy bands preserving at
the same time considerably high counting rates. From eq. (2) we can immediately
obtain a rough estimate of the energy resolution:

(4)
$$\frac{\Delta\omega_2}{\omega_2^{max}} = \frac{1}{1+2\lambda}(\gamma\theta_c)^2 \quad,$$

where $\theta_c$ is the collimator half-angle, and $\Delta\omega_2 = \omega_2(\theta_c) - \omega_2^{max}$.

As far as the photon polarization is concerned, for very relativistic electrons their elicity is a good quantum number.

Therefore, the electron spin flip amplitude is negligeable and the backward
scattered photons retain their original polarization. At other angles the photon polarization can change due to the role of the orbital angular momentum. If we use a Laser with Brewster windows which produce plane polarized light, the high energy photons produced by a head-on Compton scattering at $\theta = 0$ will have similar polarization.

For head-on collision, and $\theta \ll 1$ the linear polarization behaves as[4]:

(5)
$$P \simeq \begin{cases} (1+n^4)^{-1} & n \leq 1, \\ n^4(1+n^4)^{-1} & n \geq 1, \end{cases}$$

and in particular for $\Delta\omega_2/\omega_2^{max} \ll 1$ its mean value turns out to be:

(6)
$$\langle P \rangle \simeq 1 - (\frac{\Delta\omega_2}{\omega_2^{max}})^2 \quad.$$

Several attempts have been carried out during the past 15 years in order to
study this process experimentally.

The major attempts have been made at the Lebedev Physical Institute (E $\simeq$
$\simeq$ 600 MeV; $\omega_2 \simeq$ 7 MeV)[5] and at the Cambridge Electron Accelerator (E $\simeq$ 6 GeV;
$\omega_2 \simeq$ 0.4 GeV)[6] where the 1.78 eV Ruby laser line was always used. Without en-
tering into any details of these experiments (see Figs. 3 and 4), we can say that no-
ne of these could provide an intense photon beam for experimental research either

152

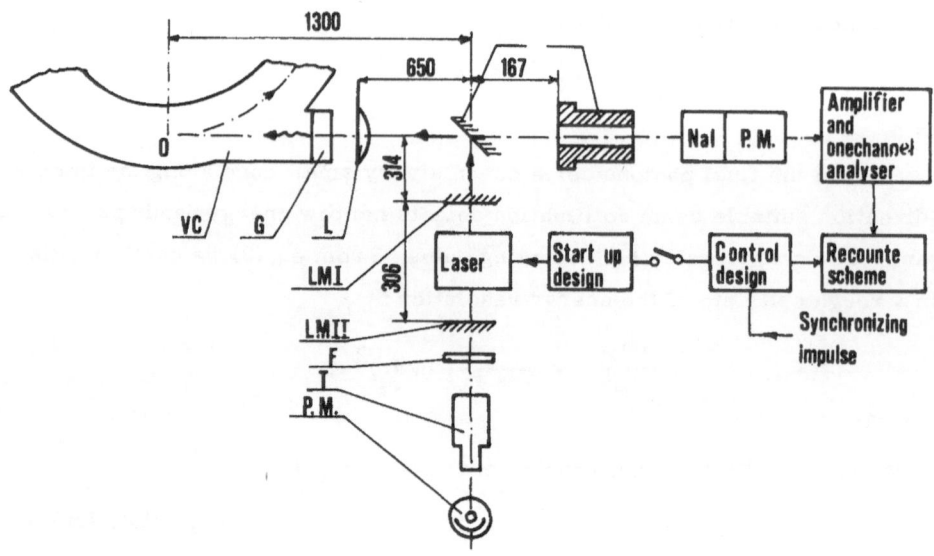

Fig. 3 - Layout of the Lebedev Institute experiment.

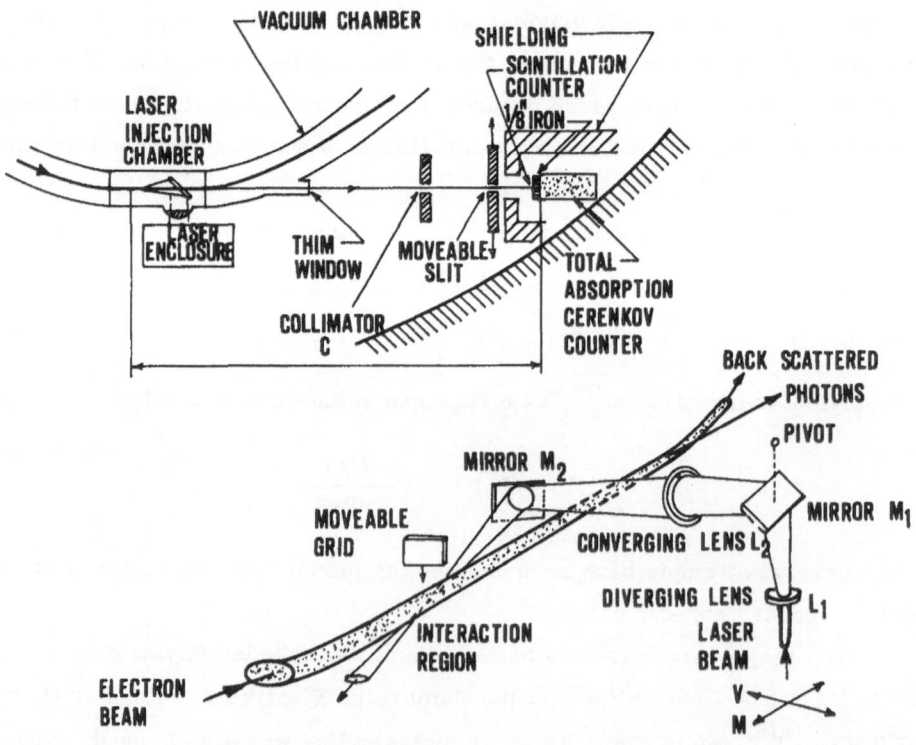

Fig. 4 - a) Schematic diagrams of the CEA experimenatl set-up; b) Laser injection geometry at CEA (see ref. (6)).

in nuclear or in particle physics.

On the contrary, a more successful attempt has been carried out at SLAC
($E \simeq 20$ GeV; $\omega_2 \simeq 5$ GeV)[7] where experiments on vector meson photoproduction
by polarized gammas have been performed with a hydrogen bubble chamber (see
Fig. 5).

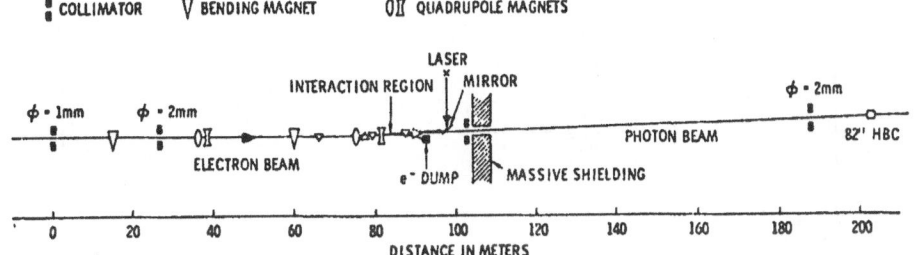

Fig. 5 - Conceptual schematic of the back scattered laser beam at SLAC
(see ref. (7)).

## 2. - THE FRASCATI PROJECT.

### 2.1. - Generalities.

While the previous attempts were characterized by the output laser beam
directly sent against the electrons, in our plane, extensively described elsewhere[4],
the laser effect is supposed to take place in a cavity as long as the electron machine
(ADONE) straight section. A schematic sketch of the cavity arrangement on the machine pipe is shown in Fig. 6.

The improvement we represent lies in the fact that we use the 2.54 eV line
of an Argon Ion laser and that, in addition, the power stored in the cavity can be up
to $\sim 100$ times as intense as the standard 3.5 Watt output power given by our CR8
(Coherent Radiation) Argon Ion laser at 4880 Å.

On the other hand this solution raises the problem to install the laser optics
on the machine pipe kept at the high vacuum of the storage ring. Furthermore it needs a careful allignment to be done between both the electron and laser lines. Another important question to bear in mind is that in our set-up the cavity mirror must
be protected against the damage caused on their coatings by the intense flux of synchrotron radiation impinging upon it.

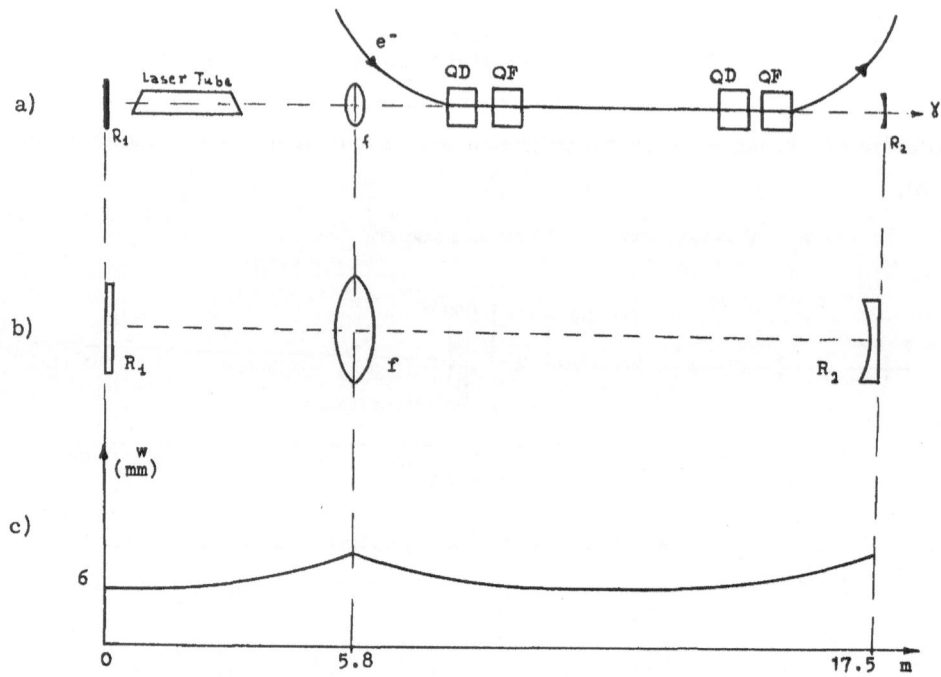

Fig. 6 - a) Schematic arrangement of our laser cavity on the Adone vacuum
pipe (QF and QD are focusing and decofusing quadrupoles respectively);
b) laser cavity optics - $R_1$, $R_2$ are mirrors and f is a lens; c) beam profile in the cavity.

All these technical problems have been carefully considered and their solutions are now in the throes of being set up.

The design of the cavity dimensions (curvature radii, focuses, and distances) has been studied with great accuracy to guarantee high stability, low diffraction
losses and maximum overlap both in the active medium and in the interaction region
with the electron beam[8]. The complete layout of the whole experiment is sketched
in Fig. 7.

## 2.2. - Beam characteristics and counting rates.

Since the circulating electrons in the storage ring have a finite angular divergency it is, of course, useless to collimate with an angle smaller than this electron divergency. This quantity is proportional to the electron energy:

$$\theta_e = \sigma \gamma ,$$

where $\sigma$ is a characteristic parameter of the machine. Therefore, the minimum
energy resolution one can have is:

155

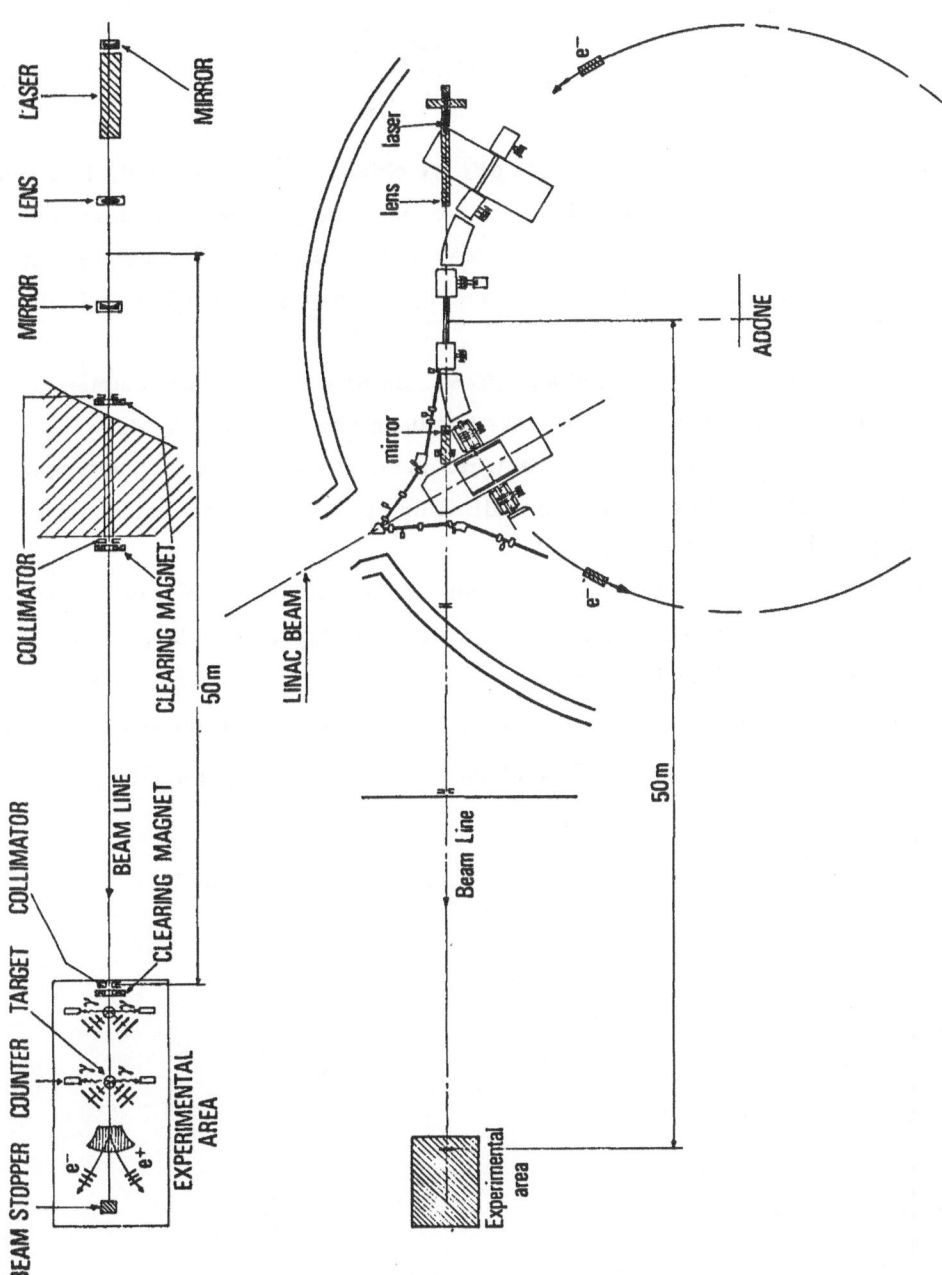

Fig. 7 – a) Sketch of the complete set-up of the experiment; b) Detailed view of the beam line.

$$\Delta\omega_2 \simeq \omega_2^{max}(\gamma\,\theta_e)^2 = \sigma^2 \frac{(\omega_2^{max})^3}{16\,\omega_1^2} \;. \tag{7}$$

In our case, for E = 1.5 GeV, $\omega_1$ = 2.54 eV, we obtain $\omega_2^{max}$ = 82 MeV. Moreover by optimizing the Adone electron optics, one can push the parameter $\sigma$ down to a value of $\sim 10^{-8}$ rad[9] and obtain an energy resolution $\Delta\omega_2/\omega_2^{max} \sim 1\%$. Furthermore with such a value, from eq. (6), the photon polarization at E = 1.5 GeV comes out to be $P \approx 0.99$.

In order to take into account any effect raising from the finite extension of the source, the Compton scattering process has been simulated with a Monte Carlo calculation reconstructing the energy spectrum and counting rate of the photons scattered inside the solid angle defined by a collimator placed at a fixed distance from the centre of the interaction region.

In particular, since at     each     end  of the straight section two quadrupoles are placed which worsen the electron angular divergency of a large factor ($\sim 5$) (see Fig. 8), we will distinguish two cases:

a) quadrupoles out, the collisions take place in the region between the two quadrupoles;
b) quadrupoles in, the collisions take place in the whole straight section.

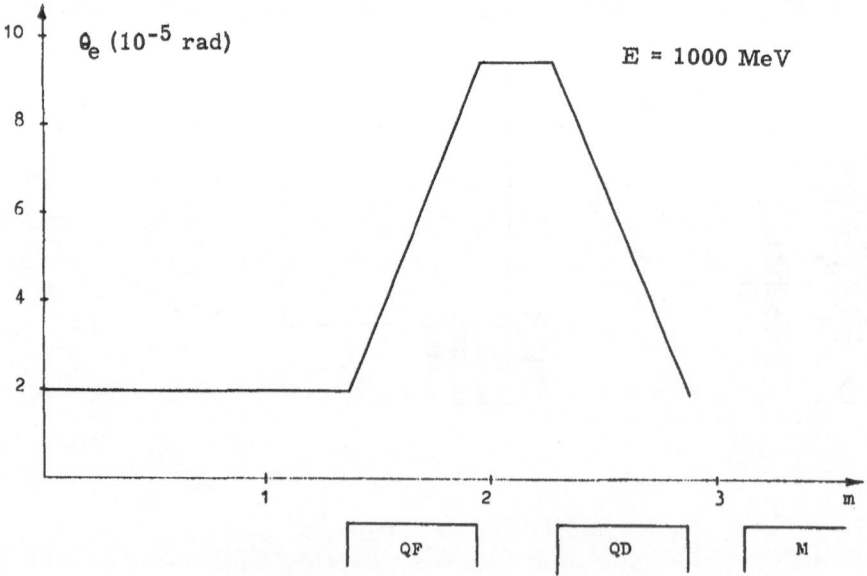

Fig. 8 - Electron angular divergency in the straight section for $\sigma = 10^{-8}$ rad (QF, QD are focusing and defocusing quadrupole, M bending magnet).

For both cases the following set of parameters has been used ($\sigma = 10^{-8}$ rad):

| E (GeV) | $\theta_e$ (rad) | $\omega_1$ (eV) | $\varrho$ (m) | $\xi$ (m) | $\eta$ (m) | $\chi$ (m) | P (watt) | I (mA) |
|---|---|---|---|---|---|---|---|---|
| 0.75 | $1.5 \times 10^{-5}$ | 2.54 | $7.1 \times 10^{-4}$ | $1.8 \times 10^{-3}$ | $7.5 \times 10^{-4}$ | $7.1 \times 10^{-4}$ | 250 | 100 |
| 1.5 | $3.0 \times 10^{-5}$ | 2.54 | $7.1 \times 10^{-4}$ | $9.1 \times 10^{-4}$ | $3.8 \times 10^{-4}$ | $7.1 \times 10^{-4}$ | 250 | 100 |

where:

$\theta_e$ = electron angular divergency,

$\omega_1$ = laser photon energy,

$\rho$ = laser spot radius,

$\xi$ = radial electron spot dimension (1 standard deviation of a gaussian like distribution),

$\eta$ = vertical electron spot dimension (1 standard deviation of a gaussian like distribution),

P = laser power stored in the cavity,

I = electron circulating current.

Since the position of the central trajectory of the electron beam shows random fluctuations this effect has been simulated assuming that it has a gaussian distribution both in the radial and vertical directions. The parameter $\chi$ represents one standard deviation of these distributions.

The Monte Carlo calculations have been performed for different values of the radius of the collimator defining the gamma ray beam and its distance from the center of the interaction region. The results have indicated that over a large range of values for these two parameters the energy resolution and counting rate depend only on the solid angle defined by the collimator and not indipendently on its position and bored radius.

Histograms showing typical energy distributions at E = 1.5 GeV and E = 0.75 GeV are reported in Figs. 9-12: the maximum energies are $\sim 82$ and $\sim 21$ MeV respectively. Provided that the peak is sufficiently narrow one can neglect the details of the distributions and define the parameter

$$R = \frac{1}{\bar{\omega}_2} \left[ < (\omega_2 - \bar{\omega}_2)^2 > \right]^{1/2} ,$$

as a good estimate of the energy resolution.

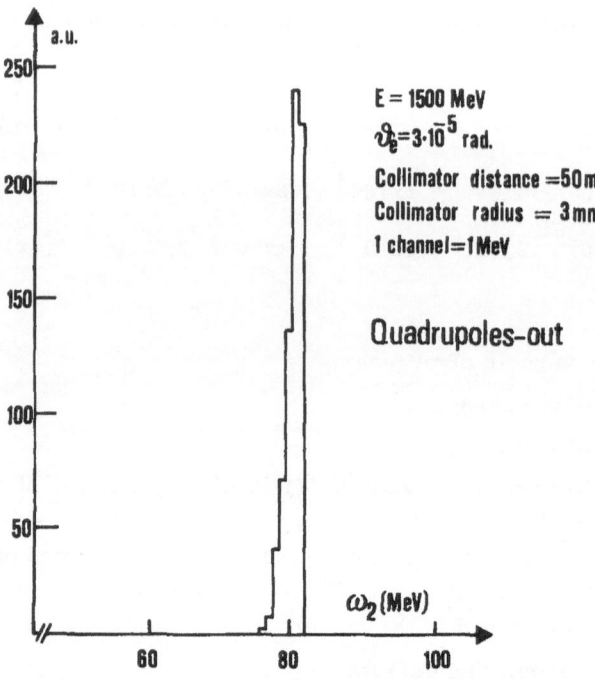

Fig. 9 - Calculated energy distribution for E = 1. 5 GeV-quadrupoles out.

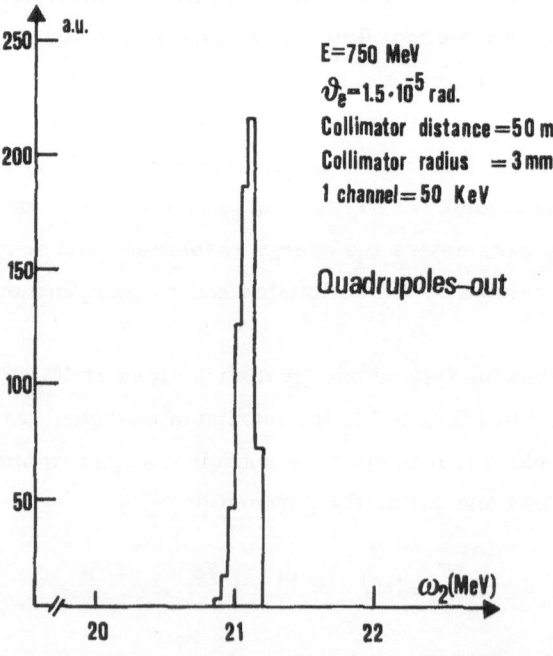

Fig. 10 - Calculated energy distribution for E = 0. 75 GeV-quadrupoles out.

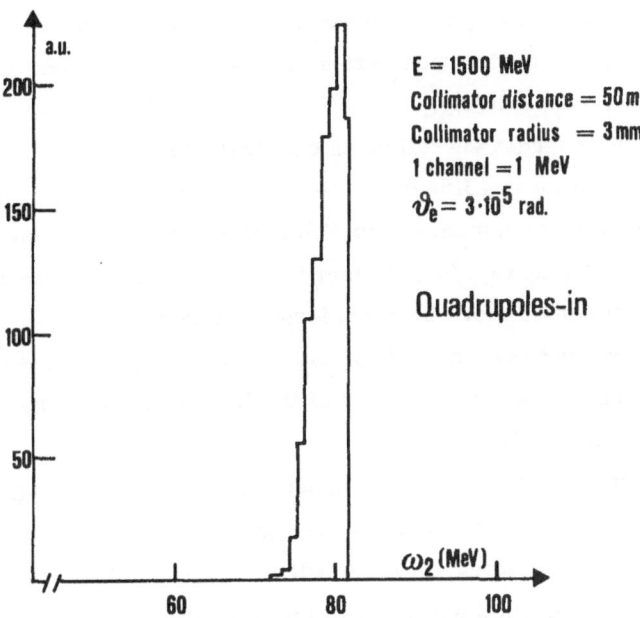

Fig. 11 - Calculated energy distribution for E = 1.5 Gev-quadrupoles in.

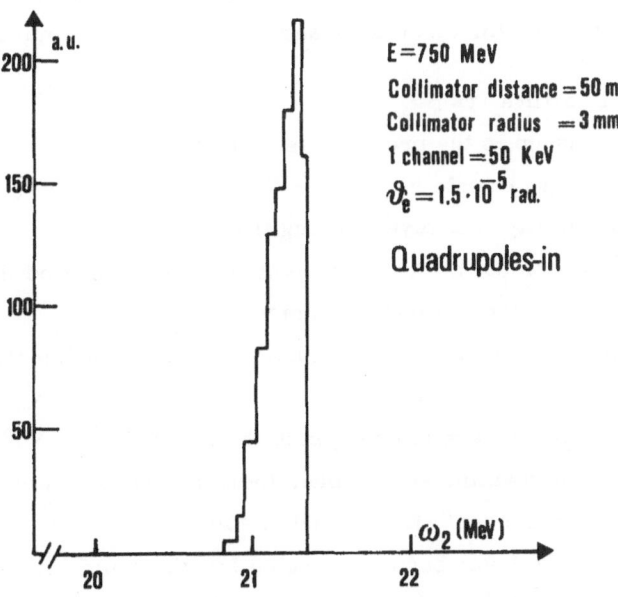

Fig. 12 - Calculated energy distribution for E = 0.75 GeV-quadrupoles in.

Curves of R as well as photon intensities ($\overset{\bullet}{N}_\gamma$) versus the accepted solid angle ($\Delta\Omega$) around the $\theta = 0$ direction, are reported in Figs. 13 and 14 for E = 1.5 GeV and E = 0.75 GeV respectively.

These results suggest the following considerations:

a) The counting rate is almost a linear function of the solid angle as it could be expected from geometrical considerations. This, of course, can only be true for values of the solid angles up to $\pi/\gamma^2$. In fact, for correspondingly larger values of $\theta$, the cross section can no more be considered constant.

b) The ratio of the counting rate in the two configurations, quadrupoles in and quadrupoles out, is consistent with the ratio of the lengths of the respective interaction regions as shown in Fig. 8.

c) The energy resolution R shows a linear dependence on the solid angle only for large values of it, where the effect of the divergence of the electron beam and finite beam size can be neglected. In the quadrupole out configuration the minimum values of R which can be obtained for zero solid angle are consistent with the values given by eq. (7).

d) The minima values of R which can be obtained in the quadrupole in configuration are much larger than those obtainable in the other configuration. This indicates the importance to avoid the electron-laser interaction in the quadrupole region.

## 2.3. - Recent developments.

As shortly discussed in the previous section, the energy resolution can be improved by avoiding the quadrupole regions.

There are essentially two ways of doing that:

a) to make the electrons pass out of the quadrupole axis so as to bend them out of the laser beam line inside the quadrupole regions;

b) to bunch the laser beam in such a way that photon and electron bunches collide only in the middle of the straight section.

At the present time we are studying both the possibilities. While the first solution looks more simple than the second one, there are serious worries with regards to the stability and reproducibility of such a solution.

Should we bunch the laser beam, the pulse length should not exceed $\sim 16.7$ nsec that corresponds to the straight section length without quadrupoles. Two possible ways to face this problem are:

a) a mode-locked operating laser;

b) a dumping of the laser cavity.

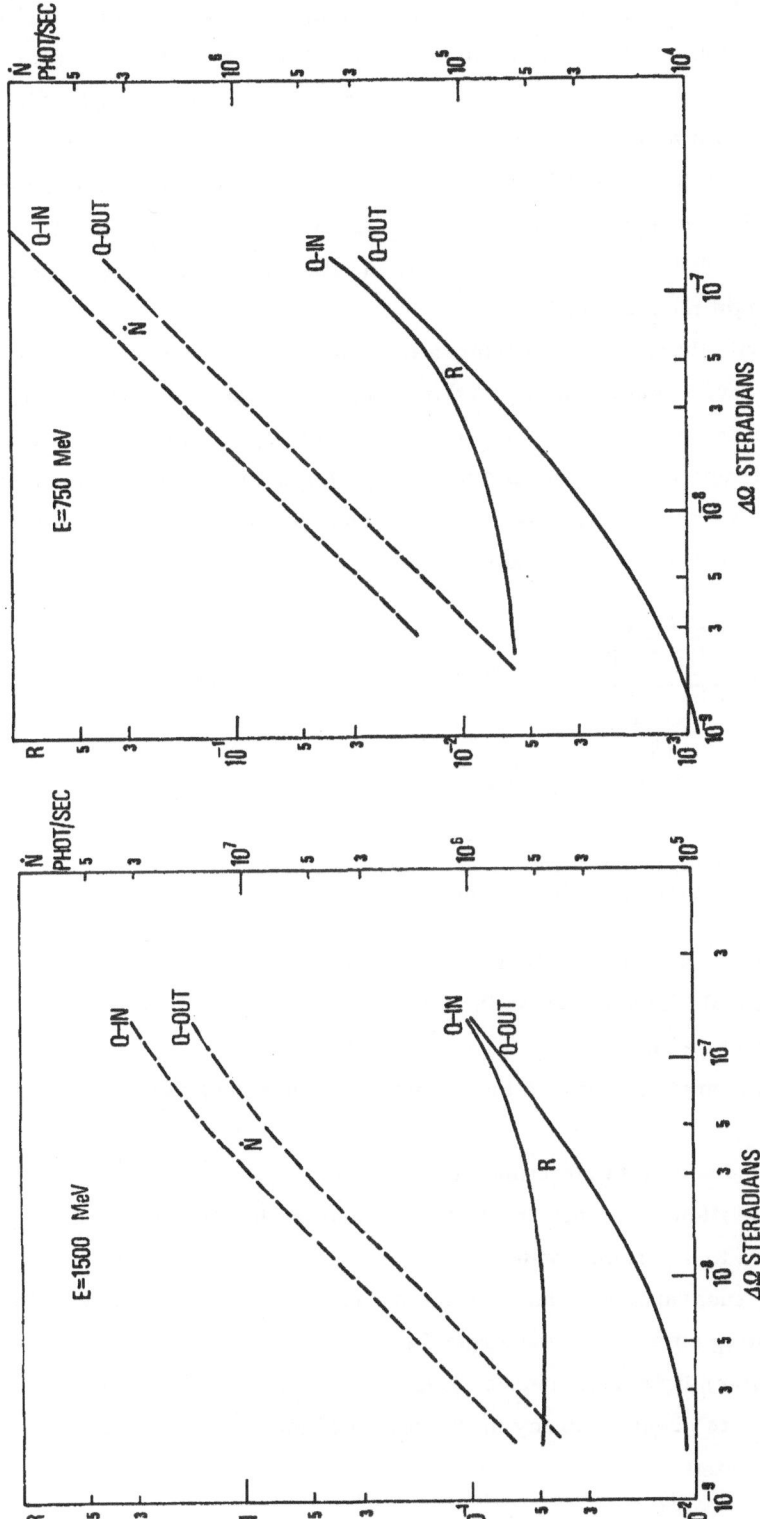

Fig. 13 - Energy resolutions (R) and photon counting rates ($\dot{N}_\gamma$) versus solid angle ($\Delta\Omega$) for the quadrupole in and out configurations at E = 1.5 GeV.

Fig. 14 - Energy resolutions (R) and photon counting rates ($\dot{N}_\gamma$) versus solid angle ($\Delta\Omega$) for the quadrupole in and out configurations at E = 0.75 GeV.

The mode locking technique allows to get train pulse whose repetition frequency is related to the cavity length by the usual relation $f = c/2L$. In our case, by imposing the repetion frequency to be equal to the Adone frequency of 8.56 MHz, we are forced to work with a cavity $\sim$ 17.5 m long.

The minimum pulse-length is mainly controlled by the line width of the laser transition and for an Argon Ion laser it is nowadays easy to achieve pulses of 800 W, 1 nsec wide. The usual methods are based either on the acousto-optic or on the electro-optic light modulation[11].

Laser cavity dumping is a technique that makes use of a fast light deflector to periodically remove some of the light from an optical resonator that serves as a temporary storage device (see Fig. 15). Ideally, the cavity dumper should be able to divert a large fraction of the circulating light within a time less than the round trip travel time of light within the cavity. In addition, the dumper must exhibit very low optical losses when in an "off" state.

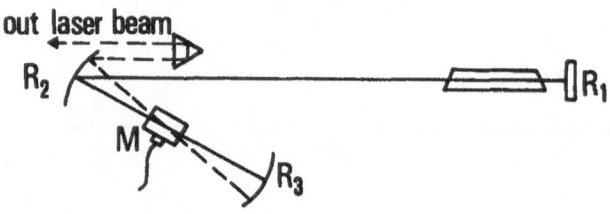

Fig. 15 - Typical cavity dumping arrangement. $R_1$, $R_2$, $R_3$ are mirrors and M stands for the Bragg cell modulator.

Acousto-optic devices can be fabricated on high-quality substrates such as fused quartz that should introduce less than 0.2 percent loss in the cavity: further details on the subject can be found in the literature[12].

The main operating features that would make this method very attractive are the following:
- high peak power pulses up to 30 times the CW power,
- choice of the repetition frequency from single pulse to 20 MHz,
- short pulses down to 15 nanoseconds.

Recently, measurements have been performed in our laboratory with a cavity dumper operating on a 2 Watt Ion Argon laser.

In particular, its performances at the 514.5 n m wavelength (2.41 eV), have been studied with great accuracy in the region of the Adone RF frequency in which we are interested.

The typical light pulse one can get is shown in Fig. 16 and the best operating features of the whole device can be sumarized as follows:

- repetition frequency    8.56 Mz
- pulse rise time    ~7    nsec
- pulse peak power    30    W
- FWHM    15    nsec.

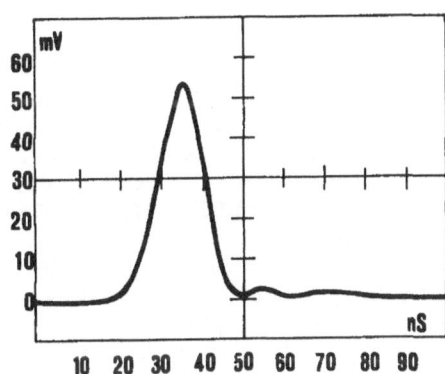

Fig. 16 - Typical light pulse shape
obtained with the cavity dumping
technique.

In such a situation the obtainable photon intensity drops down by a factor
~8 with respect to the numbers quoted in Figs. 13 and 14 making desirable more
laser power.

Therefore further possibilities to extend this technique to more powerful
lasers have been explored.

On the basis of the available laser performances provided by the manufac-
turing firms, the extrapolation of our present results yields the following picture:

| Argon Ion Laser Type | Peak power (W) pulse width = 15 ns | $\dot{N}_\gamma$ ($\Delta\Omega = 5 \times 10^{-8}$) E = 1.5 GeV, Q-OUT | Comments |
|---|---|---|---|
| Model 166-03 (2W) Spectra Physics | 30 | $1.7 \times 10^6$ | present and operating in our laboratory |
| CR8 - Coherent Radiation (5W) | 75 | $4.2 \times 10^6$ | present but not opera-ting yet |
| Model 170-03 (9W) Spectra Physics | 135 | $7.6 \times 10^6$ | ----- |
| Model 921 (18W) Spectra Physics | 270 | $1.5 \times 10^7$ | ----- |

As a final remark to conclude with, we must emphasize that the use of this
technique enables us to bypass the difficulty of the mirror damage due to the synchrotron radiation.

As a matter of fact the coated quartz mirror indicated in the sketch of
Fig. 7a, can be now substituted with a metallic one (platinum, gold) far more re-

sistent to the X-ray radiation. More over its reflectivity ($\sim 90\%$) is no longer such crucial parameter as it was in the case of the laser cavity.

REFERENCES. -

(1) - R. H. Melbourne, Phys. Rev. Letters 10, 75 (1963).
(2) - F. R. Arutyunian and V. A. Tumanian, Phys. Letters 4, 176 (1963).
(3) - V. N. Bayer and V. A. Khoze, Sov. J. Nuclear Phys. 2, 238 (1969).
(4) - L. Casano, A. Marino, G. Matone, M. Roccella, C. Schaerf and A. Tranquilli, Frascati report LNF-74/60 (1974); Laser and Unconventional Optics Journal 55, 3 (1975).
(5) - O. F. Kulikov et al., Phys. Letters 13, 344 (1964).
(6) - C. Bemporad et al., Phys. Rev. 138B, 1546 (1965).
(7) - J. Ballam et al., Phys. Rev. Letters 23, 498 (1969).
(8) - A. Tranquilli, Frascati report LNF-75/10 (1975).
(9) - M. Bassetti, Memorandum Adone E-12 (1974).
(10) - A. Fubini et al., Frascati report LNF-74/12 (1974).
(11) - G. Matone and A. Tranquilli, Frascati report LNF-76/7 (1976).
(12) - D. Maydan, J. Appl. Phys. 41, 1552 (1970).

# LATEST FROM THE DALINAC[*]

A. Richter

Institut für Kernphysik der Technischen Hochschule Darmstadt,
6100 Darmstadt, Germany

Recent developments and results from the Darmstadt electron linear accelerator (DALINAC) are summarized. The DALINAC produces electron beams with energies up to 70 MeV and electron scattering experiments are possible within the momentum transfers ranging from $q = 0.12$ fm$^{-1}$ to $q = 0.70$ fm$^{-1}$. With the newly installed energy loss system an overall energy resolution (FWHM) of $\Delta E \approx 30$ keV has been reached. The main research interest at the DALINAC has focussed at the following four topics: (i) atomic inner shell ionization by relativistic electron impact; (ii) E2-giant resonances in heavy spherical and deformed nuclei; (iii) M1 transition strength in light and heavy nuclei and (iv) inelastic electron scattering on low lying states - isospin-forbidden $\Delta T = 0$ E1 transitions in the selfconjugate nuclei $^{16}$O and $^{40}$Ca and E0 transitions in the Ca-isotopes. These topics are discussed by way of specific examples, the motivation for their investigation and many results are given.

## Introduction

In this talk I will report on latest developments and results from our laboratory. The Darmstadt electron linear accelerator (DALINAC) operates at incident beam energies between 20 and 70 MeV and hence allows momentum transfers up to $q \lesssim .7$ fm$^{-1}$. The accelerator had been installed already in the early sixties[1] and has since been in operation almost continuously. Recently, a high resolution energy loss spectrometer[2-4] similar to the one at MIT (ref. 5) has been built with an energy resolution of $\Delta E/E \approx 5 \times 10^{-4}$, thereby allowing second generation inelastic electron scattering experiments on nuclei with a first generation accelerator. Our main research interest at the DALINAC has focussed onto the following 4 topics:

A. Atomic inner shell ionization by relativistic electron impact
B. E2 giant resonances in spherical and deformed nuclei-$^{142,150}$Nd and $^{144,150,154}$Sm
C. M1 transition strength in light and heavy nuclei $^{14}$N, $^{28}$Si, $^{58}$Ni, $^{90}$Zr and $^{208}$Pb

---

[*] Supported in part by Deutsche Forschungsgemeinschaft

D. Selected topics in inelastic electron scattering on low lying states

    a) Isospin-forbidden $\Delta T = 0$  E1 transitions in selfconjugate nuclei - $^{16}$O, $^{40}$Ca

    b) EO transitions in the Ca isotopes

There is no time in this talk to cover these subjects completely. I will there-fore illustrate them by way of specific examples. Topic A and B make use of electron beams with moderate energy resolution, while C and D are being investigated by means of our high resolution facility. Since this is a new facility, I will make some re-marks on the experimental set-up.

Before dealing with those topics I should like to introduce the name of my col-laborators in the laboratory and acknowledge their contributions to this talk:

R. Frey, A. Friebel, H. Genz, H.D. Gräf, D. Hoffmann, G. Kühner, W. Löw,
W. Mettner, D. Meuer, H. Miska, R. Schneider, G. Schrieder, D. Schüll,
A. Schwierczinski, E. Spamer, H. Theissen, O. Titze and Th. Walcher

Five of these experimentalists have recently left the laboratory but have been involved in the investigation of the above subjects at various stages of the work.

## Experimental apparatus

The schematic layout of the electron scattering system at Darmstadt is shown in fig. 1. The electron beam from the linac ranging in energy between 20 and 70 MeV can be directed into 3 beamlines, a line where it may be converted into bremsstrahlung-photons, a line with the old 120$^{\circ}$ spectrometer of moderate energy resolution ($\Delta E \simeq 200$ keV) and our new beam line with the 169$^{\circ}$ spectrometer as its heart-piece. The research topics introduced above are being investigated at the beam lines label-led SF3 and SF4, respectively, with the magnetic spectrometers at the end. While constructing and adding our new energy loss system, the injector of the linac has been improved as well as vacuum chambers have everywhere been enlarged along the beam lines. This has resulted in a generally very clean beam without beam halos and has led to a low room background. This background is measurable in the 120$^{\circ}$ spectrometer by moving the counter system out of the focal plane of the spectrometer. We now find that it is only a few percent of the calculated bremsstrahlung background and is about constant at the energies available from the accelerator.

The newly added beam transport system starts after the pre-analyzer - 3 magnets, which bend the beam by 40$^{\circ}$ - and consists of two 70$^{\circ}$ bending magnets M1 and M2 and various beam transport elements. Great instrumental care has been applied to the con-struction of these elements (i.e. to the mirror plates between the jokes of the mag-nets and the pole tips, the shapes of the pole face etc.) and to the long time stabi-

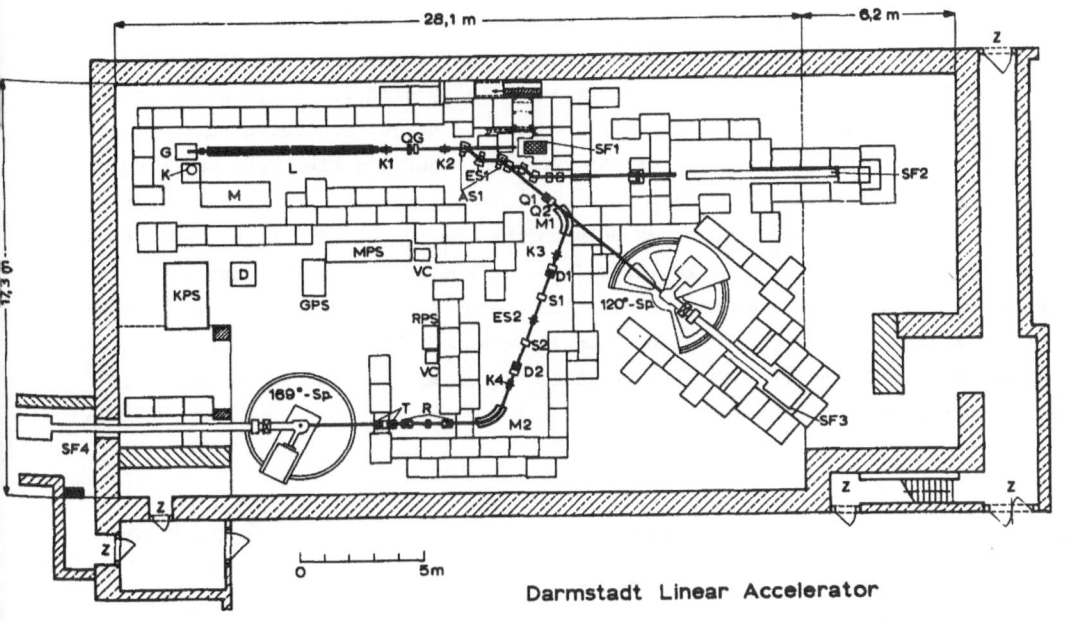

Darmstadt Linear Accelerator

Fig. 1.    Schematic layout of the Darmstadt electron scattering system.

lity (< 2 x 10$^{-5}$) of the power supplies of the bending magnets. The beam transport
system is symmetric with respect to a pair of slits ES2. The beam has three waists,
one at K3, a second one at ES2 and a third one at K4. An enlarged image of the beam
at K3 is   created at ES2 with a dispersion of 5 cm/% and then again a reduced image at
K4. The disperision of the beam is produced by the quadrupole singulet S2 and is matched
to the dispersion of the magnetic spectrometer with the help of the quadrupole trip-
let T in front of the spectrometer. Since the plane of dispersion is still horizontal
when the beam leaves the bending magnet M2 - the vertically erected spectrometer,
however, accepts only a vertically disperse beam - the beam has to be rotated by 90$^{\circ}$
by means of 5 quadrupoles R.

The geometrical extension of the beam spot on the target is typically 1 mm x 10 mm
and depends of course on the excitation energy of the nuclear level to be investigated. Up to 30 μA of beam current are used in experiments. A very simple sketch of
the principle of the spectrometer operated in the energy loss mode as compared to the
conventional mode is given in fig. 2. In the conventional mode, all electrons are
focussed onto one point of the target. The scattered electrons appear then in the
focal plane of the spectrometer at momenta $p_o$ and $p_o \pm \delta p$. A reduction of the momentum

spread of the initial beam can of course be achieved but only at the expense of reduc-
ing the beam intensity. This disadvantage can be overcome when the spectrometer is
operated in the energy loss mode which is essentially a mirror image of the conventional mode. If the electrons are positioned at the target according to their deviation $\delta p$ from the nominal value $p_o$ of the momentum, then all electrons independent of
their primary momentum $p_o \pm \delta p_o$ are focussed onto one point in the focal plane. The
position of this image point depends only on the energy loss $E_x$ of the inelastically
scattered electrons in the target. Therefore, in the course of the experiment the
dispersion has to be changed with excitation energy $E_x$. In this procedure an accuracy
of $\pm$ 15% in the setting of the dispersion is accepted. It should be noted, that clos-
ing the slit ES2 (fig. 1) such that the beam has an energy resolution of $2 \times 10^{-4}$,
yields in the conventional mode a point-like beam spot on the target with a diameter
$\leq$ 1 mm.

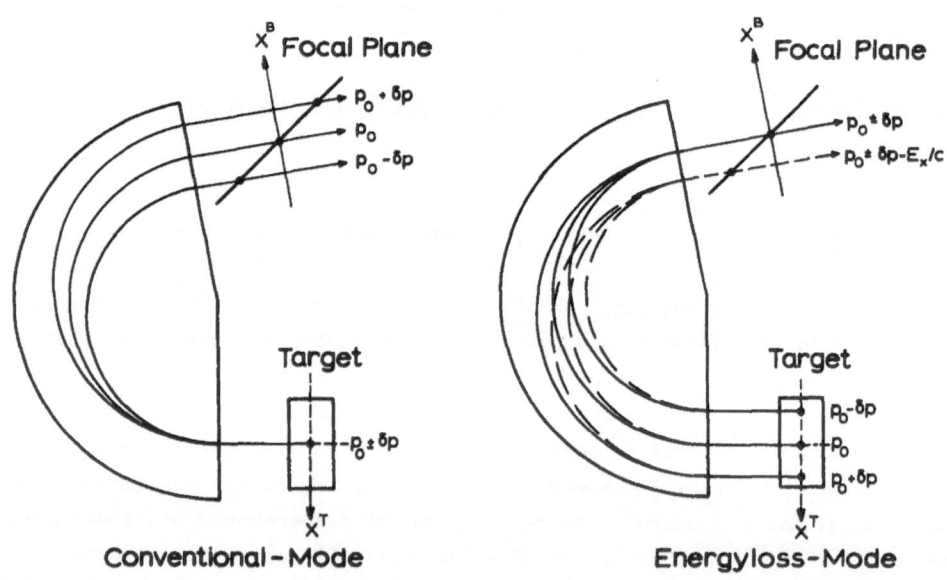

Fig. 2.    Schematic representation of the magnetic spectrometer operated in the conventional mode and in the energy loss mode, respectively.

The detector system in the focal plane of the $169°$-spectrometer is an array of
36 overlapping plastic scintillators and is a refined version of the Amsterdam system[6].
Through coincidence-anticoincidence requirements between adjacent scintillators 69
detection channels are formed - each with a relative momentum width of $\Delta p/p = 3 \times 10^{-4}$.
For background suppression, a Čerenkov counter placed behind the scintillators is
put in coincidence with these. This rather elaborate system is connected on line to
a computer system for data aquisition.

Setting-up the various beam transport elements in order to work in the energy loss mode takes about 3 h. Since the total momentum width of the counter system in the focal plane of the spectrometer is only about 2%, and the solid angle of the spectrometer is 2.5 msr, accumulating a spectrum covering 3 MeV in excitation energy with reasonable statistics requires about 80 h of run time. Hence there are extreme requirements on the long-time stability of the whole system. Typical target thicknesses for optimum resolution are thereby between (10-20) $mg/cm^2$ and the collected charge in general is 2mC per measured point in the spectra.

The full capability of the new system will be demonstrated later in conjunction with the experiments. But before doing that, let us discuss briefly the research topics A and B for which we use the beam line with moderate resolution.

## A. Atomic inner shell ionization by relativistic electron impact

The measurement of atomic inner shell ionization cross sections has become possible only due to the improvements of the beam quality at the DALINAC since an extremely low background is necessary for the detection of x rays. The experimental information on these cross sections is still extremely scarce[6-10]. So far, systematic studies exist for electron energies below 2 MeV and have recently become available above 70 MeV (ref. 11). Theoretically, these cross sections are interesting for a variety of reasons. The various nonrelativistic theories have to be extended into the realm of relativistic effects, i.e. in the description of the cross section in terms of a fully relativistic calculation, relativistic wave functions should be employed for the incident, outgoing, bound and ejected electron, respectively, and the interaction responsible for the process must also be of relativistic form. So far, semirelativistic theories exist and most of the calculations are done in Born approximation (see e.g. refs. 6, 7, 12-17).

Let us briefly look at the process where the incident electron knocks out a K-shell electron. In an impact parameter treatment the cross section can be divided into two parts

$$\sigma_K = \sigma_{b<a_K} + \sigma_{b>a_K}$$

where the first term results from close collisions, i.e. where the impact parameter b is smaller than the K-electron shell radius $a_K$, and the second term from a distant collision. The close term is treated in terms of a relativistic Møller interaction (Møller electron-electron scattering) while the distant term is usually expressed by the virtual photon spectrum of the moving projectile interacting with the bound electron. Of course, the latter method is entirely consistent with the way inelastic electron scattering on a nucleus is treated. In this connection it is also interesting to ask if the cross section for x-ray production exhibits a scaling behaviour and to find out the scaling variables when atoms with electrons as point-like constituents are being bombarded with relativistic projectils.

Figure 3 shows the experimental set-up for the measurement of the x rays (above). Since absolute cross sections have to be measured the detector efficiency and solid angle, the fluorescence yield, the target thickness and the number of eletrons incident on the target have to be known. In order to monitor the product of target thickness and electron beam through the target we use the 70 MeV/c spectrometer under a forward scattering angle and measure nuclear elastic electron scattering and the x rays simultaneously.    The lower part of this figure displays an x-ray spectrum obtained with 50 MeV electrons on Au. It is almost free of background and exhibits various K- and L-shell transitions.

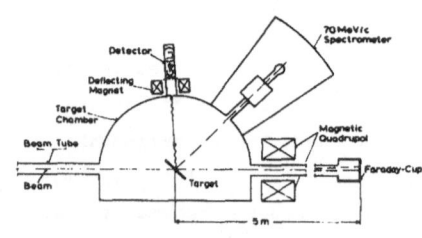

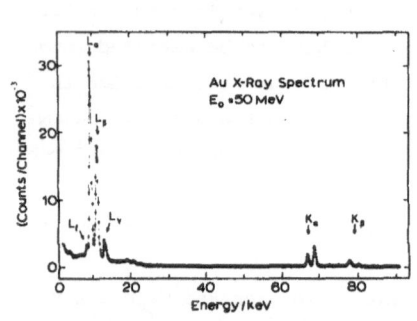

Fig. 3.    Experimental set-up for the measurement of x rays after relativistic electron impact (upper part) and a typical x-ray spectrum induced by 50 MeV electrons on a thin Au target (lower part).

Our first results of cross section measurements[19,20] of various elements between Y and Bi are given in fig. 4. This figure displays the Z-dependence as well as the energy dependence and is compared to theoretical predictions. Existing data on Au at 2 MeV and 150 MeV are also plotted (some more data from Sendai between 70 and 150 MeV have been reported since this figure was prepared), and it is hoped that a very precise determination of the cross section will serve as a severe test for the theoretical models I have mentioned earlier.

It is interesting to speculate whether the cross sections for inner shell ionization obey a scaling law at relativistic electron energies. If the present data for all measured elements at the various incident electron energies $E_o$ are plotted in the form $\sigma \cdot I = f(E_o^2/I)$ in a double logarithmic representation (this scaling behaviour is suggested by the theory of Kolbenstvedt (ref. 14) compared to our data in the previous figure) where I is the K-shell ionization energy, fig. 5 is obtained. The cross sections indeed cluster around a general curve, both for the K- and L-shell ionization cross section. At the present stage, however, it is not clear why this particular scaling describes the data rather well and we investigate this question together with measurements also on higher shell ionization further.

After this brief excursion into the field of photoatomic reactions let us now discuss photonuclear experiments at our laboratory.

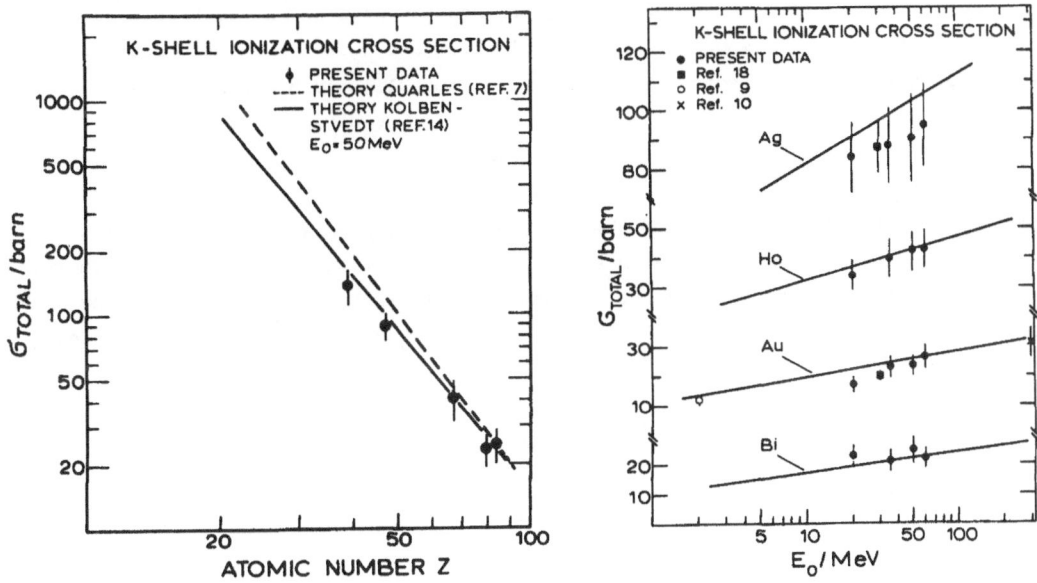

Fig. 4. K-shell ionization cross sections by 50 MeV electron impact for Y, Ag, Ho, Au and Bi targets (left part) and energy dependence of the cross sections on Ag, Ho, Au and Bi (right part). The lines drawn are various theoretical predictions.

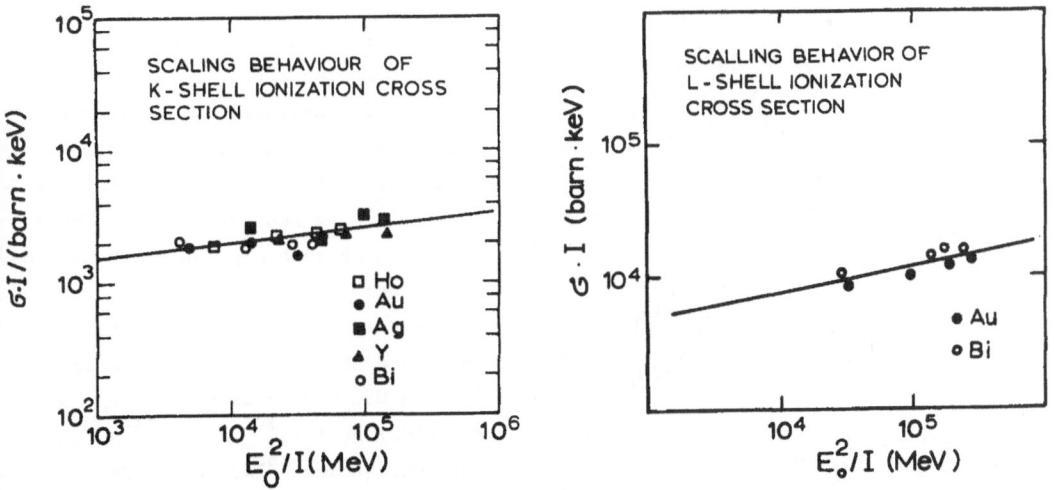

Fig. 5. K- and L-shell ionization cross sections in the $\sigma I$ vs $E_0^2/I$ representation of the present data.

## B. E2-giant resonances in spherical and deformed nuclei

The existence of giant resonances other than the well known electric dipole giant resonance is       supported both by electron and hadron inelastic scattering (see e.g. ref. 21). The isoscalar E2 giant resonance lies in heavy nuclei (A > 100) at

$$E_x \simeq 63 \ A^{-1/3} \text{MeV},$$

and hence is lower in energy than the E1 giant resonance at $E_x \simeq 77 \ A^{-1/3}$ MeV. Its strength is usually concentrated in a broad structure of width $\Gamma \simeq 3 - 5$ MeV. Therefore it is possible to investigate these resonances in inelastic electron scattering with moderate energy resolution ($\Delta E \simeq 200$ keV). We have studied recently the following systems:

$$^{142,150}\text{Nd and } ^{144,150,154}\text{Sm}.$$

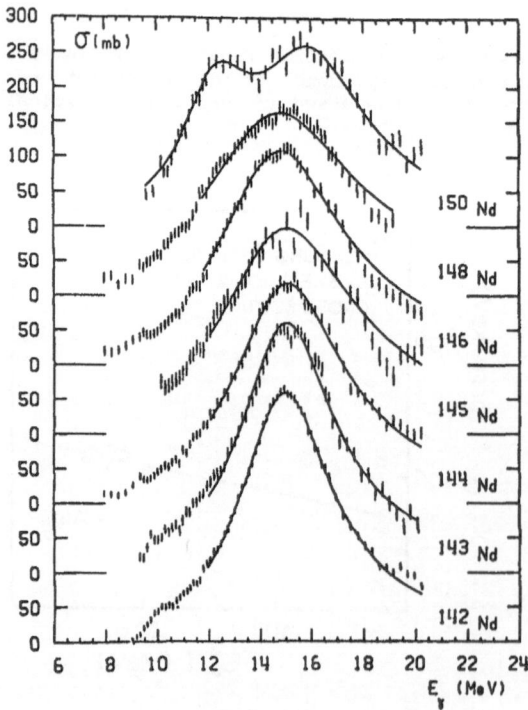

Fig. 6.   Total photoabsorption cross section of the Nd-isotopes (from ref. 22).

It is well known, that the giant E1 resonance splits into two components by going from the spherical nucleus to the deformed nucleus. The beautiful and very precise measurements by Bergere and his group at Saclay have demonstrated this recently also for the Nd isotopes[22]. As a reminder, fig. 6 shows the total photoabsorption cross section. The question which interested us mostly is: Does the isoscalar E2 giant resonance show also a splitting or at least a broadening in going from the spherical nuclei $^{142}$Nd and $^{144}$Sm to the deformed nuclei $^{150}$Nd and $^{154}$Sm, respectively, and how much of the E2 sum rule is exhausted?

Figure 7 displays two inelastic electron scattering spectra at an incident energy $E_o$ = 65 MeV and a scattering angle $\theta$ = 129° on the Nd isotopes. The spectra extend from about 5 MeV in excitation energy to a little over 25 MeV. The raw spectra as they come out directly during the run (upper part) clearly show structure superimposed onto a generally smooth background. This background is the main source of uncertainty in determining reliable transition strengths and widths of the giant resonances. (As we know, a sizable background, but of different origin, in scattering experiments with hadronic projectiles (see e.g. ref. 23) causes similar uncertainties in these quantities). The main reasons for the background are the following:

1. Production of bremsstrahlung during elastic scattering
2. Production of bremsstrahlung after inelastic scattering
3. Double processes
4. Møllerscattering
5. General room background

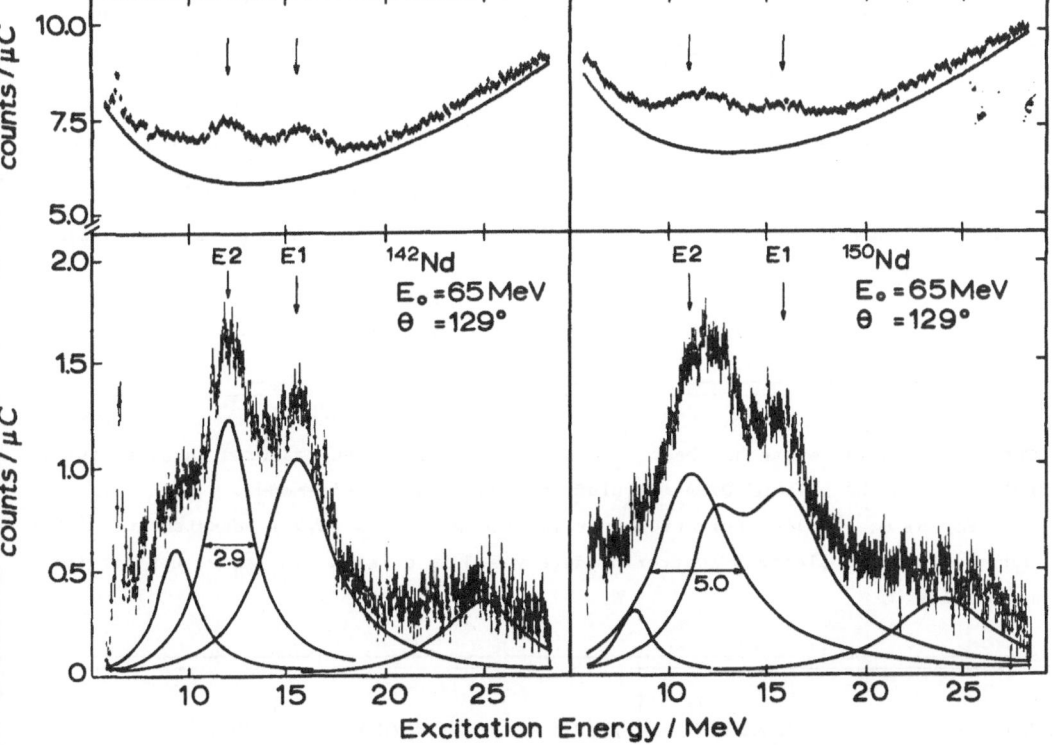

Fig. 7.    Spectra of inelastically scattered electrons on $^{142}$Nd and $^{150}$Nd. The upper part of the figure shows the raw data together with the calculated background, the lower part the data after background subtraction together with a decomposition into various lines. The width of the isoscalar E2 giant resonance is also indicated.

There is no time here to go into the details of how we treat these different sources of the background in the analysis of our spectra. We have put a great effort into this problem and believe that we can treat points 1. - 4. theoretically and can measure the fifth point reliably. The result is the background line in the upper part of fig. 7., which once determined for $^{142}$Nd is just scaled for $^{150}$Nd by the ratio of the target thicknesses. Subtracting this background from the measured counts, and using the shape and the transition strength for the E1 resonance from the photo-neutron experiments as a further constraint we can decompose the structure in the subtracted spectra with the help of various Lorentzian-lines. While the E1 giant resonance at about 15 MeV exhibits the splitting in going from $^{142}$Nd to $^{150}$Nd the isoscalar E2 giant resonance at about 12 MeV seems to become broader. The structures at around 9 MeV correspond possibly to a clustering of M1 strength, while the broad bump at around 25 MeV excitation energy might be the isovector E2 giant resonance. It agrees fairly well in excitation energy with $E_x \simeq 130 \ A^{-1/3}$ and is presently investigated further.

Similar results, but still preliminary, have been obtained for the three Sm-isotopes studied (fig. 8). The following two tables briefly summarize what we can say at present about the properties of the isoscalar E2 resonances in Nd and Sm:

T A B L E   I

| Nucleus | $E_x$/MeV | $\Gamma$/MeV | $B(E2)/fm^4$ | % EWSR |
|---------|-----------|--------------|--------------|--------|
| $^{142}$Nd | 12.0 ± 0.2 | 2.9 ± 0.3 | 2900 ± 370 | 71 |
| $^{150}$Nd | 11.1 ± 0.2 | 5.0 ± 0.2 | 3430 ± 270 | 79 |

The transition strength has been obtained by comparing the measured angular distributions with the help of DWBA calculations using the Tassie-model.

For Sm we compare excitation energies and widths to $(\alpha,\alpha')$ measurements[23] and notice, that our electron scattering data yield in general smaller values for these quantities.

T A B L E   II

| Nucleus | (e,e') | | $(\alpha,\alpha')$ | |
|---------|--------|--|-------------------|--|
| | $E_x$/MeV | $\Gamma$/MeV | $E_x$/MeV | $\Gamma$/MeV |
| $^{144}$Sm | 11.9 ± 0.2 | 2.9 ± 0.2 | 13.2 ± 0.3 | 3.90 ± 0.19 |
| $^{150}$Sm | 11.8 ± 0.2 | 3.3 ± 0.2 | 12.5 ± 0.2 | 4.25 ± 0.16 |
| $^{154}$Sm | 10.9 ± 0.2 | 4.5 ± 0.2 | 12.4 ± 0.3 | 4.72 ± 0.25 |

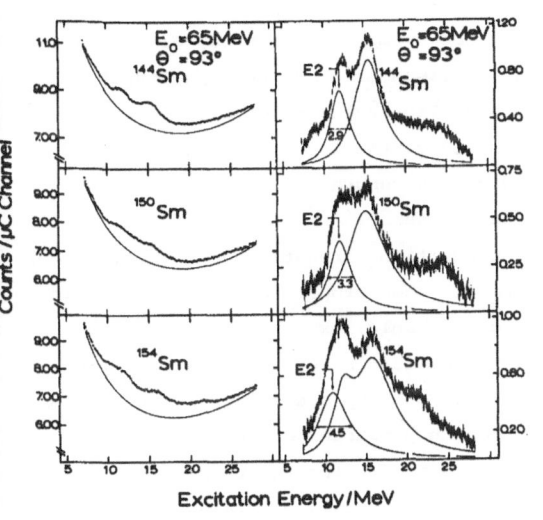

Fig. 8.   Spectra of inelastically scattered electrons from $^{144}$Sm, $^{150}$Sm and $^{154}$Sm before and after background subtraction. The variation in width of the E2 giant resonance in going from the spherical to the deformed nucleus is given in the figure.

Finally, in fig. 9 a comparison between the widths for the isoscalar E2 giant resonances observed by hadron and electron scattering has been made (upper part). The electron scattering values around mass 150 are measured at our laboratory and are in general smaller than the hadronic values. The solid line in this figure represents a recent theoretical calculation of Auerbach and Yeverechyahu (ref. 24), which employs the hydrodynamic model and the concept of viscosity. This calculation surely reproduces the overall trend of the data. The experimental data, however, may show shell effects over and above the general trend. Also for the fraction of the energy weighted E2 sum rule exhausted in the different experiments the electron

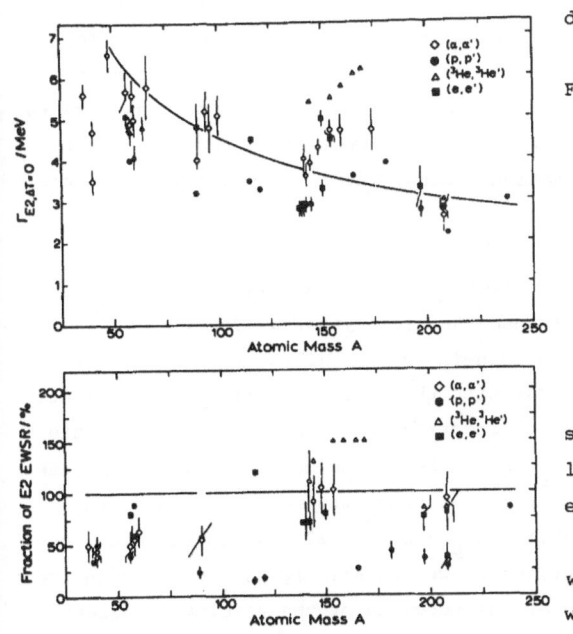

Fig. 9.   Widths (upper part) and fraction of the energy weighted sum rule (lower part) for the isoscalar giant quadrupole resonance observed in various hadronic and electronic scattering experiments. The overall trend in the widths is described by a calculation of ref. 24 (solid curve) although the data might exhibit shell effects.

scattering data from our laboratory yield lower values as compared to the hadron experiments.

In the main part of this seminar we will now discuss the experimental program where we use our newly installed high resolution facility.

## C. M1 transition strength in light and heavy nuclei

Various methods (see refs. 25 and 26) have been employed to study M1 transitions, like capture reactions $(X,\gamma)$ where X denotes a nucleon or a nucleus, photoneutron reactions $(\gamma,n)$, gamma-ray fluorescence $(\gamma,\gamma')$ and also inelastic electron scattering, mainly at $\theta = 180°$ (ref. 27). In terms of a sum rule, the majority of these transitions contains a large fraction of magnetic dipole strength in the nuclear ground state and hence often constitutes the M1-giant resonance.

Before discussing some examples of such transitions investigated with high energy resolution at the DALINAC let us briefly recall some properties of M1 transitions like excitation energy of $1^+$ states, structure, sum rule, excitation in electron scattering and related processes. The main interest for investigating these transitions is clearly the importance of determining the spin-spin and spin-isospin interactions in the nucleus. In particular, the strength parameter $g_o'$ in the effective interaction[28] $g_o' \, \tau_1 \cdot \tau_2 \sigma_1 \cdot \sigma_2$, which is also of interest in the theory of pion condensation in nuclear matter (see e.g. ref. 29), might eventually be evaluated by locating the excitation energy and strength of magnetic dipole states in nuclei.

Location: As summarized by Hanna[26] the giant resonances of lowest multipolarity are located as displayed in fig. 10. The M1-strength observed experimentally lies at an excitation energy of $E_x \simeq (35 \pm .5)A^{-1/3}$ MeV, i.e. the center of gravity in $^{28}$Si, $^{58}$Ni and $^{208}$Pb would be at $E_x \simeq 11.7$, 9.2 and 6.0 MeV, respectively. We observe two important properties of M1 transitions over the range of mass numbers: (i) They are energetically lower than the E1 and E2 giant resonances whose location invert for the heavier nuclei.(ii) The excitation strength is energetically not such a monotonic function and is not concentrated in one or two broad states as is mostly the case for the E1 and E2 giant resonances. The M1 strength usually rests in a few sharp states distributed over a broader band of excitation energy. The $1^+$ states are also bound in the heavier nuclei.

Excitation mode: How can we picture the mechanism for the excitation of such states? As fig. 11 shows, we might macroscopically use the hydrodynamical collective model (see e.g.ref.30) to describe spin-spin and spin-isospin oscillations. In the former, the isoscalar mode, all neutrons and protons with the same spin direction oscillate against the nucleons with the opposite spin direction. The B(M1) strength for this mode is proportional to $(\mu_p + \mu_n)^2 \simeq .8$, $\mu_p$ and $\mu_n$ being the magnetic moments of the proton and neutron, respectively. This strength is hence weak as compared to the strength of the isovector mode, where $B(M1) \sim (\mu_p - \mu_n)^2 \simeq 22$. An experimental investigation of these types of transitions, which we essentially have already in the deuteron, hence allows the determination of the strength of the effective nuclear

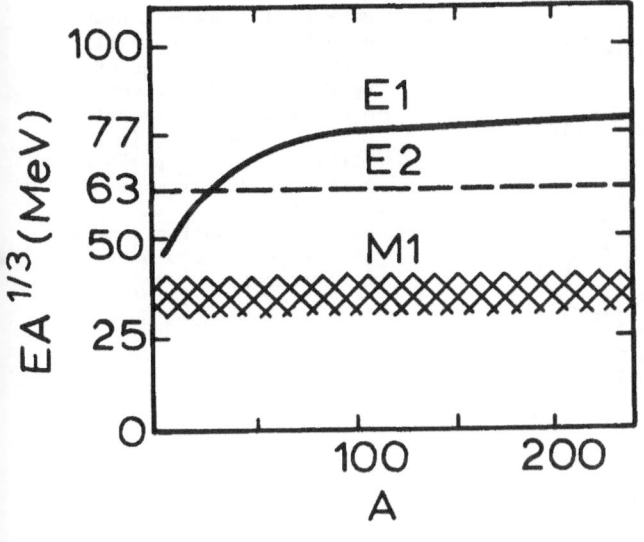

Fig. 10.    Location of giant
resonance strength
as a function of
mass number
(from ref. 26).

force causing these transitions.

A more microscopic ap-
proach to these magnetic dipole transitions utilizes the
particle-hole picture within
the frame of the shell model.
Since the magnetic dipole
operator has only matrix elements between single particle
states with the same orbital-
and principal quantum number,
spin-flip transitions can occur only between spin-orbit partners

$$j_> = \ell+1/2 \quad \text{and} \quad j_< = \ell-1/2.$$

The structure of a $|1^+\rangle$ state in general is hence of the type

$$|1^+\rangle = [(\ell_{j_>})^{-1}(\ell_{j_<})]$$

i.e. the transitions we searched for experimentally in various nuclei involve $1^+$ states
of the following structure:

$$^{28}\text{Si}: \quad |1^+\rangle = \quad [(d_{5/2})^{-1}(d_{3/2})]_{n,p}$$

$$^{58}\text{Ni}: \quad |1^+\rangle = \quad \begin{array}{l} [(f_{7/2})^{-1}(f_{5/2})]_{n,p} \\ [(p_{3/2})^{-1}(p_{1/2})]_{n} \end{array}$$

$$^{90}\text{Zr}: \quad |1^+\rangle = \quad [(g_{9/2})^{-1}(g_{7/2})]_{n}$$

$$^{208}\text{Pb}: \quad |1^+\rangle = \quad \begin{array}{l} [(h_{11/2})^{-1}(h_{9/2})]_{p} \\ [(i_{13/2})^{-1}(i_{11/2})]_{n} \end{array}$$

# Magnetic Dipole Spin Flip Transition

Spin Mode    Spin Isospin Mode

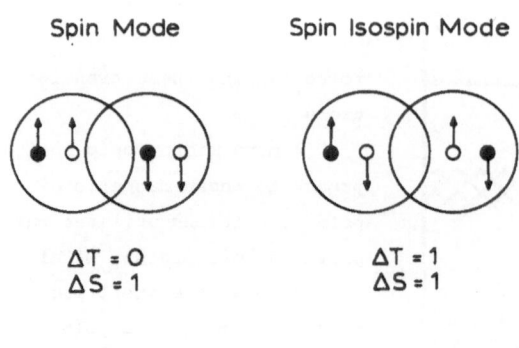

$\Delta T = 0$
$\Delta S = 1$

$\Delta T = 1$
$\Delta S = 1$

● Protons          o Neutrons

Fig. 11.  Schematic representation of magnetic dipole spin flip transitions in terms of the hydrodynamical collective model.The B(M1) strength in the isoscalar mode is proportional to $(\mu_p + \mu_n)^2 \simeq \cdot 8$, in the isovector mode to $(\mu_p - \mu_n)^2 \simeq 22$, $\mu_p$ and $\mu_n$ being the magnetic moments of the proton and neutron, respectively.

<u>Strength and sum rule:</u>  We should expect in general that the magnetic dipole strength is strongest in nuclei where the $j_>$-shell is full and the $j_<$-shell is empty. This collective effect is seen immediately with the help of a sum rule for $(J^\pi, T) = (0^+, o) \rightarrow (1^+, 1)$ transitions.This sum rule is energy-weighted and has been derived by Kurath[31] for selfconjugate nuclei and has recently been generalized by Lipparini <u>et al.</u>(ref. 32).

$$\sum_n (E_n - E_o) B(M1, 0 \rightarrow n) \simeq \frac{3}{4\pi} a(\mu_p - \mu_n + 1/2)^2 < o| \sum_k \vec{\ell}(k) \vec{s}(k) |o>$$

Here a is the spin-orbit splitting parameter. The ground state expectation value of the spin-orbit operator can easily be evaluated in terms of the shell model. Since

$$\vec{\ell} \cdot \vec{s} = [j(j+1) - \ell(\ell+1) - s(s+1)]/2$$

it follows that

$$2<o| \sum_k \vec{\ell}(k) \vec{s}(k) |o> = \ell n_{j>} - \ell(\ell+1) n_{j<},$$

where n is the number of nucleons in the respective orbits. Therefore, $n_{j>}$ as compared to $n_{j<}$ and $\ell$ need to be large for a strong transition. The same argument is of course also seen to be true if the closure sum rule[33] is used instead of the Kurath energy-weighted M1 sum rule.

The simple shell model of independent particles certainly is only the most primitive approximation for the evaluation of the ground state expectation value of the spin-orbit operator. This fact makes the study of M1 transitions extremely interesting. A very precise experimental determination of the left hand side of the sum rule expression might lead to the detection of ground-state correlations, core polarization, mesonic effects etc. The study of these M1 transitions hence supplements e.g. the measurements of magnetic moments[34].

Isospin-structure and relation to other processes: Let us consider for the sake of simplicity inelastic electron scattering from a $(J^\pi,T) = (1^+,0)$ to a $(0^+,1)$ state in the selfconjugate nucleus $^{14}$N. From fig. 12 we see that there is a well known analogy between the spin-isospin flip electromagnetic transitions, Gamow-Teller ß decays,π and μ capture reactions and charge exchange reactions involving the excitation or the decay, respectively, of states which are isobaric analogue states[35]. The various

Fig. 12. Schematic illustration of analogous spin-isospin flip transitions like Gamow-Teller ß-decay, charge exchange reactions, radiative pion capture etc. compared to an M1 transition studied by inelastic electron scattering in a self-conjugate nucleus.

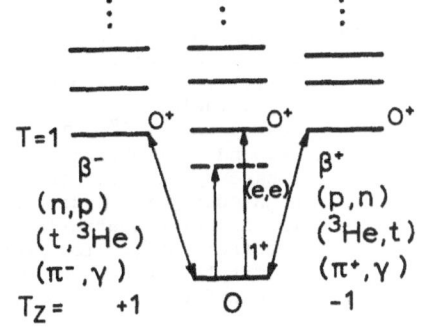

ISOSPINSTRUCTURE (e.g. N=Z nuclei)

transitions between an initial state $\langle J_i M_i|$ and a final state $|J_f M_f\rangle$ (radiative pion capture, Gamow-Teller beta decay and inelastic electron scattering) are in an impulse approximation treatment given by the following matrix elements

$$M_{(\pi\pm,\gamma)} = \langle J_f M_f|\sum_{j=1}^{A} \vec{\sigma}_j \vec{\tau}_j^{\pm} \hat{e} e^{-i\vec{k}\vec{r}_j}|J_i M_i\rangle$$

$$M_{\beta\pm}^{GT} = \langle J_f M_f|\sum_{j=1}^{A} \vec{\sigma}_j \vec{\tau}_j^{\pm}|J_i M_i\rangle$$

$$M_{e,e'} = \langle J_f M_f|\sum_{j=1}^{A} (\mu_p-\mu_n)\vec{\sigma}_j \tau_j^{(3)} e^{-i\vec{q}\vec{r}_j}|J_i M_i\rangle.$$

Here the quantities $\vec{\sigma}_j$ and $\vec{\tau}_j$ denote the spin and isospin of the $j^{th}$ nucleon, $\hat{\epsilon}$ is the photon polarization, $\mu_p$ and $\mu_n$ are the proton and neutron magnetic moments and q is the 3-momentum transfer.

Hence, if we study these processes in cases where the nuclear information is well known, i.e. where weak and electromagnetic transition rates have been well determined, e.g. pionic transitions can be predicted, as has been discussed recently by Botton[36]. Furthermore by using an elementary particle treatment of nuclei we can hope to learn something about meson exchange effects in nuclei, the renormalization of the axial vector coupling constant, the corrections for going from soft to real pions etc. (see ref. 37).

Let us illustrate this exciting analogy in fig. 13, where we compare a recently measured photon spectrum of the pion capture reaction[38] on $^{14}$N, i.e. $^{14}$N($\pi^-,\gamma$)$^{14}$C, with high resolution electron scattering data from the DALINAC. We see several M1 excitations at backward angle electron scattering with a concentration of strength in

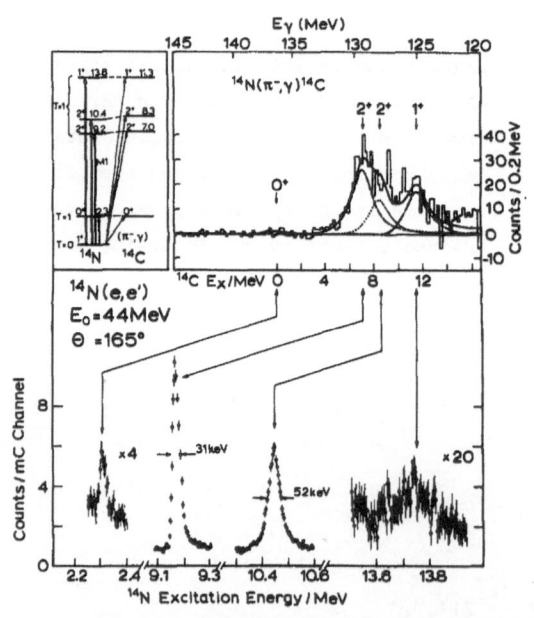

Fig. 13. Comparison between analogous transitions observed in the ($\pi^-,\gamma$) reaction and backward angle inelastic electron scattering on $^{14}$N. The corresponding transitions are linked by arrows. Note the high energy resolution in the electron scattering spectrum, which (for reasons of display) had to be cut into different energy bins.

two $2^+$ states at 9.2 and 10.4 MeV (note that our energy resolution is such that we can even determine the difference in the total widths of these states) and a weak transition to the $1^+$-state at 2.3 MeV. The M1 transitions in electron scattering have extremely similar analogous transitions in the radiative pion capture $\gamma$-ray spectrum. In fact, the counterpart of the ($\pi^-,\gamma$) transition into the 11.3 MeV state in $^{14}$C, the 13.8 MeV state in $^{14}$N, has now been measured for the first time in (e,e'). The detection of this transition is not simple since the state is highly unbound with respect to particle emission and we determine a natural width $\Gamma \simeq 85$ keV. A very careful study now is underway to determine the M1 transition strength between the ground state and the state at 2.3 MeV in addition to the formfactor of this transition for low momentum transfers because of the relation of this transition to leptonic and semi-leptonic processes in the mass 14 nuclei.

<u>Examples for M1 transitions in light and heavy nuclei:</u> We discuss now a few of our
experiments performed with high energy resolution in order to study magnetic dipole
transitions. So far,we have investigated $^{28}$Si, $^{58}$Ni, $^{90}$Zr and $^{208}$Pb. As the simple
shell model argument from above shows, these nuclei should be possible strong candidates for collective M1 transitions (fig. 14).

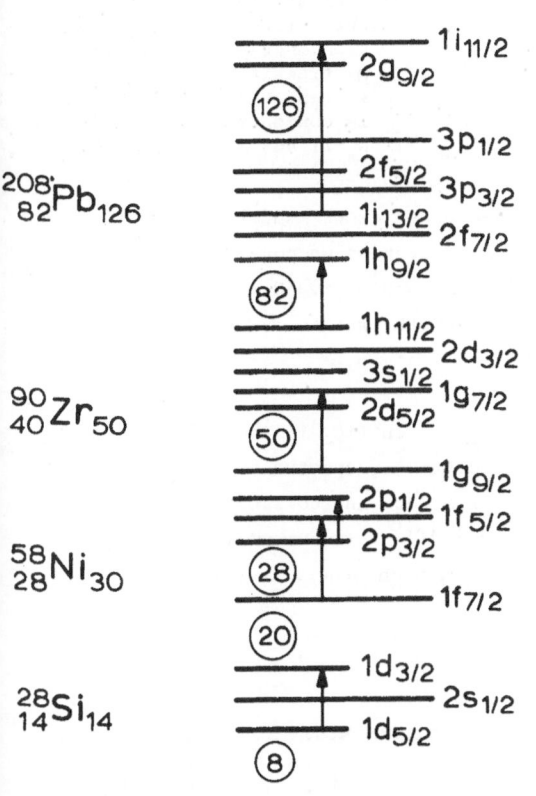

Fig. 14.  Shell model picture in order
to display possible candidates
for collective M1 transitions.
Several such transitions between spin-orbit partners are
indicated by arrows.

## Possible candidates for collective M1-transitions

$^{28}$Si: Figure 15 displays an inelastic
electron scattering spectrum from a natural Si-target. Riding on a smooth
background (mostly due to the radiative tail of the elastic peak not shown
in this spectrum)there are many lines corresponding to the excitation of states
in $^{28}$Si. The spin and parity for most
of these are known[39].The area under
the invidual peaks can be determined
with high precision and the smallest
measurable cross section with our set-
up is about 1 x 10$^{-34}$ cm$^2$· The energy
resolution in this spectrum is 34 keV (FWHM). The lower part of this spectrum shows
the region of 1$^+$, T=1 states. We observe M1 transitions to states at 10.594, 10.725,
10.901, 11.445 and 12.331 MeV and hence less strength than had been identified previously with moderate energy resolution. However, as is the case in most light nuclei
there is essentially one strong state which carries most of the isovector M1 strength.

Table III summarizes the results on isovector M1 transitions in $^{28}$Si. If the
summed strength $\Sigma\Gamma_\gamma^0$ = 41.6 eV is related to the prediction of Kurath's sum rule[31] -
here we use a representation of this sum rule due to Kuehne, Axel and Sutton (ref. 40) -

then the experimentally determined number of 7.3 ± 0.4 compares favourably with a Nilson model prediction of 7.6 where $^{28}$Si in its ground state is assumed to be of oblate shape. This result, of course, is in agreement with other, sometimes much less

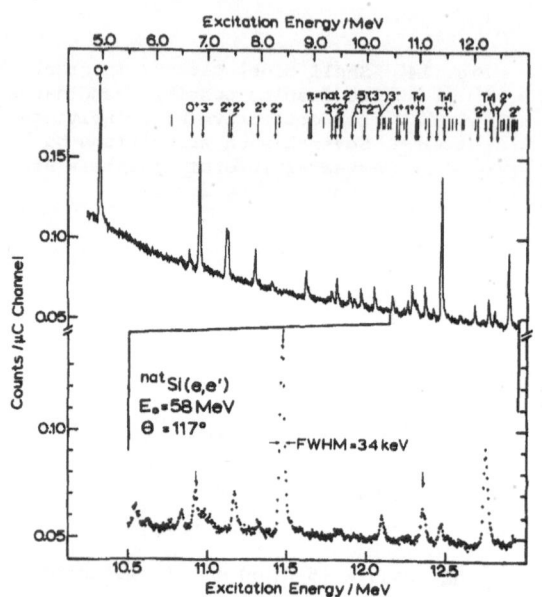

accurate experimental observations. For completeness we note, that the shell model would give 8 and the Nilsson model with prolate deformation 4.3.

Fig. 15. Inelastic electron scattering spectrum of 58 MeV electrons from Si. The lower part shows the enlarged region of excitation energy where isovector M1 transitions in $^{28}$Si are observed. Most of the strength is concentrated in the state at $E_x$=11.445 MeV.

$^{58}$Ni: In fig. 16 inelastic electron scattering spectra on $^{58}$Ni at $\theta$ = 180$^\circ$ from Fagg and collaborators[41] and at $\theta$ = 165$^\circ$ from Darmstadt are compared with each other. In the region

T A B L E   III

| $E_x$ MeV | $B(M1,q=k)$ fm$^2$ | $R_{tr}$ fm | $\Gamma_\gamma^0$ eV | $\Gamma_\gamma^0/\Gamma_W$ |
|---|---|---|---|---|
| 10.594 | $(2.75 \pm 0.46) \cdot 10^{-3}$ | 4.2±0.4 | 1.14 | 0.046 |
| 10.725 | $(2.12 \pm 0.32) \cdot 10^{-3}$ | 3.6±0.2 | 0.91 | 0.035 |
| 10.901 | $(9.47 \pm 0.59) \cdot 10^{-3}$ | 2.55±0.12 | 4.29 | 0.16 |
| 11.445 | $(5.34 \pm 0.26) \cdot 10^{-2}$ | 2.75±0.08 | 27.9 | 0.89 |
| 12.331 | $(1.12 \pm 0.05) \cdot 10^{-2}$ | 3.03±0.05 | 7.33 | 0.19 |

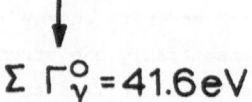

$\Sigma\, \Gamma_\gamma^0 = 41.6\,eV$

of expected M1 transitions between about 9 to 11 MeV the density of levels is already
high. Nevertheless even with moderate energy resolution some strong transverse excitations could be identified, although the high resolution work clearly points to the
fact that because of the large number of states observed in the region of interest
one has to be extremely careful which of the states are to be associated with M1 excitations.

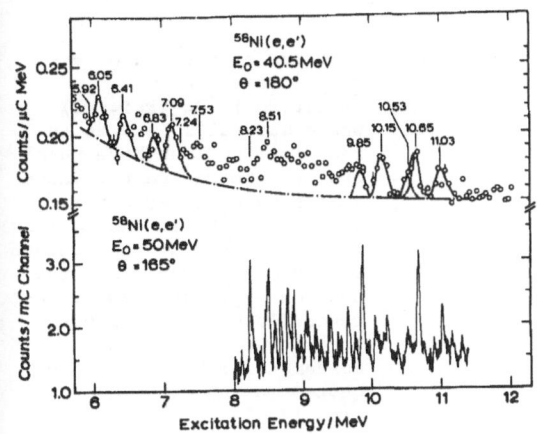

Fig. 16. Comparison between backward
angle inelastic electron scattering spectra from $^{58}$Ni measured with medium energy resolution at NRL (ref. 41) and with
high energy resolution at the
DALINAC. Note the high level
density in the energy region
of expected M1 excitations.

Figure 17 exhibits the high resolution spectra in detail. They were
taken between $\theta = 93^{\circ}$ and $\theta = 165^{\circ}$ with
varying energy resolution, the best one
being $\Delta E = 25$ keV. The decrease of the
background, which is mainly longitudinal in origin, in going from forward

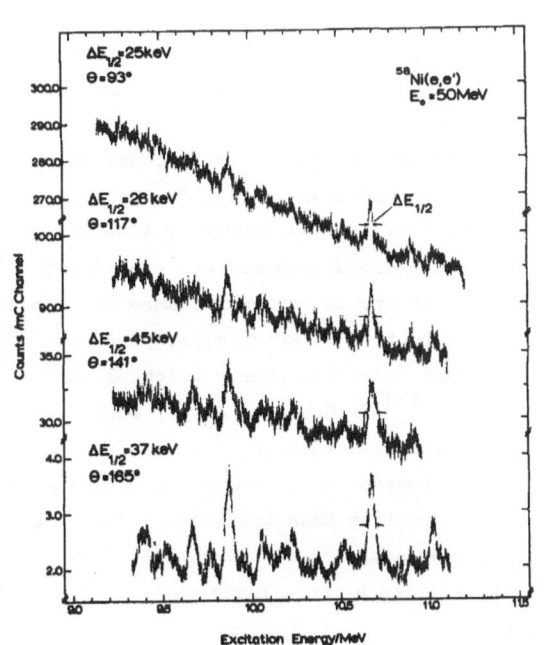

Fig. 17. Inelastic electron scattering
spectra from $^{58}$Ni at $E_o$=50 MeV
and between scattering angles
$\theta = 93^{\circ}$ and $165^{\circ}$. The energy
resolution (FWHM) in the spec-
tra is also indicated. Five
$1^{+}$ states are observed at
$E_x$ = 9.852, 10.224, 10.515,
10.676 and 11.020 MeV.

to backward angles and the corresponding
enhancement of transverse excitations
is clearly visible. There are five candidates for $1^{+}$ states at $E_x$ = 9.852,
10.224, 10.515, 10.676 and 11.020 MeV.
How can we be certain about their multipolarity? We ascertain them by two
ways. First, as fig. 18 shows, we used
(like the authors of ref. 41) the charge
exchange reaction $^{58}$Ni(t,$^{3}$He)$^{58}$Co (ref.42)
and the two-nucleon transfer reactions

$^{62}$Ni(d,α)$^{58}$Co (ref. 43) and $^{56}$Fe($^3$He,p)$^{58}$Co (ref. 44) which identified low lying 1$^+$ states in $^{58}$Co. All these states in the parent nucleus $^{58}$Co have high lying iso-baric analogue states in $^{58}$Ni, which are shifted with respect to their parent states by the Coulomb energy difference minus the neutron-proton mass difference. If we take the locations of our strong transverse excitations in the region of the expected ΔT = 1 M1 excitations and compare them to the excitation energies of the projected $^{58}$Co parent states then a Coulomb energy difference of $\Delta E_c$ = (9.193 ± 0.021) MeV is

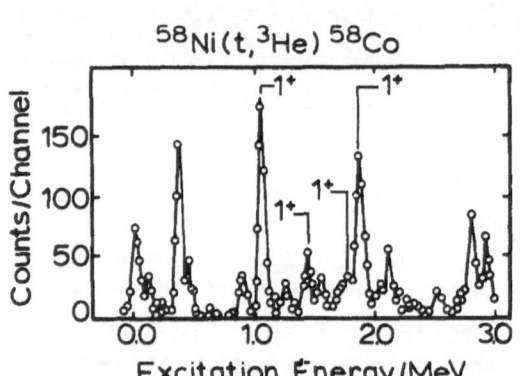

Fig. 18.  Spectrum of the charge ex-change reaction $^{58}$Ni(t,$^3$He)$^{58}$Co$_+$(ref. 42) where several 1$^+$ states in the final nucleus have been observed (upper part). These states have isobaric analogue states in $^{58}$Ni at excitation energies given by the Coulomb energy difference minus the neutron - proton mass diffe-rence (lower part). Their location is in very good agreement with the energies quoted in the caption of fig. 17.

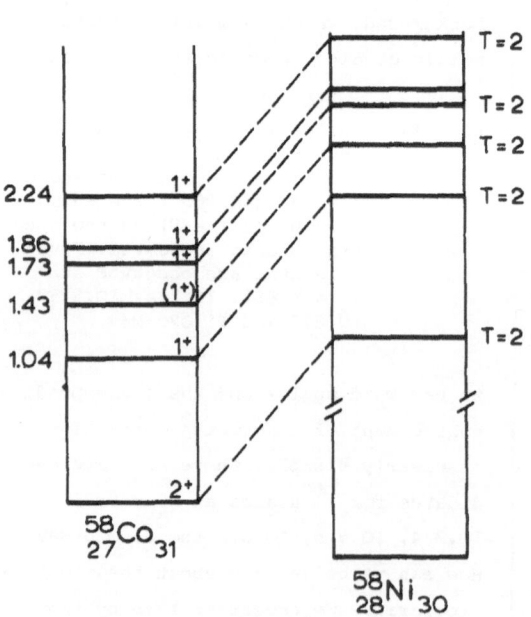

derived. This value was not known but it excellently fits into the systematics of Coulomb energy differences (ref. 45).

The second indication that we are dealing with isovector M1 excitations in $^{58}$Ni is furnished by the very good agreement of the measured angular distri-butions with DWBA calculations. This is demonstrated in fig. 19 for four of the experimental angular distributions. From a comparison between experiment and theory we derive a summed M1 strength of about 16 eV which corresponds to only about 7% of the energy weighted sum rule$^{31,32)}$. Hence we detect much less M1 strength than in $^{28}$Si. The strength we observe is approximately a factor of two smaller then the value found by the authors of ref. 41 since the M1-cross sections in the high resolution experiment are smaller by that factor as compared with the ones from the medium energy resolution work.

In order to find the missing M1 strength (ref. 46) we will extend our measurement to higher excitation energies in $^{58}$Ni. We will also search for heretofore unknown
isoscalar M1 strength at lower excitation energies.

$^{90}$Zr:  The location of giant magnetic dipole strength in this nucleus is extremely important since in the simple shell model picture only neutrons should participate in
the $g_{9/2} \rightarrow g_{7/2}$ spin-flip transition. There are 10 neutrons in the $g_{9/2}$ shell and the

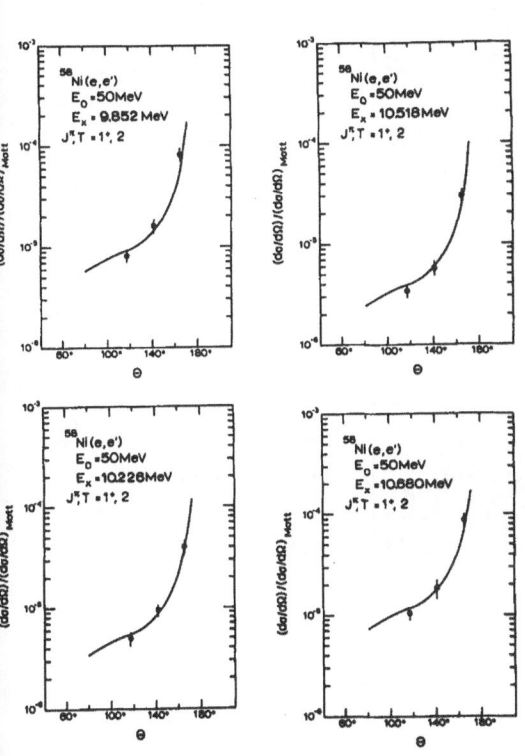

Fig. 19.  Angular distributions for four
of the experimentally determined M1 excitations in $^{58}$Ni
compared to DWBA-calculations.

$g_{7/2}$ shell is completely empty - ideal
conditions for observing a collective
M1 transition. Despite this fact a previous (e,e') experiment with moderate
resolution (ref. 47) and a (p,p') experiment (ref. 48) have not been successful in locating M1 strength. We
have searched for M1 transitions in $^{90}$Zr
between excitation energies of 7.5 and
12.5 MeV. The result of this search is
shown in fig. 20. The shell model predicts the M1 strength at $E_x \simeq 9$ MeV and
we indeed find with an energy resolution of about 30 keV (indicated on the
inserts which display the lines from
elastic scattering) transverse excitations at the predicted excitation energy in all three spectra measured so far
($\theta = 165^{\circ}$, $E_o$ = 58, 55 and 44 MeV).

These spectra were decomposed by a fitting procedure using the line shape from elastic
scattering and the resulting cross sections were compared to DWBA calculations for
M1 and M2 transitions. Our preliminary analysis yields 2 possible candidates for $1^+$-
states at $E_x$ = 8.81 and $E_x$ = 9.20 MeV with a strength of about 6.4 eV. Compared to
the simple shell model estimate of 45 eV, this value constitutes about 13% of the expected total strength and about 31% compared to a renormalized RPA calculation of Krewald
and Speth[49], who predict 19.5eV.Most of the other states are presumably M2 excitations,
i.e. states at $E_x$ = 8.86, 8.92, 8.97, 9.10, 9.13, 9.26 and 9.30 MeV. Their strength
is approximately $\Sigma B(M2) = 4.0 \times 10^2 \mu_N^2 fm^2$ which is about 8% of the total strength predicted by Krewald, Speth[49], and ~ 20% of the strength predicted around 10 MeV excitation energy.

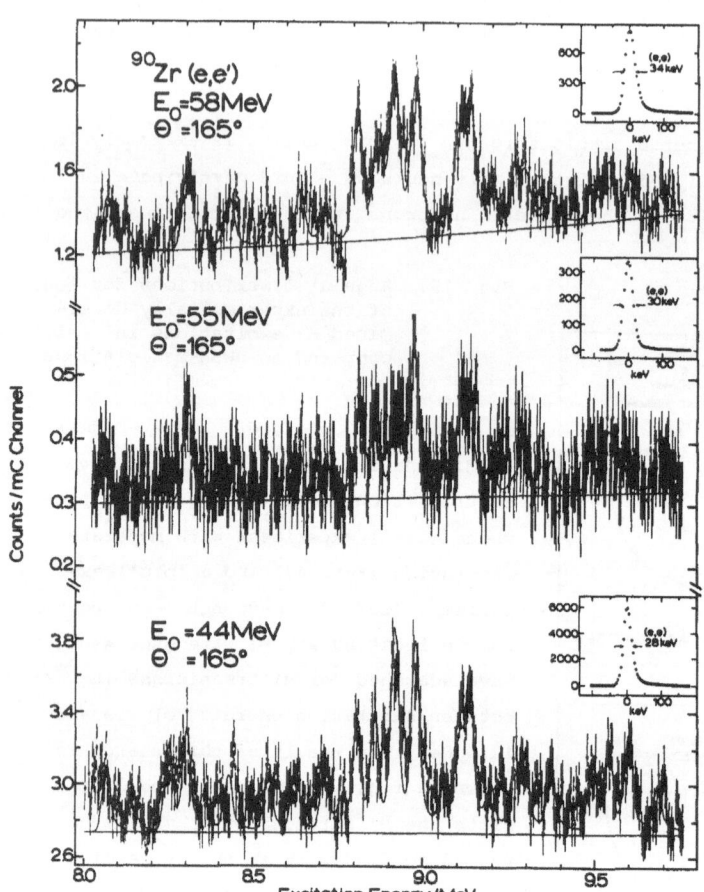

Fig. 20. Inelastic elec-
tron scattering
spectra on $^{90}$Zr
taken under $\theta$=165°
at 3 different
energies. In the
upper right hand
corner the measur-
ed elastic line
is given and the
energy resolution
(FWHM) is indi-
cated. Shown
also is the de-
composition of
the various spec-
tra into lines.
Most of the states
excited around
9 MeV excitation
energy are pre-
sumably 2$^-$ states,
candidates for
1$^+$ states are
the ones at
$E_x$ = 8.81 and
9.20 MeV.

We hence notice that
(as was the case in $^{58}$Ni)
there is very little M1
strength where it is ex-
pected and certainly the
strength one sees can
hardly be called collec-
tive, nor can it be called
a giant M1 resonance.

It is, however, gratifying, that we at least see some M1 strength. This fact
supports indirectly the observation of giant Gamow-Teller strength in the charge
exchange reaction $^{90}$Zr(p,n)$^{90}$Nb.   (ref. 50).

$\underline{^{208}Pb}$: As it is illustrated in fig. 21, collective M1 excitations in $^{208}$Pb should
result from $|h^{-1}_{11/2} \, h_{9/2}\rangle$ proton - and $|i^{-1}_{13/2} \, i_{11/2}\rangle$ neutron excitations. The respec-
tive single particle energies are 5.6 MeV and 5.8 MeV and the two configurations are
therefore expected to mix thoroughly. There is an interesting history to the problem
of M1 excitations in $^{208}$Pb, both experimentally and theoretically. Threshold photoneutron experiments on $^{208}$Pb pioneered the field, but out of the originally proposed
5 states (ref. 51) for possible 1$^+$ candidates only one survived (ref. 52) with a
strength of about 20 eV. Presently, charged particle experiments like (p,p'$\gamma$),

($\alpha,\alpha'\gamma$) and ($d,d'\gamma$) experiments are underway to detect the missing M1 strength below threshold (ref. 53). An electron scattering experiment[54] with medium energy resolution at NRL also detected sizable M2 strength near threshold. Theoretical efforts to

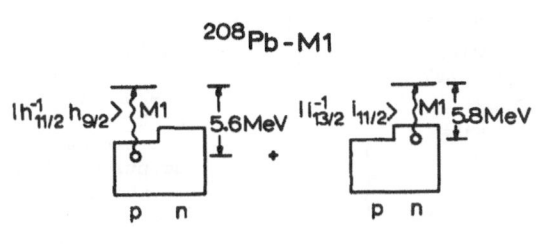

Fig. 21. Schematic representation of expected neutron and proton M1 excitations in $^{208}$Pb.

predict the location and strength of M1 excitations have been made within the shell model by Vergados[55] and Lee and Pittel[56], within the frame of a more refined RPA calculation by Ring and Speth[57] and by Grecksch, Huber and Knüpfer[58], and recently on quite general grounds by Bohr and Mottelson[59]. While the calculations in refs. 55, 56 and 59 give a total M1 strength of roughly 70 to 80 eV, the RPA calculations yield about 30 - 40 eV. Since $^{208}$Pb is crucial for the determination of the spin-isospin dependent part of the interaction we set out to study the region of expected M1 transitions with high resolution inelastic electron scattering. So far, we still have preliminary results only.

Figure 22 shows there are indeed strong transverse excitations observed in inelastic scattering. There we have compared backward angle electron scattering spectra taken at medium and high resolution. Lindgren et al. (ref. 54) assign two of the peaks (at 7.40 and 7.91 MeV) of the four outstanding ones to be due

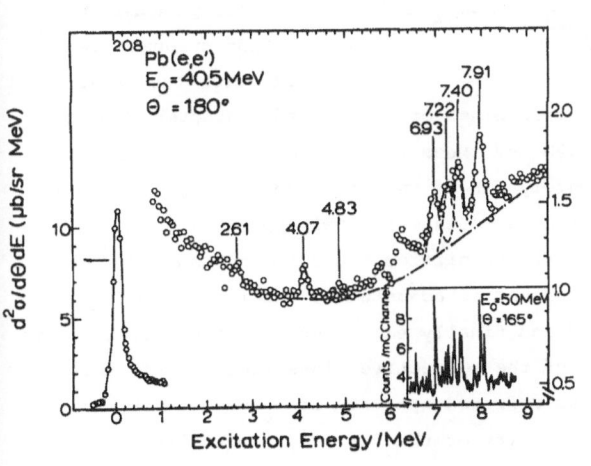

Fig. 22. Comparison between backward angle electron scattering spectra from the NRL (ref. 54) taken with medium energy resolution and from the DALINAC (insert in the lower right hand corner) with high resolution. There are about 40 states excited in the energy region common to both experiments. The strength at the centers of gravity 7.40 and 7.91 MeV has been identified by the authors of ref. 54 as being due to M2 excitations.

to M2 excitations.    The clustering of strength between about 6 and 8 MeV, however,
is only the envelope of about 40 states. This is proved in detail by our high resolution experiments performed at $E_o$ = 50 and 63.5 MeV.

Figure 23 displays the four measured spectra at 50 MeV and scattering angles
θ = 93°, 129°, 141° and 165°. The best energy resolution (FWHM) achieved was ΔE = 24 keV

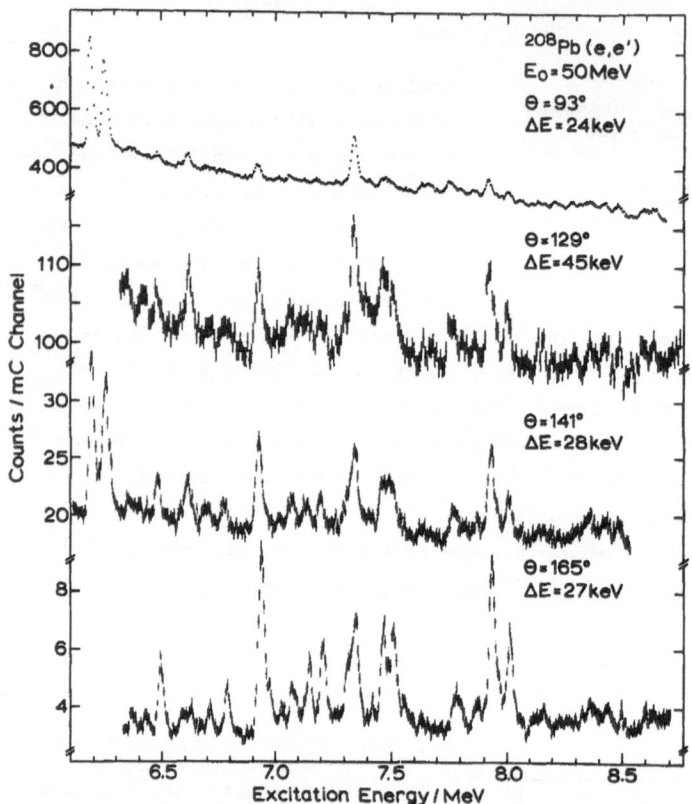

Fig. 23.    High resolution
inelastic elec-
tron scattering
spectra on $^{208}$Pb
at $E_o$ = 50 MeV.
The energy re-
solution (FWHM)
in the four spec-
tra varies be-
tween 24 and
45 keV.

Several possible candidates
for $1^+$ or $2^-$ states stand
out above the background.
In this region of excita-
tion energy in $^{208}$Pb, the
level density is already
high. In fact, the
energy resolution in our
experiment is not suffi-
cient to even resolve the
selectively excited states.
A similar picture of ex-
citation strength re-
sults from the measurement
at 63.5 MeV  (fig. 24), except that some more lines of transverse character show up.
The neutron threshold lies at $Q (\gamma,n)$ = 7.38 MeV. Measuring one of the 8 spectra displayed in figs. 23 and 24 takes approximately 80h of run time and hence puts extreme
boundary conditions on the stability of the whole experimental apparatus.

The analysis of the very complex spectra is difficult and time-consuming. The
procedure is the following: First, a background is adjusted to the raw data of fig. 23
and 24, the background is then subtracted and finally each spectrum is decomposed into
individual lines. Thereby the line shape of the respective lines from elastic scattering is used as a guide. Figure 25 shows a fitted spectrum at $E_o$ = 63.5 MeV and
θ = 165° together with a measure for the quality of the fit. There are altogether
44 lines hidden under the envelope given by the experimental points and one constraint
of the fit is, of course, that the value of the absolute energy of each line comes out

consistent from the analysis of all 8 measured spectra. A precise determination of
the cross section is extremely important since it is hoped, that then some of the

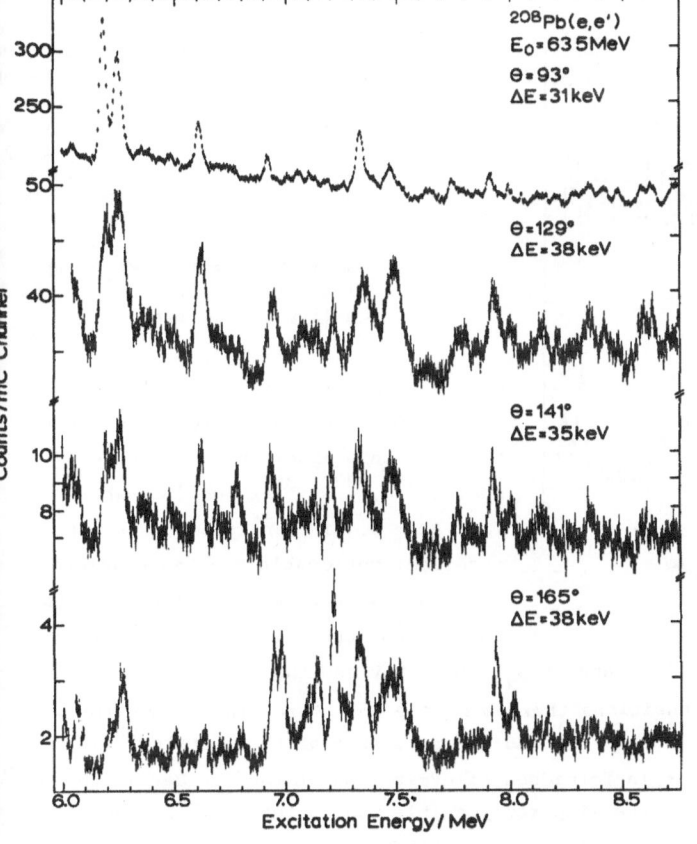

Fig. 24. Same as fig. 23,
but at
$E_o$ = 63.5 MeV

model dependence in the extraction of excitation strength (by way of DWBA calculations with microscopic formfactors compared to the data) might be removed. We are presently in the process of producing absolute cross sections and angular distributions from which we hope to infere the multipolarity and strength of the states excited in $^{208}$Pb between 6 to 8 MeV.

We also plan to search for the new kind of giant M1 resonances proposed recently by Speth et al.[60] RPA calculations using a configuration space which includes $2\hbar\omega$ excitations predict very collective $1^+$-resonances in $^{208}$Pb around an excitation energy between 19 and 25 MeV.

## D. Selected topics in inelastic electron scattering on low lying states

In this brief section I should like to emphasize the importance of studying also weak transitions. These transitions can be detected only with a high resolution facility, since only there is the peak to background ratio in the spectra sufficiently large to measure weakly excited states.

a) Isospin-forbidden $\Delta T = 0$, E1 transitions in self conjugate nuclei – $^{16}$O and $^{40}$Ca

Electric dipole transitions in selfconjugate nuclei have to proceed with a change of isospin ($\Delta T = 1$) if they are isospin-allowed. The systematics of E1 transitions[61]

in light and medium heavy nuclei shows that the average ground state radiation widths of $\Delta T = 1$ transitions vary between $\Gamma_\gamma^o \sim 10^{-2}$ to $10^{-4}$ W.u., while the isospin-forbidden

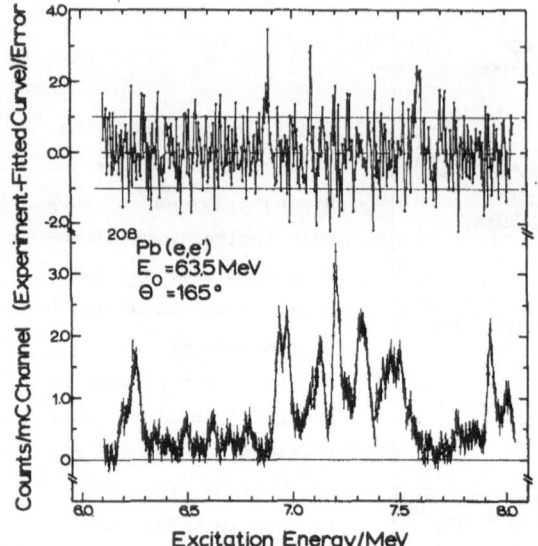

Fig. 25. Inelastic electron scattering spectrum on $^{208}$Pb at $E_o$=63.5 MeV and $\theta$=165°, after the background has been subtracted. The experimental spectrum has been decomposed into individual lines by line shape fitting procedure and the result is compared to the experimental points in form of a full line (lower part). The upper part of the figure shows a measure of the quality of the fit.

ones have a strength between about $10^{-4}$ - $10^{-6}$ W.u. There are a few exceptions to this observation in the 4n nuclei $^{16}$O, $^{32}$S, $^{36}$Ar and $^{40}$Ca where fast $\Delta T = 0$, E1 transitions have been observed. Noticeable are the transitions between the state at $E_x$ = 7.12 MeV and the ground state in $^{16}$O with a strength of $4 \times 10^{-4}$ W.u. and between the 1$^-$ state at $E_x$ = 6.95 MeV and the groundstate in $^{40}$Ca with $\Gamma_\gamma^o \simeq 2 \times 10^{-3}$ W.u. These transitions therefore pose an interesting object for study of the mechanism of isospin-violation[62-64] and the structure of these states[65].

We have investigated the isospin-forbidden E1 excitation of the 7.2 MeV state in $^{16}$O and the 6.95 MeV state in $^{40}$Ca. In the simplest possible model, the harmonic oscillator shell model, the 1$^-$ state in $^{16}$O is formed by exciting a particle from the p-shell into the s-d-shell and in $^{40}$Ca, correspondingly from the s-d- into the f-p-shell. The formfactor in this p-h model

$$F(q) = (A_o q^3 b^3 + \Phi A_1 qb) \exp(-q^2 b^2/4)$$

then consists of two parts, an isoscalar contribution ($\Delta T = 0$) which is proportional to the third power of the momentum transfer q and will hence dominate the cross section at high q and an isovector contribution resulting from the orthogonal part in the wave function which is almost linear in q (b is the harmonic oscillator parameter). Since the measured cross section is proportional to the square of this formfactor, i.e. $d\sigma/d\omega \propto |F(q)|^2$, the two terms should interfere. A comparison of this simple model should then give the ratio of the amplitudes $A_1/A_o$ and the phase of the interference.

Figure 26 shows the formfactor for the 7.12 MeV, 1$^-$, T = 0 state in $^{16}$O (ref. 66). In addition to our measurements at the DALINAC, data from NBS (ref. 67) and F(q) at the photon point from the lieftime of ref. 68 are also plotted. The experimental data can only be described when the isoscalar and the isovector contributions interfere

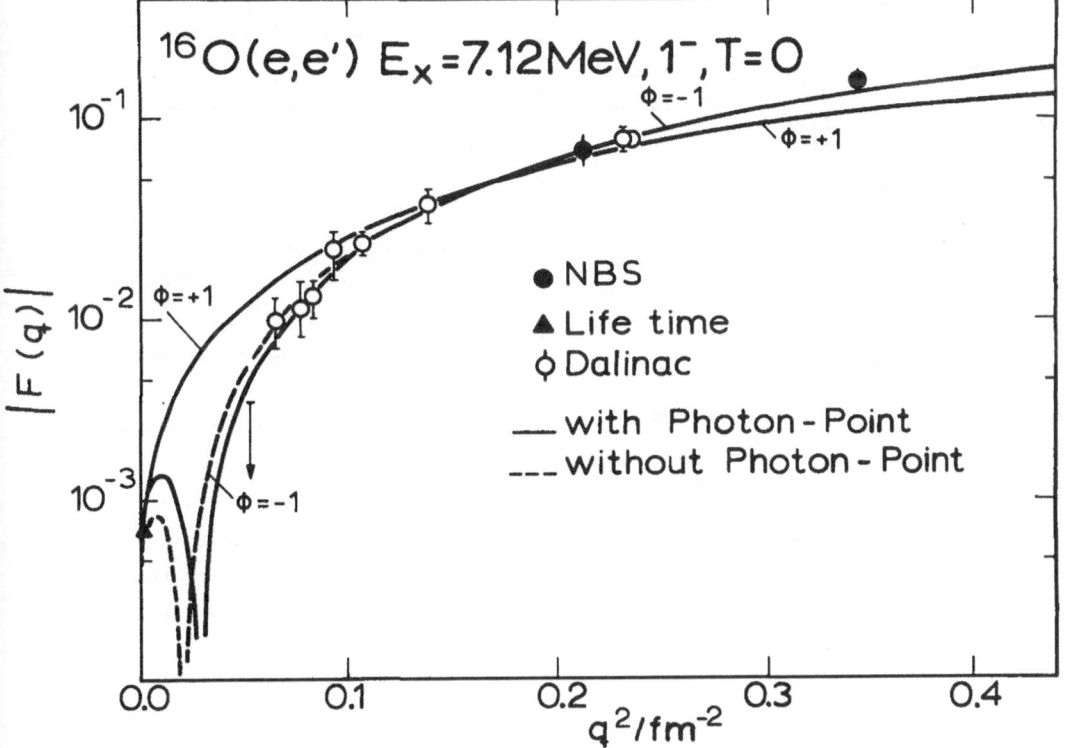

Fig. 26.    Formfactor F(q) as a function of $q^2$. Besides the measured points from the DALINAC, data from the NBS (ref. 67) and from the lifetime (ref. 68) are shown.

destructively, i.e. if the phase $\Phi = -1$. This fact leads to a minimum of the formfactor at very low q. The ratio of the amplitudes $A_1/A_0$ amounts to about 0.5% and is a rough measure of the isospin violation. There are different sources of this violation[62-64]. A recent extensive continuum shell model calculation[69] reproduces as well the minimum of the formfactor as about its absolute magnitude.

The situation in $^{40}$Ca is very similar, except that the $1^-$, T = 0 state is only separated by 41 keV from a $2^+$, T = 0 state at $E_x = 6.910$ MeV. As fig. 27 shows high resolution is needed in order to determine a reliable transition strength. Our preliminary results of the measured formfactor are displayed in fig. 28. This formfactor still contains both the longitudinal and the transverse contribution to the cross section but again reveals a minimum at very low q. If the longitudinal part and the transverse part are separated by measuring the angular dependence of the cross section at constant momentum transfer, the lifetime of the state[70] at the photonpoint

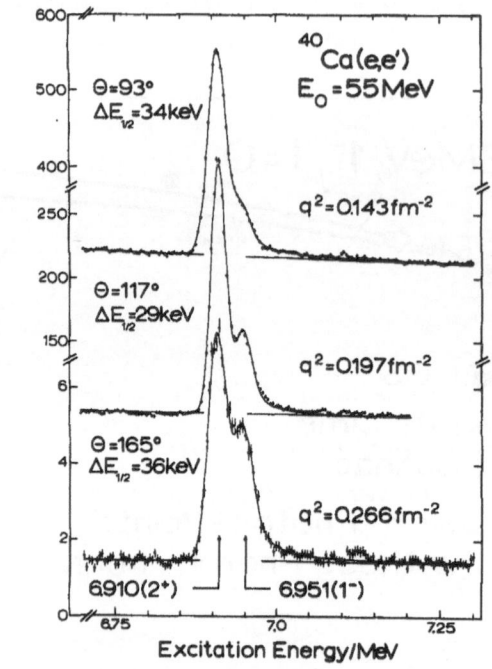

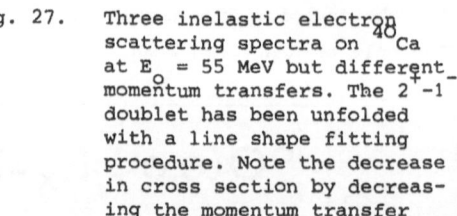

Fig. 27. Three inelastic electron scattering spectra on $^{40}$Ca at $E_o$ = 55 MeV but different momentum transfers. The $2^+$-$1^-$ doublet has been unfolded with a line shape fitting procedure. Note the decrease in cross section by decreasing the momentum transfer

$q = k = E_x/\hbar c$ is reproduced with the fit of the simple harmonic oscillator model described above. Both, the E1 transition in $^{16}$O and in $^{40}$Ca might have also some importance for the determination of exchange currents.

b) E0 transitions in the Ca isotopes

Core-excited deformed states and shell model states seem to coexist in the excitation spectrum of the Ca-isotopes. If $0^+$-states above the

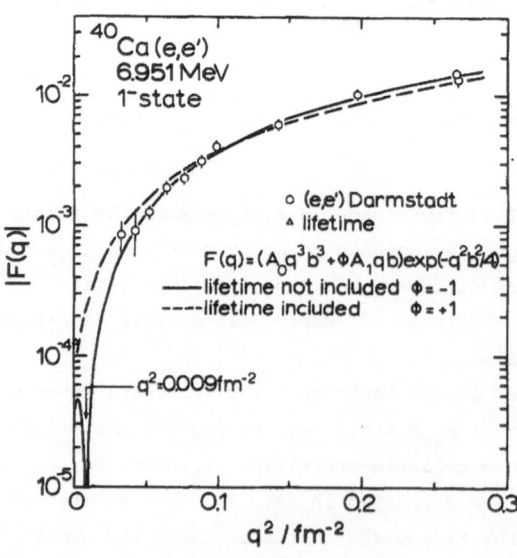

Fig. 28. Formfactor for the isospin-forbidden electroexcitation of the $1^-$, T = 0 state at $E_x$ = 6.951 MeV in $^{40}$Ca. The formfactor still contains both the longitudinal and the transverse contribution and is described by the simple harmonic oscillator model (see text) only if the isoscalar and the isovector part interfere destructively.

ground states in the even Ca isotopes are excited in inelastic electron scattering then proton core excitations have to be present. We have investigated the lowest $0^+$ states above the ground states in $^{40,42,44,48}$Ca (see fig. 29) in order to determine their monopole strength and hence a measure of the core excitation directly.

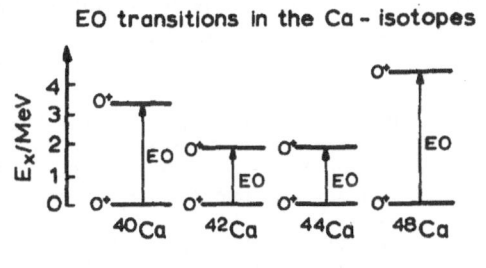

EO transitions in the Ca - isotopes
$E_x$/MeV
0⁺
EO
$^{40}$Ca
$^{42}$Ca
$^{44}$Ca
$^{48}$Ca


Fig. 29. Location of lowest $0^+$ states above the ground states in the even-even $^{40,42,44,48}$Ca-isotopes.

Figure 30 shows characteristic spectra on all four Ca isotopes at excitation energies where the monopole transitions are expected. We observe them indeed with about the same strength in $^{40}$Ca-$^{48}$Ca and in $^{42}$Ca-$^{44}$Ca, a statement which is made quantitative in fig. 31. There the monopole matrix elements determined in the present experiment are compared with the ones obtained by the measurement of the pair

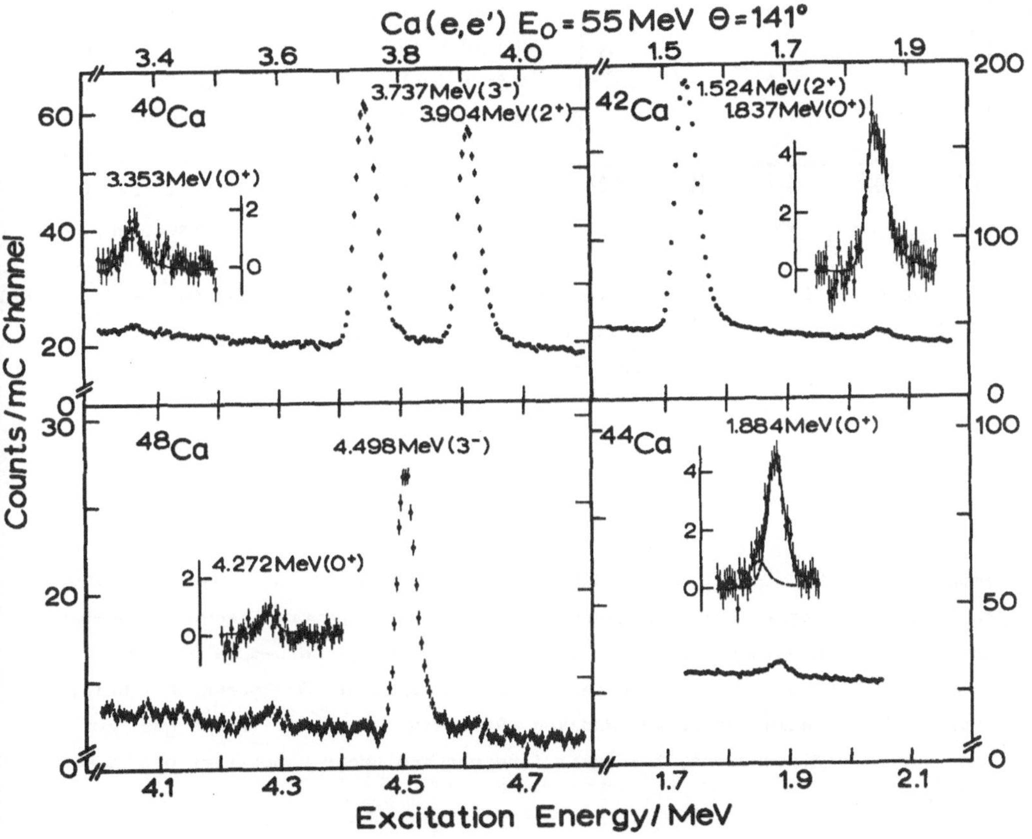

Ca(e,e') $E_0$ = 55 MeV Θ = 141°
$^{40}$Ca
3.737MeV(3⁻)
3.904MeV(2⁺)
3.353MeV(0⁺)
$^{42}$Ca
1.524MeV(2⁺)
1.837MeV(0⁺)
$^{48}$Ca
4.498MeV(3⁻)
4.272MeV(0⁺)
$^{44}$Ca
1.884MeV(0⁺)
Counts/mC Channel
Excitation Energy/MeV


Fig. 30. Inelastic electron scattering spectra on the Ca isotopes in the vicinity of the lowest $0^+$ state above the ground state. The $0^+$-states are appreciably excited and the corresponding lines are enlarged in the inserts of the spectra.

decay and the lifetime. Since for the latter method[71], the pair decay branching ratio (which is small) has to be known in the case of $^{42}$Ca and $^{44}$Ca, the high resolution

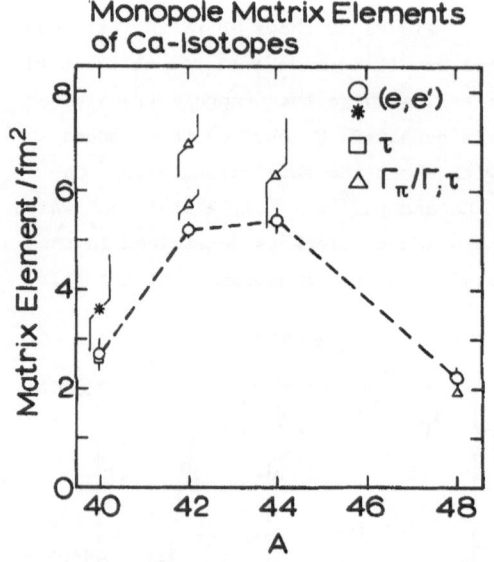

## Monopole Matrix Elements of Ca-Isotopes

Fig. 31.     Monopole matrix elements for the excitation of the $0^+$-states in the Ca isotopes shown in fig. 30. The new and precise measurements from the DALINAC are compared to and old measurement with medium energy resolution (ref. 73) and to results from lifetime measurements and pair decay (refs. 71,74, 75,76)

electron scattering results are much more accurate. These monopole matrix elements should serve as a very stringend test for models including core excitation. If, e.g. the interaction used by Sakakura, Arima and Sebe (ref. 72) for $^{40}$Ca is used for a description of the monopole matrix elements also for the other even Ca isotopes, not only does the 12 particle-4 hole state in $^{48}$Ca lies very low in energy (the interaction favours many particles in the $f_{1/2}$ shell) but also is the behaviour of the matrix elements found experimentally (see fig. 31) so far in complete disagreement with the theory.

## Acknowledgement

Among my collaborators I have thanked in the introduction of this talk I would like to acknowledge again the help given to me by Dr. E. Spamer. I have had numerous discussions with him on the subject of this talk and he has helped me patiently with the manuscript. I should furthermore thank the theorists Dr. de Forest, Dr. Heil, Dr. Knüpfer, Dr. Krewald, Professor Manakos, Professor Speth, Dr. Stock and Prof. Strottman for many discussions, for their interest and their support of our experimental work.

## References

1) F. Gudden, G. Fricke, H. G. Clerc and P. Brix, Z. Physik 181 (1964) 453

2) R. Frey, H.D. Gräf, H. Miska, R. Schneider, D. Schüll, A. Schwierczinski, E. Spamer, H. Theissen, O. Titze and Th. Walcher, Nucl. Instr., to be submitted

3) H. Miska, H.D. Gräf, A. Richter, R. Schneider, D. Schüll, E. Spamer, H. Theissen and O. Titze, Phys. Lett. 58B (1975) 155

4) A. Sçhwierczinski, R. Frey, A. Richter, E. Spamer, H. Theissen, O. Titze, Th. Walcher, S. Krewald and R. Rosenfelder, Phys. Rev. Lett 35 (1975) 1244

5) S. Kowalski, W. Bertozzi and C.P. Sargent, in "Medium energy nuclear physics with electron linear accelerators", MIT 1967 Summer Study, USAEC report no. TID-24667 (1967), p. 39

6) D.H. Madison and E. Merzbacher, in Atomic Inner-Shell Processes, B. Crasemann, ed., Acad. Press, Inc., New York (1975), p. 11

7) C.A. Quarles, Phys. Lett. 39A (1972) 375; and Electric and Atomic Collisions, IX ICPEAC, J.F. Risley and R. Geballe, ed., University of Washington Press, Seattle (1975) p. 898

8) C.A. Quarles, The Use of Small Accelerators in Research and Teaching, CONF-741040-P1, Nat. Tech. Inf. Service, US Dept. Com., Springfield, Va, (1974) p. 351

9) W. Scholz, Angela Li-Scholz, R. Collê and I.L. Preiss, Phys. Rev. Lett. 29 (1972) 761

10) L.M. Middleman, R.L. Ford and R. Hofstadter, Phys. Rev. A2 (1970) 1429

11) For a summary of recent work see, e.g., S. Morita, in Proceedings of the Second International Conference on Inner Shell Ionization Phenomena, Freiburg 1976, ed. R. Brenn and W. Mehlhorn, Fakultät für Physik, Universität Freiburg, Germany

12) H. Kolbenstvedt, J. appl. Phys. 38 (1967) 4785

13) B.L. Moisewitsch and D.M. Davidovic in ref. 7, p. 906

14) H. Kolbenstvedt, J. appl. Phys. 46 (1975) 2771

15) W. Lotz, Z. Physik 206 (1967) 205; Astrophys. J. Suppl. 14 (1967) 207

16) W. Hink, J. Jessenberger and A. Ziegler, Z. Physik 226 (1969) 463

17) V.M. Pessa and W.R. Newell, Phys. Scripta 3 (1971) 165

18) G.R. Dangerfield and B.M. Spicer, J. Phys. B., Atom. Molec. Phys. 8 (1975) 1744

19) H. Genz, D.H.H. Hoffmann, A. Richter and E. Spamer, ref. 11, abstracts p. 229

20) H. Genz, D.H.H. Hoffmann, W.Löw and A. Richter, Contribution to the International Conference on the Physics of X-ray Spectra, Aug. 29 - Sept. 2, 1976, Washington, D.C., U.S.A.

21) G.R. Satchler, Phys. Rep. 14 (1974) 97

22) P. Carlos, H. Beil, R. Bergere, A. Lepretre and A. Veyssiere, Nucl. Phys. A172 (1971) 437

23) D.H. Youngblood, J.M. Moss, C.M. Rozsa, J.D. Bronson, A. D. Bacher and D.R. Brown, Phys. Rev. C13 (1976) 994

24) N. Auerbach and A. Yeverechyahu, Ann. Phys. (N.Y.) 95 (1975) 35

25) S.S. Hanna, in Isospin in Nuclear Physics, edited by D.H. Wilkinson (North-Holland, Amsterdam), 1969, p. 593

26) S.S. Hanna, in Proc. Int. Conf. on nuclear structure and spectroscopy, 1974, vol. 2, ed. H.P. Blok and A.E.L. Dieperink (Scholar's Amsterdam, 1974) p. 249

27) L.R. Fagg, Rev. Mod. Phys. 47 (1975) 683

28) A.B. Migdal, O.A. Markin and I.I. Mishustin, Sov. Phys. JETP 39 (1974) 212

29) G.E. Brown and W. Weise, Phys. Rep., in press

30) H. Überall, Electron Scattering from Complex Nuclei (Academic Press, New York, 1971)

31) D. Kurath, Phys. Rev. 130 (1963) 1525

32) E. Lipparini, S. Stringari, M. Traini and R. Leonardi, Nuovo Cimento 31A (1976) 207

33) S. Yoshida and L. Zamick, Ann. Rev. of Nucl. Science 22 (1972) 121

34) J. Speth, Magnetic moments and magnetic giant resonances, in Proc. Int. Symposium on Nuclear Structure; Coexistence of Single Particle and Collective Types of Excitations, Balatonfüred, Hungary, 1975, in press

35) The interconnections between electromagnetic and weak interactions with nuclei have been emphazised by various authors. For a resent summary sea T.W. Donelly and J.D. Walecka, Ann. Rev. of Nucl. Science 25 (1975) 329

36) N. de Botton, in Proc. Int. Symposium on "Effets mesoniques dans les noyaux diffusion d'electrons a energie intermediaire", Saclay, Mai 12 - 16, 1975 (Comissariat a l'Energy Atomique, Department de Physique Nucleaire), p. 51

37) M. Ericson and M. Rho Phys. Rep. 5C (1972) 57

38) H.W. Baer, J.A. Bistirlich, N. de Botton, S. Cooper, K.M. Crowe, P. Truöl and J.D. Vergados, Phys. Rev. C12 (1975) 921

39) P.M Endt and C. van der Leun, Nucl. Phys. A214 (1973) 1

40) H.W. Kuehne, P. Axel and D.C. Sutton, Phys. Rev. 163 (1967) 1278

41) R.A. Lindgren, W.L. Bendel, E.C. Jones, Jr., L.W. Fagg, X.K. Maruyama, J.W. Light-body and S.P. Fivozinsky, preprint

42) E.R. Flynn and J.D. Garrett, Phys. Rev. Lett. 29 (1972) 1748

43) M.J. Schneider and W.W. Daehnick, Phys. Rev. C5 (1975) 1330

44) T. Caldwell, O. Nathan, O. Hansen and H. Bork, Nucl. Phys. A202 (1973) 225

45) J.A. Nolen and J.P. Schiffer, Ann. Rev. of Nucl. Science 19 (1969) 519

46) That some of the missing isovector M1 strength might be at higher excitation

energies is the result of a shell model calculation of D. Strottman, private communication.

47) L.W. Fagg, W.L. Bendel, E.C. Jones, N. Ensslin and F.E. Cecil, in Proc. of the Intl. Conf. on Nucl. Phys., Munich, 1973, ed. J. de Boer and H.J. Mang (North Holland, 1973), vol. 1, p. 631

48) F.E. Cecil, G.T. Garvey and W. J. Braithwaite, Nucl Phys. A232 (1974) 22

49) S. Krewald and J. Speth, Phys. Lett. 52B (1974) 295

50) R.R. Doering, A. Galonsky, D.M. Patterson and G.F. Bertsch, Phys. Rev. Lett. 35 (1975) 1691; Phys. Rev. Lett. 36 (1975) 344(E)

51) C.D. Bowman, R.J. Baglan, B.L. Berman and T.W. Phillips, Phys. Rev. Lett. 25 (1970) 1302

52) R.J. Holt and H.E. Jackson, Phys. Rev. Lett. 36 (1976) 244

53) G.T. Garvey, private communication

54) R.A. Lindgren, W.L. Bendel, L.W. Fagg and E.C. Jones, Jr., Phys. Rev. Lett. 35 (1975) 1423; Phys. Rev. Lett. 36 (1976) 116(E)

55) J.D. Vergados, Phys. Lett. 36B (1971) 12

56) T.S.H. Lee and S. Pittel, Phys. Rev. C11 (1975) 607

57) P. Ring and J. Speth, Phys. Lett. 44B (1973) 477

58) E. Grecksch, W. Knüpfer and M.G. Huber, Lett. Nuovo Cim. 14 (1975) 505

59) A. Bohr and B. Mottelson, Nuclear Structure, vol. 2 (Benjamin, New York, 1975)

60) J. Speth, J. Wambach, V. Klemt and S. Krewald, Phys. Lett., in press

61) P.M. Endt and C. van der Leun, Nucl. Phys. A235 (1974) 27

62) D.H. Gloeckner and R.D. Lawson, Phys. Lett. 56B (1975) 301

63) A. Arima, P. Manakos and D. Strottman, Phys. Lett. 60B (1975) 1

64) F.C. Barker, Phys. Lett., submitted

65) M.N. Harakeh, J.R. Comfort and A. van der Woude, Phys. Lett. 62B (1976) 155

66) H. Miska, H.D. Gräf, A. Richter, D. Schüll, E. Spamer and O. Titze, Phys. Lett. 59B (1975) 441

67) J.C. Bergstrom, W. Bertozzi, S. Kowalski, X.K. Maruyama, J.W. Lightbody, Jr., S.P. Fivozinsky and S. Penner, Phys. Rev. Lett. 24 (1970) 152

68) C.P. Swann, Nucl. Phys. A150 (1970) 300

69) V. Heil and W. Stock, Phys. Lett., submitted

70) F.R. Metzger, Phys. Rev. 165 (1968) 1245

71) M. Ulrickson, W. Hartwig, N. Benczer-Koller, J.R. MacDonald and J.W. Tape, Phys. Rev. C13 (1976) 536

72) M. Sakakura, A. Arima and T. Sebe, Phys. Lett. $\underline{61B}$ (1976) 335

73) P. Strehl, Z. Physik $\underline{234}$ (1970) 416

74) S. Gorodetzky, N. Schulz, J. Chevallier and A.C. Knipper, J. de Phys. $\underline{27}$ (1966) 521

75) B.N. Belyaev, S.S. Vasilenko, D.M. Kaminker, Izv. Akad. Nauk. SSSR, Ser. Fiz. $\underline{35}$ (1971) 806 [Bull. Acad. Sci. USSR, Phys. Ser. $\underline{35}$ (1971) 742]

76) N. Benczer-Koller, G.G. Seaman, M.C. Bertin, J.W. Tape and J.R. MacDonald, Phys. Rev. $\underline{C2}$ (1970) 1037

<u>PRELIMINARY RESULTS ON THE ANNIHILATION PHOTON</u>
<u>BEAM AT THE FRASCATI LINAC LABORATORY</u>

G.P.Capitani* E.De Sanctis* C.Guaraldo* G.Ricco,**
M.Sanzone**R.Scrimaglio* and A.Zucchiatti**
presented by "MARCELLA SANZONE"
Istituto di Scienze Fisiche dell'Università
Viale Benedetto XV, 5 - 16132 Genova

In this talk I will present the preliminary results on the new monochromatic photon beam from positron annihilation obtained at Frascati Linac laboratory (1).

Positron annihilation in flight at energies below 100 MeV has been widely used in recent years to obtain quasi monochromatic photon beam with variable energy. The main features of the existing facilities have been already presented by professor Bergere in his lectures. On the other side we heard also during this Course the remarkable physical interest in photonuclear reactions at the energies above 100MeV. The positron energy of Frascati Linac can be continuously varied at least from 80 up to 300 MeV without a severe loss of intensity at low energies, as we will see later on.

In fig. 1 the positron beam handling from the end of the Linac to the deviation system is schematically presented. It consist of two pair of quadrupoles $Q_1$, $Q_2$, $Q_3$ and $Q_4$ followed by a pulsed bending magnet. One burst every second is deflected so that the positron spectrum can be detected, during the measurements, by a system of secondary emission monitors in the focal plane of the analysing magnet. Another pair of quadrupoles ($Q_5$ and $Q_6$) is necessary to maintain the beam dimension quite constant along the whole course up to the deviation system. This results in fact the best condition for optimizing the positron intensity transmission. We don't forget that the positron beam emittance is of the order of 1 mrad x cm and it increases at low positron energies.

In fig. 2 the calculated envelope of the beam is presented.

(*) Sezione INFN dei LNF     (**) Sezione INFN di Genova

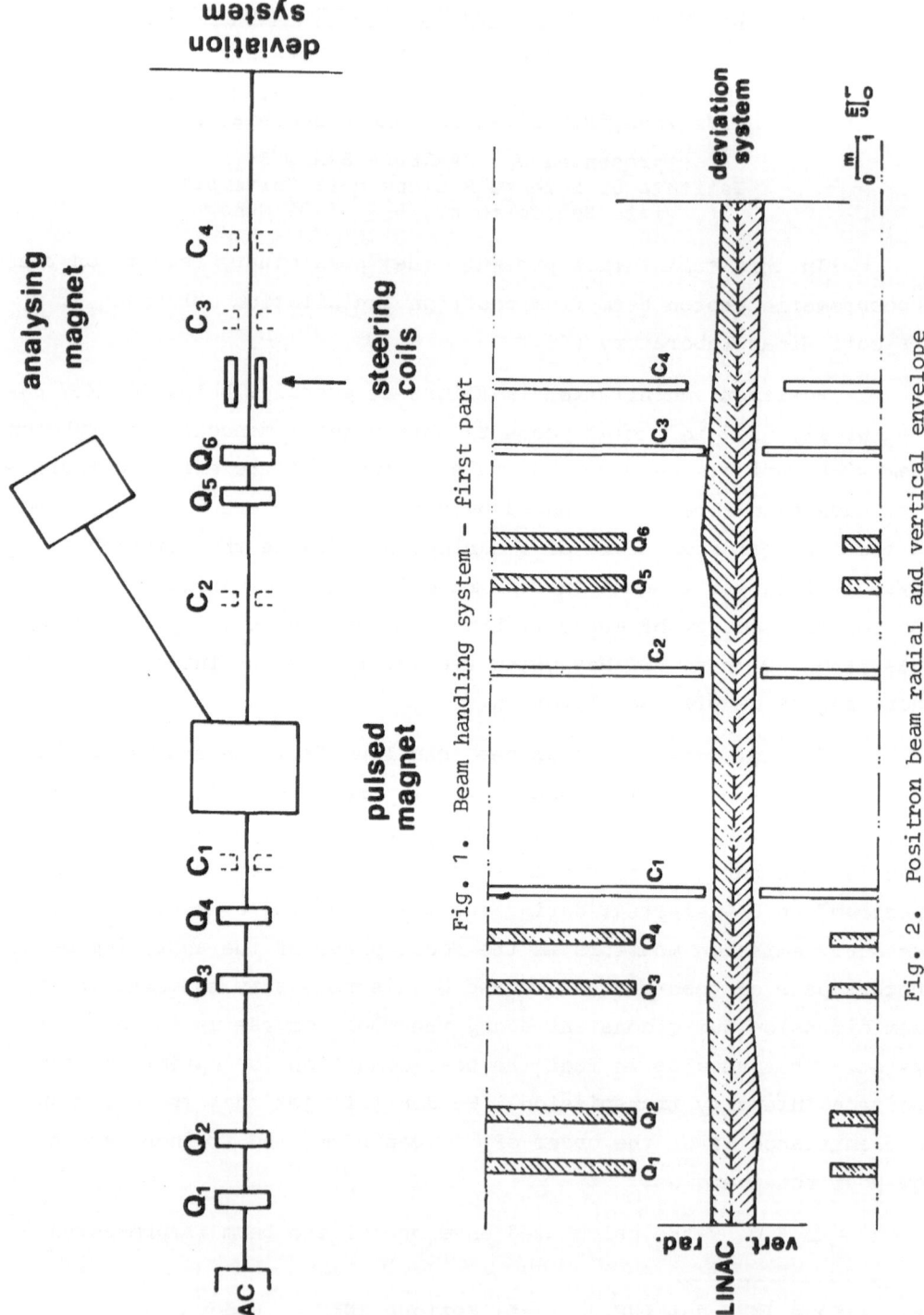

Fig. 1. Beam handling system - first part

Fig. 2. Positron beam radial and vertical envelope

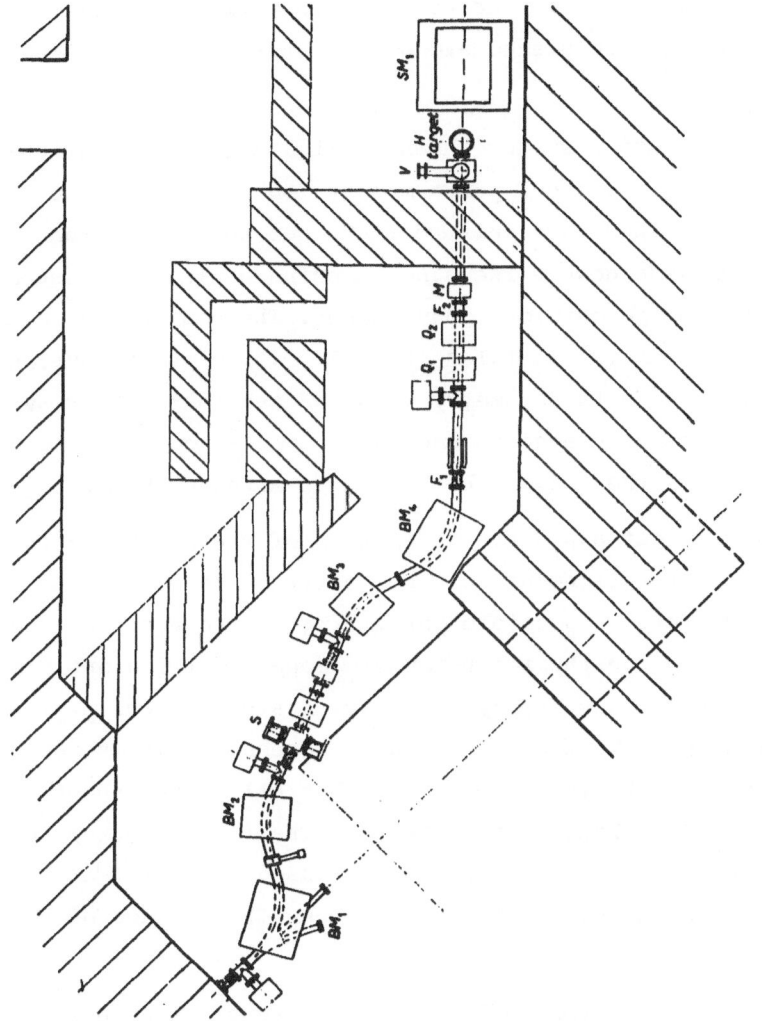

Fig. 3  Beam handling system – second part

To control the beam dimension and position four fixed collimators ($C_1$, $C_2$, $C_3$ and $C_4$) are placed along this part of beam transport system.

The 45° deviation system shown in fig. 3 is a conventional Penner type achromatic system, performed by four bending magnets $BM_1$, $BM_2$, $BM_3$ and $BM_4$. They have uniform field but nonzero intrance and exit angles to provide double focusing. Bending angles are respectively 60° and 37,5° for geometrical convenience. The beam energy can be selected by a system of two tantalum slit S, 15 mm thick each, not cooled, positioned near the symmetry plane of the four magnet system. The slits have been calibrated in energy by comparison with the positron spectrum, measured at the end of the Linac, as already shown.

In order to get a precise definition of the photon emission angle a very accurate alignment of the positron beam along the optical axis is required. This is possible optimizing on the positron monitors the beam intensity after the two removable copper collimators $F_1$ and $F_2$ ($\emptyset$ 7 mm and 6 mm respectively) by two pair of steering coils. The beam spot is observed on a plastic scintillator screen V, which can be inserted at the end of the vacuum pipe. It has elliptical shape with semiaxes 5 mm and 7 mm long respectively. The final quadrupole dublet $Q_1$ and $Q_2$, normally turned off, is only used to test the correct alignment verifying on the scintillator screen the absence of beam steering effect.

The intensity of the positron beam along the transpot channel is continuously monitored and visualized by a toroidal charge monitor system like to that described by Gardiner et al (2). A toroidal mu--metal core placed around the beam path acts as a current transformer, the primary of which is the electron beam. The signal from the seconda ry windings feeds a low impedance preamplifier placed very near the monitor to minimize the differentiation of the pick-up signal and to preserve its proportionality to the positron beam pulse. In the subsequent electronics the undershoot introduced in this differentiation is removed by a high precision linear gate opened for a short interval, spanning each beam pulse. The output pulses from the gate are d.c.

coupled to a standard current integrator whose response is proportional to the charge carried by the beam. The beam pulse is continuously compared on a oscilloscope screen with that due to a calibrator pulser induced in one wire.

After the annihilation target the positrons are deflected by a damping magnet SM1 into a shielded Faraday cup, used as a beam catcher as shown in fig. 4

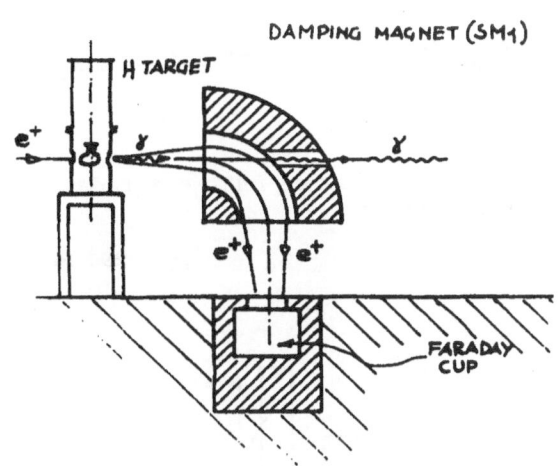

Fig. 4  Positron beam catcher

The Faraday cup signal, integrated by a standard current digitizer, provides an absolute charge monitor.

Positron peak current, as transmitted by the transport system without any energy definition, measured on the Faraday cup, in the energy range between 65 and 300 MeV is shown in fig. 5  Electron-positron converter used in this measurement was tungsten. The repetition rate was 5 Hz.

In fig. 6  the positron spectrum at  200 MeV is reported. The F.W.H.M. turns out to be ∼ 1.5 %

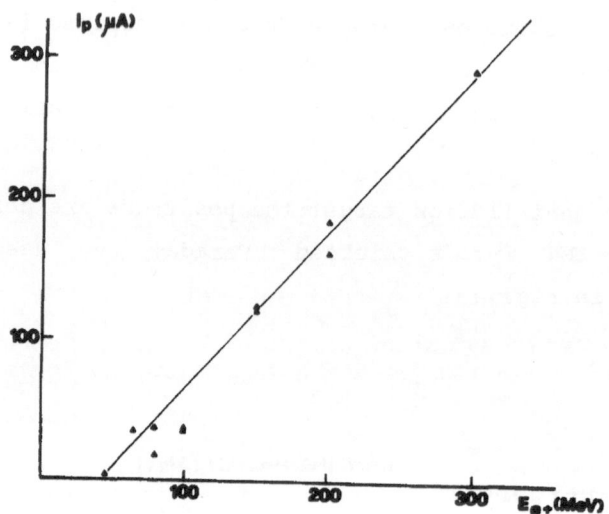

Fig. 5  Positron peak current measured on the Faraday cup, relative to the tungsten converter

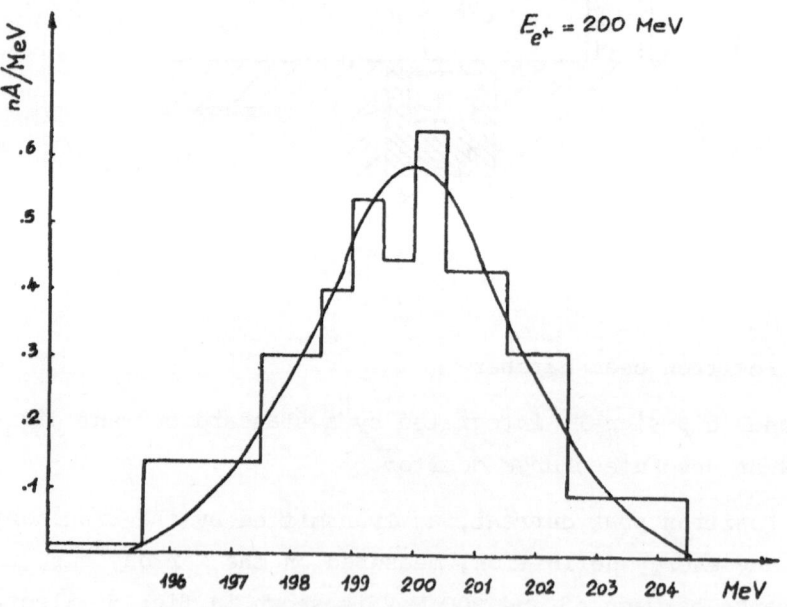

Fig. 6  Positron energy spectrum: the istogram is obtained at the end of the Linac, the solid line at the annihilation target point.

In table 1 mean features of positron beam between 80 and 300 MeV are reported

TABLE   1

| $E_{e^+}$ (MeV) | $N_{e^+}$/sec (copper)* | $\Delta E_{e^+}/E_{e^+}$ F.W.H.M. | $\Delta E_{e^+}/E_{e^+}$ 50 % ip |
|---|---|---|---|
| 80 | .3 $10^{11}$ | 2.2 % | 1.8 % |
| 100 | .3 $10^{11}$ | 1.8 % | 1.5 % |
| 150 | 1. $10^{11}$ | 1.5 % | 1.25% |
| 200 | 1.5 $10^{11}$ | 1.5 % | 1.25% |
| 250 | 2. $10^{11}$ | 1.4 % | 1.2 % |
| 300 | 2.4 $10^{11}$ | 1.4 % | 1.2 % |
| * at 100 Hz repetition rate | | | |

Two electron — positron converter are available.  The tungsten one gives a better positron yield but it is not possible to use it  at high repetition rate for vacuum security reason.  The copper one gives positron yield a factor  2.3  minus but it is possible to use it at high repetition rate.  During theese measurements we usually worked at 100 Hz, but it is possible to work at least at  150 Hz  without appreciable intensity reduction.

Positrons annihilate in a cylindrical liquid Hydrogen target (0.7 gr/cm$^2$ or 0.35 gr/cm$^2$ tick).

Photon beam monitoring has been performed by a standard NBS P2 duraluminium chamber filled with air and by a Komar type quantameter (3) with equivalent results.  Komar quantameter is essentially a Wilson multiplate ionization chamber filled with air whose dimensions, number and distances between the plates are choosen so that the sensitivity is constant in the energy range  5 MeV  to  5 GeV, as shown in fig. 7

Photon spectra have been measured collecting photons, from 200 MeV positron annihilation, at two different angles $\theta_1$ = 0. mrad and $\theta_2$ = 17.5 mrad (1).  In both cases the total angular acceptance

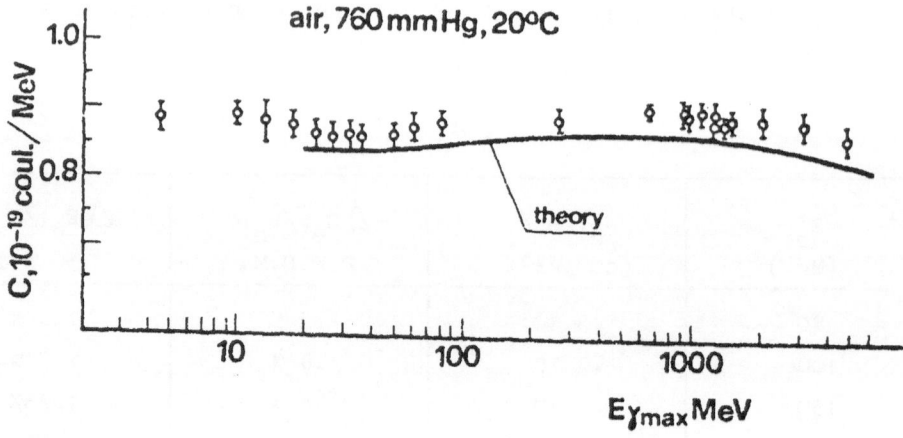

Fig. 7   Sensitivity of the Komar quantameter

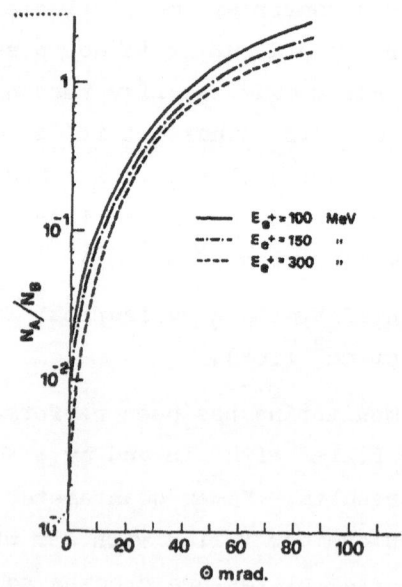

Fig. 8   Angular dependence of the ratio between annihilation and
bremsstrahlung photon yields relative to hydrogen target

was 1 mrad. It is well known that ratio between annihilation and bremsstrahlung photon intensities strongly depends from positron-photon collection angle, as shown in fig. 8. Unfortunately intensity and energy resolution strongly decrease as the angle increases.

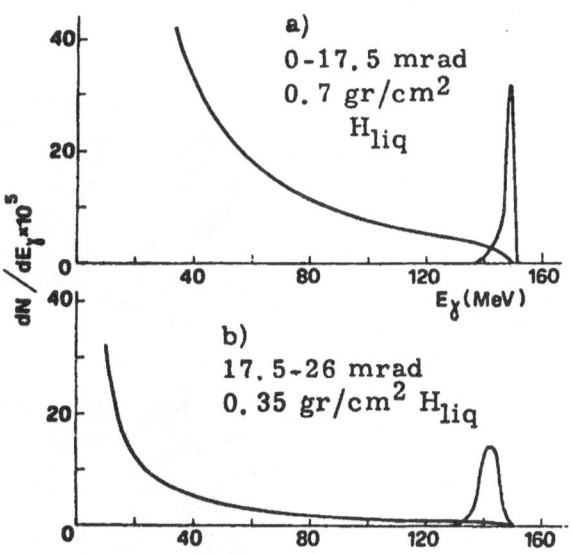

Fig. 9    Spectra of bremsstrahlung and annihilation photon for positron energy 150 MeV. (a) For 0.7 gr/cm$^2$ hydrogen target and photons collected between 0.and 17.5 mrad; (b) for 0.35 gr/cm$^2$ hydrogen target and photons collected between 17.5 and 26.mrad

As a matter of fact for some experiments (when low threshold reactions are involved) it is necessary to choose a good annihilation - brems strahlung ratio condition , for some other experiments (when very low cross sections are involved) it is necessary to choose good intensity condition. For this reason photon spectrum measurements have been performed at two different collection angles.

Photons have been collected by a set of four lead collimator $C_1$, $C_2$, $C_3$, and $C_4$ followed by three small sweeping magnets $SM_2$, $SM_3$ and $SM_4$ as shown in fig. 10a.

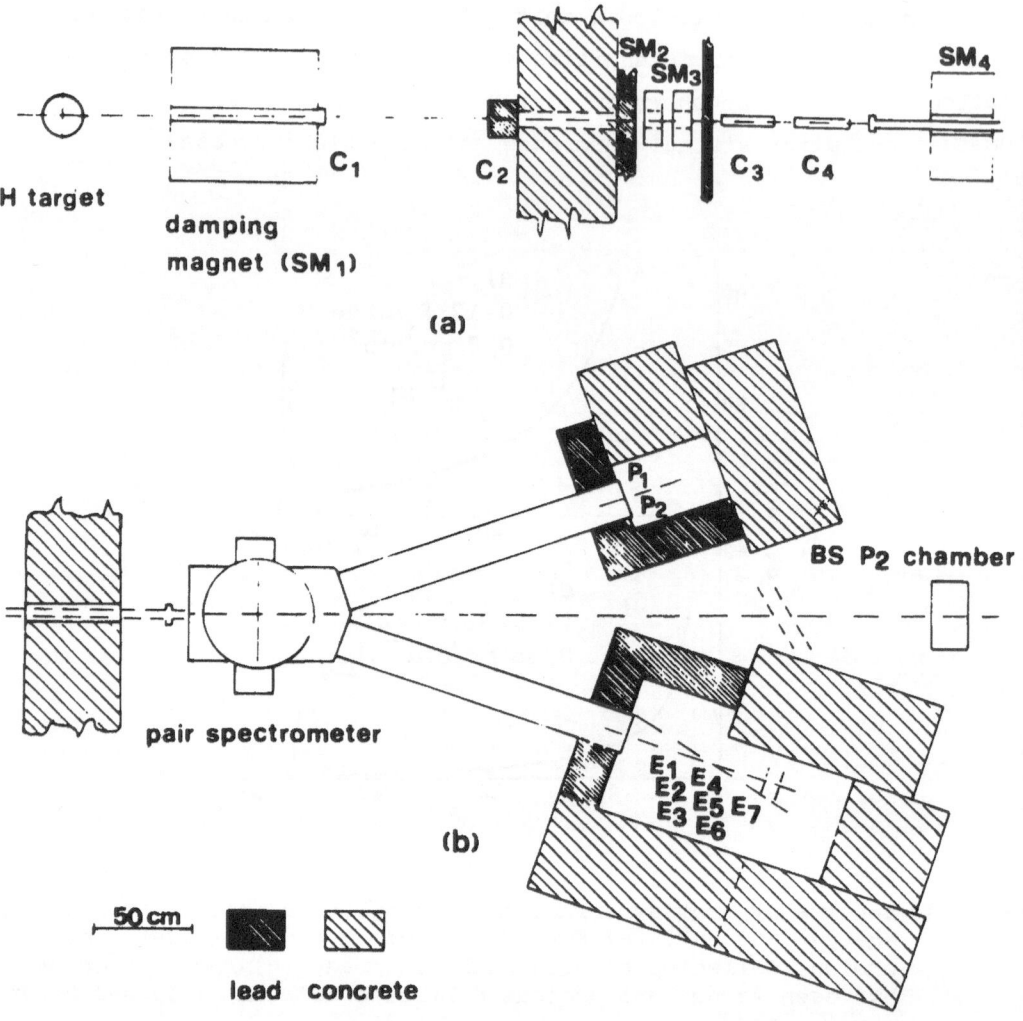

H target

damping
magnet (SM₁)

C₁

C₂

SM₂

SM₃

C₃ C₄

SM₄

**(a)**

P₁
P₂

BS P₂ chamber

pair spectrometer

E₁ E₄
E₂ E₅ E₇
E₃ E₆

**(b)**

50 cm

lead  concrete

Fig. 10  a) Layont of the photon beam collecting system  b) pair
spectrometer : experimental set up

Preliminary measuremants of photon spectrum have been performed using the only at that moment available pair spectrometer, schematically shown in fig. 10b, with flat rectangular poles and a gap 1.5 cm high. The magnet small gap and the optical properties limit the photon collimation angle to $\sim 8 \ 10^{-6}$ sterad. An Hall probe , permanently inserted at a fixed position between the poles, enables the continuous testing of the magnetic field value with 1 ‰ accuracy.

Electron - positron pairs are created in an Aluminium converter which thickness can be varied. The pairs are deflected of 19° and detected in coincidence by a system of scintillation counters. In the electron arm five energy channels are defined by five scintillation counters $E_1$ - $E_5$ (1 cm wide, 2 cm thick and 10 cm high each) in coincidence with two large scintillators $E_6$ and $E_7$ (20 x 1 x 20 cm$^3$) in order to decrease the background. In the positron arm only one energy is selected by the coincidence between the scintillation counters $P_1$ (0.6 x 2 x 10 cm$^3$) and $P_2$(2 x 0.6 x 10 cm$^3$).

The conventional electronic block diagram apparatus is presented in fig. 11

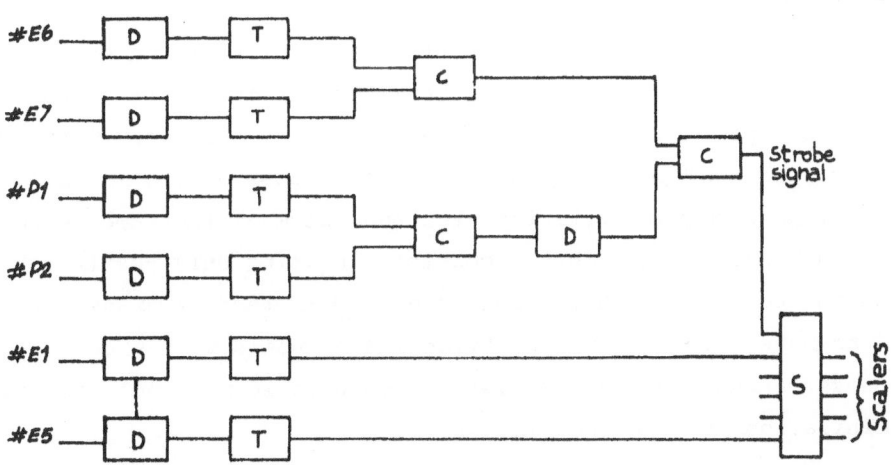

Fig. 11  Block diagram of the electronics (D delay,  T trigger discriminator, C coincidence, S strobed coincidence)

The photon spectra obtained at the positron energy of 200 MeV are reported in figs.12 and 13. Data at $\theta_1$ = 17.5 mrad (fig. 12) have been collected using pair converters of two different thickness (0.4 mm and 0.8 mm). The measurement at $\theta_2$ = 0. mrad (fig. 13) has been performed with a 0.06 mm thick converter, in order to reduce the multiple scattering effects.

Both the spectra show a peak at the correct annihilation energy with a bremsstrahlung continuous tail. The annihilation - bremsstrahlung ratio as well as the peak resolution are very sensitive to the collectio angle, as expected. In this respect during the measurements care has been taken to keep stable the ratio between positron and photon intensities .

Different background sources have been accurately investigated: annihilation target off background, pair converter off background and random coincidence between the positron and electron arms. In figs.12 and 13 the total background has been subtracted: the large experimental errors are due to poor statistics because the strong collimation, required by the spectrometer gap, cut down drastically the photon beam intensity.

To extract annihilation peak from the total photon spectrum two methods are available. Electron beam accelerated in the Linac, when the electron - positron converter is on, presents the same emittance and energy spread of positron beam. So it is possible to measure bremsstrahlung spectrum utilizing electron beam at the same energy of positron beam. Another method consists in using only positron beam but changing positron - photon converter. Different Z radiators having the same thickness in radiation lenght units will produce the same bremsstrahlung spectrum as well as comparable multipole scattering but annihilation peaks decreasing as $1/Z$ .

We have performed measurements using H and Cu annihilation target, each 0.011 radiation lenght tick. The photon spectra from the Cu target figs.12b and 13b show a tipical bremsstrahlung energy dependence with a small amount of annihilation on the head. The H

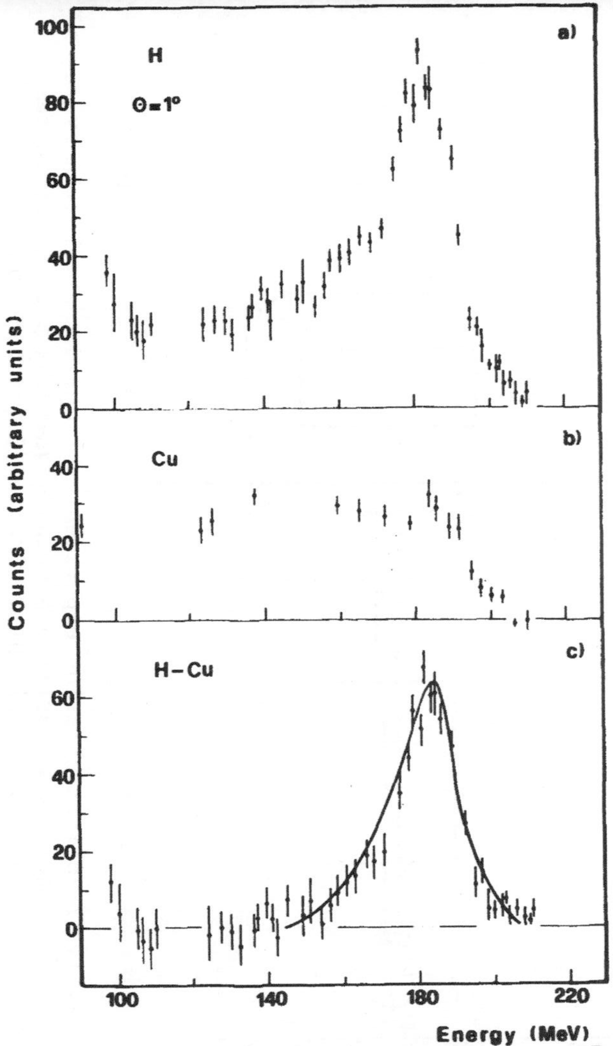

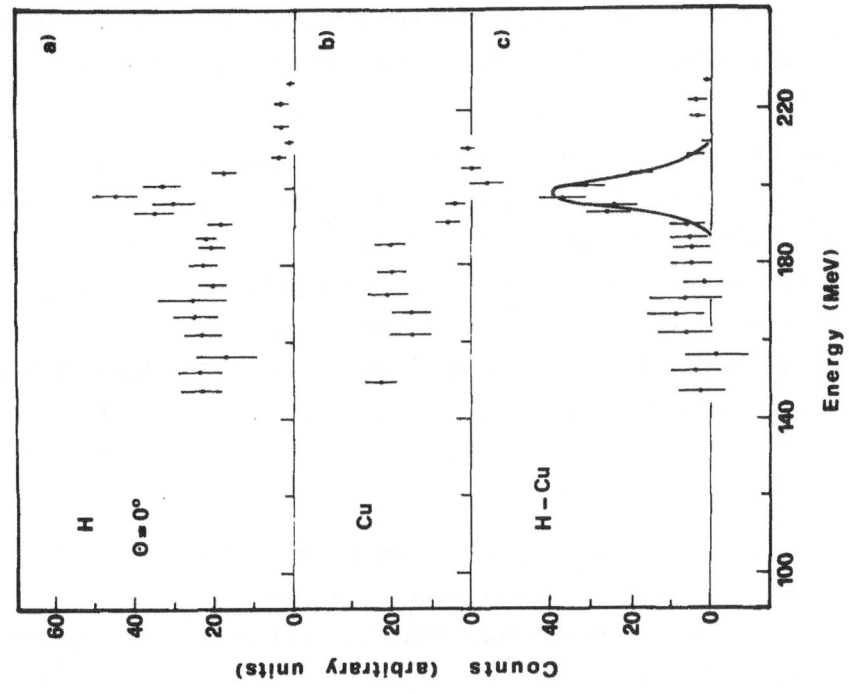

Fig. 12 Photon energy spectrum at $_1$ = 17.5 mrad

Fig. 13 Photon energy spectrum at $_2$ = 0. mrad

minus  Cu  difference spectra (fig. 12c and fig. 13c) correspond there
fore to the contribution, only sligtly understimated, of the annihilatio
photons in Hydrogen.

The annihilation photon spectrum from  H  can be evaluated
starting from the electromagnetic cross section and taking into account
all the target thickness effects (4).  The theoretical spectrum,
corresponding to the geometrical set up of our experiment, has been
folded with a gaussian resolution function of F.W.H.M. $\sim$ 9 MeV to
take into account the finite spectrometer resolution.  The final result,
plotted as a cuntinuous curve in figs.12c and 13c, shows a fairly good
agreement between the theoretical and the experimental peaks also for
what concerns the absolute value.

The intensity of annihilation photon peaks can be evaluated
starting from  N.B.S.  ionizzation chamber or quantameter response and
taking into account the bremsstrahlung and annihilation photon spectra
calculated as said before.

This intensity turns to be  $2 \ 10^6$ photons/sec  at  0. mrad
and  $2 \ 10^5$ photons/sec at  17.5 mrad.Obviously the low intensity is due

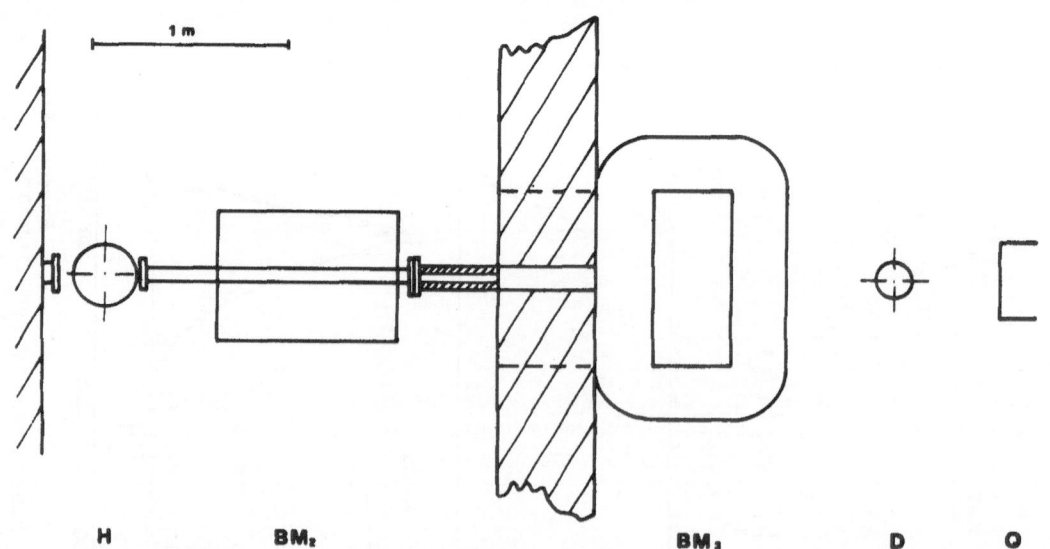

Fig. 14 Photodisintegration experiment set up. H annihilation hydrogen
        target – $BM_2$ positron damping magnet–$BM_3$ pair spectrometer
        magnet – D experimental target – Q quantameter.

I think it could be interesting to know the photon beam intensity available in some future experimental conditions. In fig. 14 the experimental set up,as presently in progress, is reported. The photon beam passes through the new pair spectrometer yoke, which will work also a sweeping magnet. The angular acceptance, the gap dimension and the bending angle of the magnet and the pair converter thickness should allow an energy resolution of the order of 1%. In this configuration it shall be possible to have the photon spectrum measure during the photo-disintegration experiments.

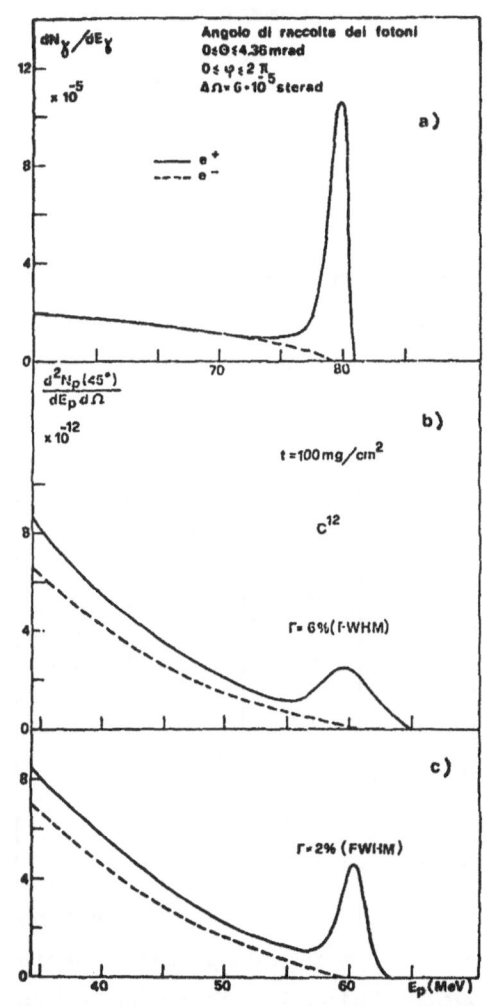 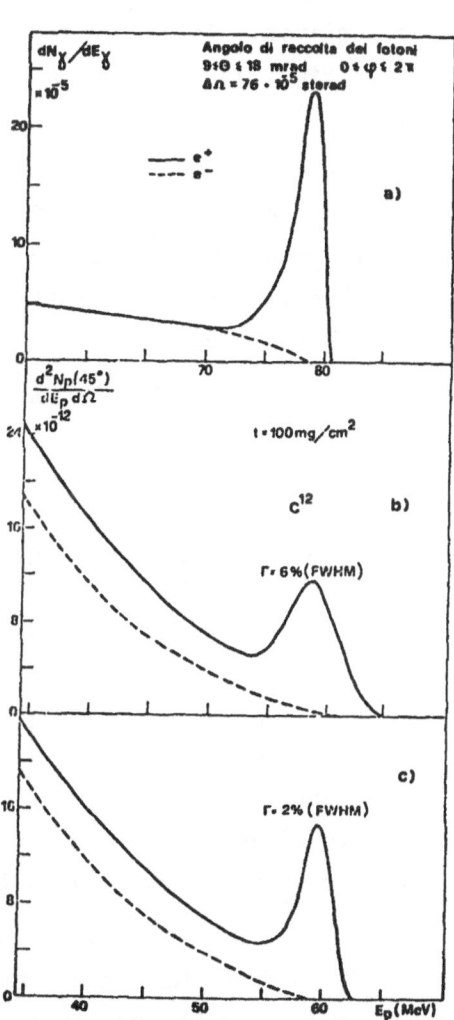

Fig. 15 Photon spectrum at 0. mrad and relative proton spectrum

Fig.16 Photon spectrum at 13.5 mrad and relative proton spectrum

For what concerns the ($\gamma$, p) experiments, in figs. 15 and 16
I reported the calculated photon spectra at 80 MeV in two collimation
condition and the relative proton spectra from the ($\gamma$, p) reaction
on $C^{12}$. The two proton spectra in each figure are relative to different
proton detector resolution (plastic scintillator figs. 15b and 16b
and NaI( ) scintillator with fast pulse analysis (5) figs. 15c and 16c).
Obviously in these experiments it is not possible to use magnetic
spectrometer for intensity reason.

The photon and proton intensity in figs. 15 and 16 are relative
to one positron. To obtain the photon and proton number per second the
values of figs. 15 and 16 have to be multiply for a factor $3 \ 10^{10}$
(see table 1).

The second evaluation I will present concerns the photodisintegration of Deuteron between 80 and 300 MeV. This experiment will
start very soon. The evaluation is relative to $E_{\gamma}$ = 150 MeV that
is the worse condition for what concerns energy resolution. In table 2
photon beam features and proton beam features in two different annihi-
lation-bremsstrahlung ratio $N_A/N_B$ conditions are reported

TABLE 2

| Photon features $E_{e+}$ = 150 MeV   0.7 gr/cm$^2$ Hydrogen | | | | | | Proton features 110 mg/cm$^2$ Deuteron | | |
|---|---|---|---|---|---|---|---|---|
| (mrad) | | $N_\gamma/N_{e+}$ | $N_\gamma/sec$ (*) | $\Delta E_\gamma/E_\gamma$ | $N_A/N_B$ | $\Delta\Omega$ (sterad) | $N_p/sec$ | $\Delta E_p/E_p$ (**) |
| 0. | 5. | $7 \ 10^{-4}$ | $.7 \ 10^8$ | 1 % | 3 % | $4 \ 10^{-3}$ | $4 \ 10^{-2}$ | 2.5 % |
| 13.5 | 5. | $7 \ 10^{-5}$ | $.7 \ 10^7$ | 5 % | 30 % | $4 \ 10^{-3}$ | $4 \ 10^{-3}$ | 6. % |

(*)  relative to Copper converter and 100 Hz repetion rate
(**) due to cinematic condition, target thickness effects, photon
     beam resolution and proton detector resolution

Finally I will show in fig. 17 photon beam features in the energy range
between 100 and 300 MeV in a cinematic condition which optimizes annihilation and bremsstrahlung ratio . Obviously photon intensity and energy
resolution in these condition are poor.

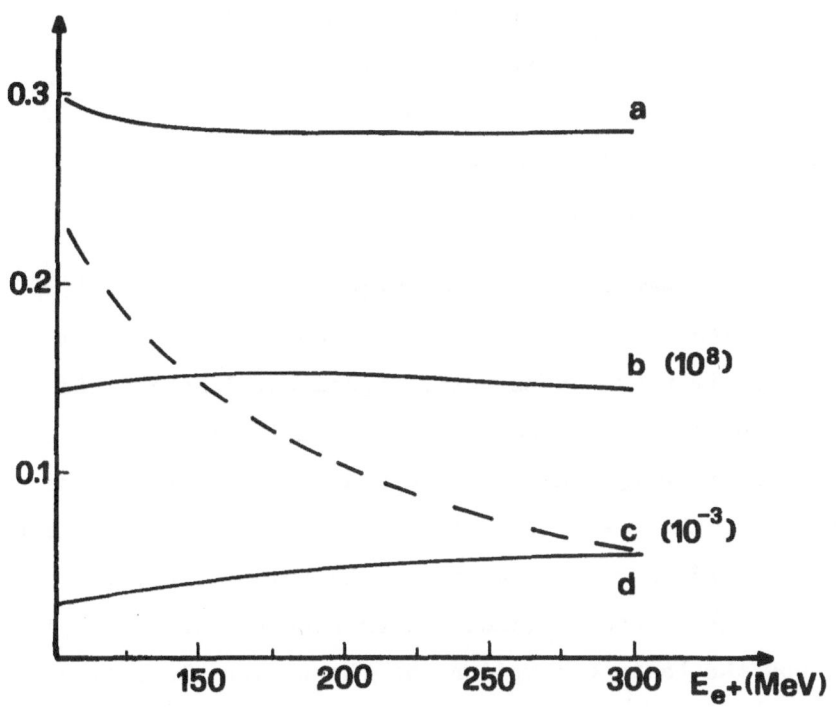

Fig. 17   Photon beam features for 0.35 gr/cm$^2$ Hydrogen target  and
          = 22 ± 6 mrad photon emission angle.  a) $N_A/N_B$, b) N /sec,
       c) N /sec, c) N /$N_{e+}$ , d)    E /E

Also in this case the photon intensity  N /sec  is relative to Copper

electron - positron converter and  100 Hz repetition rate.

## R E F E R E N C E S

1)  G.P.Capitani, E.De Sanctis, S.Faini, C.Guaraldo, R.Scrimaglio,
    G.Ricco, M.Sanzone and A.Zucchiatti - Lett. Nuovo Cimento
    16 (1976) 453 .
2)  S.N.Gardiner, J.L.Matthews and R.O.Owens - Nucl. Instr. and
    Meth. 87 (1970) 285 .
3)  A.P.Komar, S.P.Kruglov and I.V.Lopatin - Nucl. Instr. and
    Meth. 82 (1970) 125 .
4)  E.Mancini and M.Sanzone - Nucl. Instr. and Meth. 66 (1968) 87.
5)  A.Zucchiatti, M.Sanzone and E.Durante - Nucl. Instr. and Meth.
    129 (1975) 467.

# SOME EXPERIMENTAL RESULTS ON THE MEASUREMENT OF THE TOTAL PHOTOABSORPTION CROSS SECTIONS.

G.V.Solodukhov

Institute for Nuclear Research, Moscow.

The measurements of the gamma-quanta total absorption cross-section are of great interest, because they allow us to interprete unambiguously the experimental data. This value gives general characteristics of the nuclei without complicating the picture by taking into account partial reactions probabilities, which are not yet clear enough. The comparision of the experimental data on total absorption cross-section with the theoretical estimations enables us, in principle, to check the basic conceptions of the photon absorption mechanism.

In the dipole giant resonance energy region and a little above it, we can speak of two methods of total absorption cross-section measurement which give comparable results:

1/ the summing up of the cross-sections of main partial reactions which are possible at the energy under investigation.

2/ direct measurement of the $\gamma$ -beam attenuation by the sample being investigated.

The first method has been developed thanks to the monochromatic beam experiments in Saclay and Livermore.

In such experiments one can obtain the total photoneutron cross-section

$$\sigma_n = \left[ \sigma(\gamma,n) + \sigma(\gamma,pn) \right] + \sigma(\gamma,2n) + \dots \qquad (1)$$

which with good accuracy one can consider as $\sigma_{tot}$ in a wide range of A. However, there are certain constrains in the use of this method. This method is most effectively used for the nuclei within $100 < A < 209$ region. For $A < 100$ the photoproton channel becomes significant. For the nuclei with low fission barrier ($Z \sim 90$) the correct calculation of the fission neutron multiplicity and the contribution of such reactions as ($\gamma$, nf), ($\gamma$, 2nf) etc. becomes difficult. Besides, the investigation of giant resonance isospin splitting by means of this method proves difficult due to the damping of $\Delta T = 1$ transitions in the photoneutron channel. In the photoneutron experiments the contribution of ($\gamma$, $\gamma$) and ($\gamma$, $\gamma'$) reactions is not taken into account, but they may have certain influence on giant resonance shape, especially in low-energy region. However, in spite of these shortcomings the best results in the medium and medium-heavy nuclei region were obtained by means of photoneutron technique.

It is necessary to mention here the great amount of data from bremsstrahlung spectrum experiments. The errors above ($\gamma$, 2n) threshold due to nucleon multiplicity correction, as it follows from experiments, do not exceed 10%. Much more significant is the fact that due to unfolding procedure the experimental errors increase sharply with energy.

The direct method of total photoabsorption cross-sections measurement seems to be most universal, because it can be applied to any nuclei and allows us to take into account the contribution of all the partial reactions to the total cross-section. The procedure used in this method includes registration of the direct photon spectrum from the accelerator and the same spectrum, attenuated by thick sample under investigation in "good geometry". The experimental cross-section represents the sum of: The interaction of the radiation with

atomic electrons and its nuclear part which is of interest
to us

$$\sigma_{exp} = \sigma_{atom} + \sigma_{nucl} = \frac{1}{A} \ln \frac{J_o(E)}{J(E)}$$

where $J_o(E)$ and $J(E)$ -

   - direct and attenuated spectra of
     radiation intensity;

   A - factor determining number of nuclei per $cm^2$
     of the sample.

The main problem here is the extremely disadvantageous
ratio of the nuclear and atomic parts of the cross-section
measured: even for the light nuclei this ratio in the maximum
of dipole resonance does not exceed 5-10%.

As Z increases, this value is supposed to drop approximately as $1/Z$ . However,          the fact that the dipole
strength in heavy nuclei is concentrated in a rather narrow
energy range (6-8 Mev) leads to the increase of this ratio.
The following facts may have contributed to this:

The nuclear cross-section $\sigma_{nucl} \sim \frac{NZ}{A}$  . As $\frac{N}{A}$
value grows when Z increases (from $\sim$ 0.5 for the light
nuclei to $\sim$ 0.6 for uranium), $\sigma_{nucl}$ grows more rapidly than
Z, while the atomic cross-section grows, in general, more
slowly than $\sigma_{pair} \sim Z^2$ due to considerable contribution
of $\sigma_{compt} \sim Z$. As a result, $\dfrac{\sigma_{nucl}}{\sigma_{atom}}$ descreases more
slowly than $1/Z$ with the increase of Z. For example, this
ratio is of the order of 1.5-2.0% for $^{165}$Ho and 2.0-2.5%
for U. This fact makes the total absorption experiments possible over the whole range of A.

Let us consider now some experimental results obtained
by the methods mentioned above.

*DIPOLE GIANT RESONANCE REGION*

Some years ago Semenko /1/ suggested the possibility of
a "phase transition" in certain nuclei induced by dipole
exitations that could change the selfconsistent potential.
To test this hypothesis,following Semenko's proposal the measurements of the photoneutron cross sections for transition
nuclei with $N \sim 90$ were made at Saratov University. For these
nuclei abrupt changes in low-lying spectra properties and nuclear shapes are observed when a small number of neutrons are
added.

At present,measurements on $^{148}$Nd, $^{147,148,150,152,154}$Sm
and $^{152, 154, 156, 157, 158, 150}$Gd at energies from 7,5 to
20 Mev have been made. /2-5/. The results of these investigations may be briefly summarized as following.

In general the shape of the measured photoneutron reaction cross sections is in agreement with the predictions of
the dynamic collective model. Passing from $N = 88$ to $N = 90$
an abrupt change in the general shape of the cross sections
observed, which is in agreement with the change in low lying
collective exitations of these nuclei. For nuclei with $N = 88$
whose collective exitation spectra are closed to vibrational
ones, the shapes of photoneutron cross sections are similar
to these of spherical nuclei, with the exception of $^{152}$Gd
(fig. 1). At the same time for nuclei with $N = 90$ collective
exitations have a rotational form and the photoneutron cross
sections are characterized by two maxima (fig. 2).

The case of $^{152}$Gd is quite different. Although low lying
collective spectrum of this nucleus is practically similar
to $^{148}$Nd and $^{150}$Sm (vibrational nuclei) the shape of the giant
resonance in $^{152}$Gd is similar to that of $^{154}$Gd (a rotational

nucleus) (fig. 3). It is difficult to explain this result
in terms of the dynamic collective model using existing models
describing low-lying collective spectra and assuming that
the characteristics of the giant resonance are essentially
determined by that spectra. In this case one can suggest that
these effects may be connected with the above mentioned phase
transition, although  this suggestion should be verified by
more experimental evidence and theoretical investigations.

Similar results connected with a change in the shape
of the giant resonance in the nuclei under investigation with
N increasing have been obtained at Saclay /6/ (see Fig.2)
and Livermore /7/ (fig.4). But one must note that for some
nuclei studied at these laboratories, the structure of the
cross sections is less pronounced. Besides, the width of
the transvers maximum from the Saratov data for $^{160}$Gd is 50%
smaller than that of Livermore data.

At Saratov the change of the giant resonance shape for
the transition nuclei with A $\sim$ 190 was also studied.

According to modern conceptions one can expect a smooth
change in the surface characteristics of the nuclei from
prolate (isotopes of W) to an oblate spheroidal shapes
(isotopes of Pt). Intermediate nuclei (isotopes of Os) seem to
have non axial shape. According to the optical  anisotropy
theory such changes in the shape of the nuclei should be accom
panied with a respective change in the giant resonance
cross-section. Measurements of photoneutron cross sections
for a number of odd and even isotopes of W, Re, Os, Ir and Pt
have demonstrated good agreement with these theoretical
considerations (fig. 5 /8/). One should  also note that as
the nonaxiality and oblateness of these nuclei is not large,

the corresponding effects in the structure of the giant
resonance are very small.

The accumulation of the experimental data on gamma-quantum
total absorption cross-sections in $50 < A < 100$ range seems
to be of great interest from the point of view of the validity
limits of the models estimation, inasmuch as this range is
intermediate between the light nuclei $A < 40$ which are
described in microscopic models framework and the $A > 100$
nuclei, which are described well enough by the collective
models. The data available untill recently for this region
are obtained procticaly only by the summing up of the photoneutron cross-sections. Meanwhile the ( $\gamma$ , $p$ ) data show
a significant contribution of this channel to total cross-
seotion (of order of 20% for $^{64}$Zn /9 /). Therefore one has
to be very careful in making interpretations on the basis
of the photoneutron data analysis. One may illustrate this
conclusion on the selenium isotopes example. For some selenium isotopes there are $\sigma_n$ data /10, 11/ as well as the
direct $\sigma_{t.t}$ measurements data obtained in Institute for
Nuclear Research, Moscow /12/ by L.Lazareva, G.Gurevich
and G.Solodukhov. The $\sigma_n$ data /10, 11/ (the curves of
these two experiments are practically identical) for even-
even Selenium isotopes show the prominent evolution of the
shape and absolute value of the cross-section when A
grows from 76 to 82. The $^{76}$Se curve (fig. 6) is "out" at
100 mb level (about 2/3 of $^{82}$Se cross-section in maximum)
and obviously tends to splitting into two maxima which is
characteristic for the deformed nuclei. At the same time
$^{82}$Se $\sigma_n$ curve can be well approximated by one Lorenz curve

with maximum of ~ 150 mb. Such a behavior has not reasonable explanation either from low-lying spectra or from giant resonance isospin splitting formalism.

The total absorption cross-sections for $^{76}$Se and $^{82}$Se obtained in the attenuation experiments /12/ are shown in fig. 7,8. The errors indicated are statistical ones. Table 1 shows the best Lorentz fit parameters as well as some data of photoneutron experiments (in brackets). When all the partial reactions are taken into account the situation changes radically. As one can see the difference between the $^{76}$Se and $^{82}$Se $\sigma_{tot}$ curves is not profound. This character of the curves for the two isotopes is in a better agreement with the dipole-quadrupole interaction model predictions. This clearly indicates the importance of channels other than ( $\gamma$ , $n$ ) in the mass region discussed.

The study of nuclei in this region by means of total absorption technique seems to be useful for getting some additional information. As is shown in fig.8, there is some evidence that additional maximum with the width of 1,5-2,0 Mev should exist. The area of this maximum is about 1.5-2,5% of the whole experimental curve. The energy position, width and integral cross section (see table 1) do not contracdict either the experimental data or the theoretical predictions of the corresponding values for isoscalar quadrupole resonance /13/. It should be noted that in literature there is no indications to the existence of this resonance in photoabsorption experiments. Although the $\sigma_o(E2)/\sigma_o(E1)$ value is of the order of 1-2%, but as the concentration of $\sigma(E2, T = 0)$ cross section takes place in a narrow energy region, it becomes possible to observe it in the photoabsorption experiments.

If other photoneutron results /14/ are analysed from this view-point, similar structure in $(55-65).A^{-1/3}$ energy region can be noted for many nuclei. This resonance in curves should be most prominent for nuclei with fairly low ( $\gamma$ , $n$ ) threshold.

Therefore, measurements in the $\Delta T = 1$ giant resonance branch for selenium and other nuclei of this A range by means of attenuation technique seem to be interesting because the study of this branch evolution in photoneutron experiments is difficult. Now we have plans to make such experiments in the Institute for Nuclear Research.

Until recently photonuclear reactions for the low fission-barrier nuclei have not been studied satisfactorily even in the giant resonance region. Total photoabsorption data for these nuclei are especially scanty. The comparison of the data obtained in photoneutron experiments carried out in Saclay /15/ and Livermore /16/ using a quasi-monochromatic photon beam shows the significant difference of absolute cross-section values. This difference can perhaps be explained by the fact that partial reaction contributions to total photoabsorption cross section cannot be unambiguously taken into account, especially in the energy region where a competition of ( $\gamma$ , 2n), ( $\gamma$ , nf), ( $\gamma$ , 3n), ( $\gamma$ , 2nf) etc. takes place.

From this point of view the use of the absorption method which allows to obtain unambiguous results irrespective of partial channel competition seems to be preferable to the photoneutron technique for nuclei with low photofission thresholds. We therefore undertook a systematic investigation of total photoabsorption cross sections for $^{232}$Th, $^{235}$U, $^{238}$U and $^{239}$Pu in the E1 giant resonance energy region /17/. The comarision of the Moscow results (see fig.9, table 2) with that of Saclay and Livermore shows a good agreement between our and Bergere's data for $^{232}$Th and $^{238}$U. As to $^{235}$U, our cross-section value is remarkable less than that of Livermore, although the widths and energies of maxima values are in a good agreement. Deformation parameters and quadrupole momenta calculated within the framework of the Okamoto-Danos model are given in table 3. Root-mean square deformations $\bar{\beta}$ from /17/ as well as quadrupole moments from various papers are presented for comparison. These data are in reasonable agreement with our values.

The study of the giant resonance for this group of nuclei is interesting in another aspect. Since many nuclear properties depend similarly on the proton and neutron numbers, that follows of the charge independence of nuclear forces,it is reasonable to expect that the transition effects analogous to N $\sim$ 90 case should be observed as well for Z $\sim$ 90 nuclei. This effect would lead to the evolution of the photoabsorption cross-section shape.

In table 4 the simplest surface motion characteristics are given for several transition nuclei with Z near 90. Excitation energies EJ+ in table 4 are taken from reviews /18, 19/ .The subscribts g, $\beta$ , $\gamma$ correspond to the levels of the ground-state, $\beta$ -and $\gamma$ -vibrational state rotational bands respectively. $\mu$ and $\gamma$ are parameters of Davydov-Chaban low-energy collective excitation model /20/ describing the nuclear softness with respect to longitudinal vibrations and the non-axiality of the nucleus respectively. The root-mean square deformations $\bar{\beta}$ were obtained from the data of ref. /17/ by taking into account $\wedge$ and $\gamma$ values.

Using data of table 4 it is possible to explain some features of our photoabsorption cross sections. Thus, the giant resonance broadening observed corresponds to an increase of the root-mean square deformations. The longitudinal maximum in $^{232}$Th photoabsorption cross section is not so pronounced as those for Z= 92 and 94 nuclei. This fact could be explained by greater softness of Z=90 nuclei with respect to $\beta$ -vibrations which are responsible for the preferential broadening of the longitudinal maximum /21/.

As stated above the sharpest change of the surface properties of N ~ 90 transition nuclei occurs for N between 88 and 90. It should be interesting, in this connection, to investigate the optical anisotropy change of the uranium region nuclei for Z between 88 and 90, e.g. for a transition from $^{226}$Ra to $^{232}$Th. Unfortunately, existing spectroscopic data concerning low-energy excited states of $^{226}$Ra (see table 4) are too poor to make certain conclusions about cerface motion characteristics of this nucleus, while direct measurements of the $^{226}$Ra total photoabsorption cross section does not seem practicable at present.

We attempt to estimate the total photoabsorption cross-section for $^{226}$Ra using the photofission cross-section data obtained by Jagrov, Nikitina et al /22/. This experiment is carried out with the same accelerator and the same beam energy and intensity control system as those of our work. The estimation is based on a comparison of the energy dependence of the $\sigma_{tot}/\sigma_{f}$ ratios for several nuclei of the uranium region. The total photoabsorption cross section of $^{226}$Ra thus calculated is given in fig.10 together with the $\sigma$ ( $\gamma$ , tot) curves for $^{232}$Th and $^{238}$U from our work. The photoabsorption cross section for the nearest double-magic nucleus $^{208}$Pb /23/ is also shown in fig.10 for comparison. As can be seen from this fig., the $^{226}$Ra photoabsorption cross-section curve has a single-humped shape characteristic for nuclei with relatively small deformation. There are two facts in favour of this conclusion:

1. The width of the $^{226}$Ra photoabsorption curve which depends only slightly on the assumptions made during the calculation is equal to 5.1 ± 0.9 MeV. This relatively small value could be considered as an indication of a rather small nuclear deformation of $^{226}$Ra.

2. The $^{226}$Ra deformation parameter calculated from the probability of the first 2+-level excitation /17/ is equal to 0.19±0.01 which is almost 1.5 times smaller than that for $^{232}$Th.

These data seem to be convincing enough to make a conclusion that Z ∼ 90 nuclei form a transition region similar to that for N ∼ 90 nuclei. It should be noted however that the similarity of the transition effects for these two groups of nuclei is not complete. Thus, for example, for Z between 88 and 90 the nuclear equilibrium deformation and photoabsorption cross-section shape change approximately in the same manner as for the corresponding transition in the N ∼ 90 region while the low-lying collective spectra at Z ∼ 90 (see table 4) do not change so radically as those at N ∼ 90. This result seem to be connected with greater softness of N ∼ 90 nuclei. The reason for this difference is not quite

clear yet and requires additional investigations for its
explanation.

As was shown by Strutinsky /24/, shell effects are
not necessarily connected with spheroidal nuclei, but they
can manifest themselves in deformed potential as well. So-called 'deformed shells' are connected with local variations
of the single-partical level density near the Fermi surface.
The existence of the energy gap for nuclei in the vicinity of
N = 108 was theoretically predicted by Nilson et al /25/.
Experimental data analysis of low-lying spectra for the
nuclei of this region indicates some peculiarities, which (fig.11a,b)
could be explained by the effect of N = 108 closed shell.
In order to investigate the influence of the shell effects
on the giant resonance shape, the total absorption cross-sections measurements for nuclei in $165 \leq A \leq 209$ range were
made /26/ by the same group of Moscow Institute for Nuclear
Research. This influence is most pronounced when the giant
resonance width are analysed (see fig. 11).

As is well known, closed-shell spherical nuclei form
one-peaked resonance with the width of $\sim$ 4 Mev. The width
of giant resonance in deformed nuclei increases due to the
dependence of $E_2-E_1$ value on the nuclear deformation:

$$E_2 - E_1 = Const \cdot (x-1) \cdot A^{-1/3} \qquad (2)$$

where $x = a/b$

In fig.11 the $\beta$ values are indicated. For convenience,
$\beta = 0$ corresponds here to $\Gamma = 4$ Mev (spherical nuclei);
$\beta = 0.3$ corresponds to $\Gamma$ for $A = 155$ deformed nuclei.
It is interesting to note that for $160 \leq A \leq 185$ nuclei

region $\Gamma$ decreases fairly sharply in spite of the fact
that $\beta$ stays nearly constant. For deformed nuclei the
giant resonance full width is described by $\Gamma_1/2 + \Gamma_2/2 + (E_2 - E_1)$.
To explain the disagreement between $\Gamma$ and $\beta$ behavior
it should be assumed that the Lorenz lines widths decrease.
Fig.12 shows $\Gamma_1$ and $\Gamma_2$ values according to the data of
Saclay and Livermore as well as $\frac{\Gamma_1 + \Gamma_2}{2}$ curve. As one
can see, for $150 \leq A \leq 185$ region $\frac{\Gamma_1 + \Gamma_2}{2}$ value decreases by approximately 1 Mev. This fact can explain decrease
of $\Gamma$ value for this region of A. This effect has not been
discussed so far and an explanation has yet to be found. One
could suppose that it is connected with the $N = 108$ shell
filling up.

Smaller values of $\Gamma$ observed for the same deformation
parameters in A $\sim$ 230-240 range can be explained, at least
partly, by $A^{-1/3}$ dependence of $(E_2 - E_1)$ value. But it seems
very interesting to study the A $>$ 240 nuclei from this
point of view, which the $N = 152$ deformed shell effect can
manifest.

### THE REGION ABOVE GIANT DIPOLE RESONANCE

Let us turn to one more possibility of total absorption
cross-sections determination in the ( $e$ , $e'$ ) experiments.
In one-photon plane-wave approximation, the inelastic electron scattering is connected with the total hadronic cross-
section as following:

$$\frac{d^2\sigma}{d\Omega dE'}(E, E', \vartheta) = \Gamma(E, E', \vartheta)\left[\sigma_T(q^2, \omega) + \varepsilon\sigma_L(q^2, \omega)\right] \quad (3)$$

where

$$\mathcal{E} = \left[ 1 + 2 \left( 1 + \frac{\nu^2}{q^2} \right) tg^2 \frac{\vartheta}{2} \right]^{-1}$$

$$q^2 = 4 E E' \sin^2 \frac{\vartheta}{2}$$

$$\nu = E - E'$$

$$K = \frac{\omega^2 - M^2}{2M}$$ — energy of real photon, which leads to the same hadron final state, as in inelastic electron scattering case.

$M$ — nucleon mass

$$\omega = \left( M^2 + 2M\nu - q^2 \right)^{1/2}$$ — effective mass of hadron final state.

$\sigma_T (q^2, \omega)$
$\sigma_L (q^2, \omega)$ — total cross-sections of the virtual photon interraction with transversal and longitudinal polarisation

When $q^2 \to 0$
$\sigma_L \to 0$
$\sigma_T \to \sigma_\gamma$

thus $\lim\limits_{q^2 \to 0} \left( \frac{1}{\Gamma} \frac{d^2\sigma}{d\Omega\, dE'} \right) = \sigma_\gamma (\omega)$

Due to assumptions which are made the equation (3) is valid only for the light nuclei. If $E$ and $E' \gg B$ ( $B$ is coulomb barrier) one may expect that the wave function distortion of primary and scattered electrons are insignificant. As is shown experimentally, for heavy nuclei at the energies higher then 0.1 Gev, the descrepancy between experiment

and theory do not exceeds 5%. Keeping in the mind this
limit of accuracy, one can apply (3) to heavy nuclei.
If one measures the ( $e$ , $e'$ ) cross-sections for various
momentum transfered and fixed energy transfered and extra-
polates them to $q^2$ = 0 ("photon point"),one can obtain
a photoabsorption cross-section. The main advantage of this
method in comparision with direct attenuation method is more
profitable ratio of hadronic and electromagnetic cross-section parts.

By means of this technique the protoabsorption cross-
section for hydrogen and deuterium nuclei have been measured
by T.Armstrong et al /27/ .In Kharkov Physical Technical
Institute the same method was applied to a number of nuclei
in a wide A-range /29/.

Untill recently there was no data on total photonuclear
cross-sections in 150-1500 Mev region. At the same time
this energy region has the particular interest because just
here all the nucleon resonances which give remarkable contribution to total hadronic cross-section take place. Apart
from it, just in this region the collective photoabsorption
mechanism which dominates in low energy range is replaced
by single-nucleon one.

In this connection seems to be interesting to discuss
shortly the Kharkov virtual photon cross section results
for $C^{12}$, $^{27}Al$, Ni, Mo and W nuclei.

As an example, in fig. 13 the ( $e$ , $e'$ ) cross-section
values for $^{12}C$ nuclei for different momentum transfered are
given .As one can see, extrapolated "photon point" $q^2$ = 0
may be determined with a reasonable accuracy.

Fig. 14 shows $^{12}C$ Kharkov results together with that

of I.Ahrens et al and T.Armstrong et al for lower and higher
energies. /28,27/.In fig. 15, are given the total hadronic
photoabsorption cross-sections for Al, Ni, Mo, W obtained
by the same method. One must note that the experimental cross
sections exceed by factor 1.2-1.5 the sum of the single-nucleon cross-sections. The technique described of the photoabsorption cross-section extracting is connected with a number of corrections certain of which are ambiguous. Therefore,
keeping in the mind the importance of these data, the
aditional experiments seem to be useful to define the results.

A fission of the $Z \geqslant 90$ nuclei at excitation energies
higher than the fission threshold has a large probability.
Thus the measurement of the energy dependence and absolute
value of the fission cross-section for these nuclei is interesting not only for a study of the fission process itself
but for obtaining information on the behaviour of the total
nuclear absorption cross-section for fission-inducing particles.

Until recently there were no data on transuranium element
fotofission. From this point of view the results on $^{241}$Am
photofission measured by Lazareva group together with
Kharkov physical technical institute /30/ at energies 50-1300
MeV are of great interest.

$\sigma_f$ curve for $^{241}$Am is given in fig.16. First dis-
tinct maximum at 250-300 MeV corresponds to the first barion resonance $P_{33}$ (1236 Mev) as observed in the total hadron
absorption cross-section $\sigma_{\gamma p}$ at an energy 315 MeV. Second
and third barion resonances $D_{13}$ (1520) and $F_{15}$ (1688) observ-

ed at 720 and 1035 MeV respectively are smoothed considerably due to the internal motion of nucleons in the nucleus.
Second maximum in the $\sigma_f$ curve seems to be connected with
overlapping of these two maxima. The quasi-deutron absorption cross-section calculated for Am using the equation

$$\sigma_{qd} = \mathcal{L} \cdot \frac{ZN}{A} \cdot \sigma_d$$

where $\sigma_d$    - deutron photodesintegration cross
              section

    $\mathcal{L} = 10,3$    is estimated in the case of light
              nuclei

is given in fig. 17 (curve 1) together with the experimental
values of the total hadron — cross-sections (T.Armst-
rong et al. /27/), multiplied by 241 (vurve 2).and calculated
curve 3 obtained from Am data /27/ by taking into account of
the nucleon impulse distribution in the nucleus. The step-
like distribution with $P_{max} = 260$ MeV/c was used.

A comparison of the absolute values for $\sigma_f$ and $\sigma_{tot}$
curves gives a result, similar to Kharkov's one /29/: the
whole $\sigma_f$ curve within the first maximum and higher has an
amplitude which is approximately 1,6 as large as theoretical
$\sigma_{tot}$ value.

Since there are no experimental data on the total hadronic cross-section $\sigma_{tot}$ for U and Am nuclei the question
needs clarifying. It should be noted however that the $\sigma_{tot}$ calculation does not take into account such effects as short-range
nucleon correlations, interference effects etc.,which can be
essential in heavy nuclei.It should be pointed out in this connection that in the energy region of the first maximum $\sigma_{qd}$
value is as large as about 30% of $\sigma_{\gamma p}$ .

It cannot be excluded that a portion of $\sigma_{\gamma d}$ , which is not connected with a barion resonance, will contribute to $\sigma_{\gamma A}$ . For heavy nuclei a constant $\mathcal{L}$ can exceed the value used.

It is possible that an additional interaction mechanism exists whose contribution could be of several percents for light nuclei, but could increase the absorption cross sectio by factor of  at Z $\geqslant$ 90. The dependence of $\sigma_{tot} \sim A^{1,5}$ would make it possible.

From the author's viewpoint, the  errors here are rather large, and the results obtained need the definition in the experiments for other nuclei.

References

1.  S.Semenko.Yad.Phys. 1 (1965), 414.

2.  O.V.Vasilijev et al. Yad.Phys. 10 (1969), 460.

3.  O.V.Vasilijev et al. Phys. Letts. 30B (1969), 97.

4.  S.Semenko et al. Phys. Letts. 31B (1970), 429.

5.  V.A.Semenov et al.  orosi teoreticheskoi i jadernoi fiziki.
    (1976) N 6, Saratov.

6.  P.Carlos et al. Nucl. Phys. A225, (1974), 171.

7.  B.Berman et al. Phys. Rev. 185 (1969) 1576.

8.  A.M.Gorjachev et al. Yad. Phys. 17(1973). 463.

9.  G.E.Clark et al. Nucl. Phys. A218 (1973) 358.

10. A.M.Gorjachev et al. Izv. Akad. Nauk ser.phys. 39 (1975)134.

11. P.Carlos et al. Nucl. Phys. A258 (1976) 365.

12. G.M.Gurevich et al. Tezisi XXVI Soveschanija po jad.
    spectroskopii i strukture at. jadra. Baku, (1976), 433.

13. I.N.Borzov et al. Preprint Fiziko-energ.inst. N° FEI-580
    (1975).

14. B.Berman. At. Data and Nucl. Data Tables v.15 (1975), 4.

15. A.Veyssiere et al. Nucl. Phys. A199 (1973), 45.

16. C.D.Bowman et al. Phys. Rev. 133 (1964) B676.

17. G.M.Gurevich et.al. Pis(ma v JETP. 20 (1974), 741.

18. Y.A.Elis et al. Nucl. Data. 4,(1970) 543.

19. A.H.Wapstra et al. Nucl. Data B1 (1966), 5-91.

20. A.S.Davidov. Vozbujdennje sostojanija at.jader. Moskwa 1967.

21. O.V.Vasilijev et al. Yad. Phys. 13 (1971) 463.

22. E.A.Jagrov et al. Yad. Phys. 13 (1971) 934.

23. A.Veyssiere et al. Nucl. Phys. A159 (1970) 571.

24. V.M.Strutinsky. Nucl. Phys. A95 (1967) 410.

25. S.Nilsson et al. Nucl. Phys. A131 (1969), 1.

26. G.M.Gurevich et al. Pis'ma v JETP 23 (1976), 411.

27. T.Armstrong et al. Phys. Rev._D5 (1972)1640.
                       Nucl. Phys._B41 (1972) 445.
28. J.Ahrens et al. Nucl. Phys._A251 (1975) 479.
29. V.G.Vlasenko et.al. Yad.Phys._23 (1976) 504.
30. V.M.Alexandrov et al. III Int.Seminar of electromagnetic
    interactions of nuclei, Moscow, 1975.

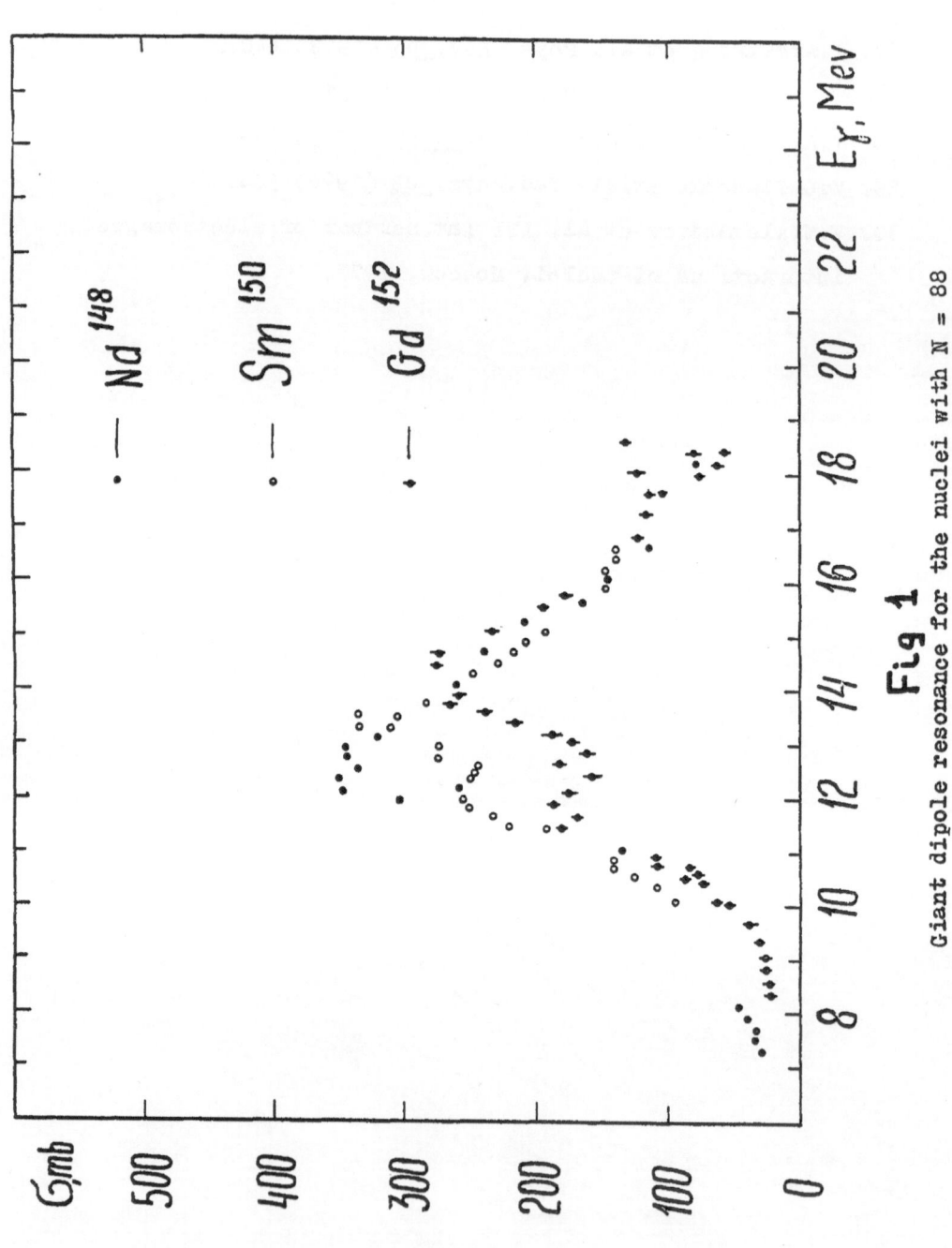

**Fig 1**

Giant dipole resonance for the nuclei with N = 88

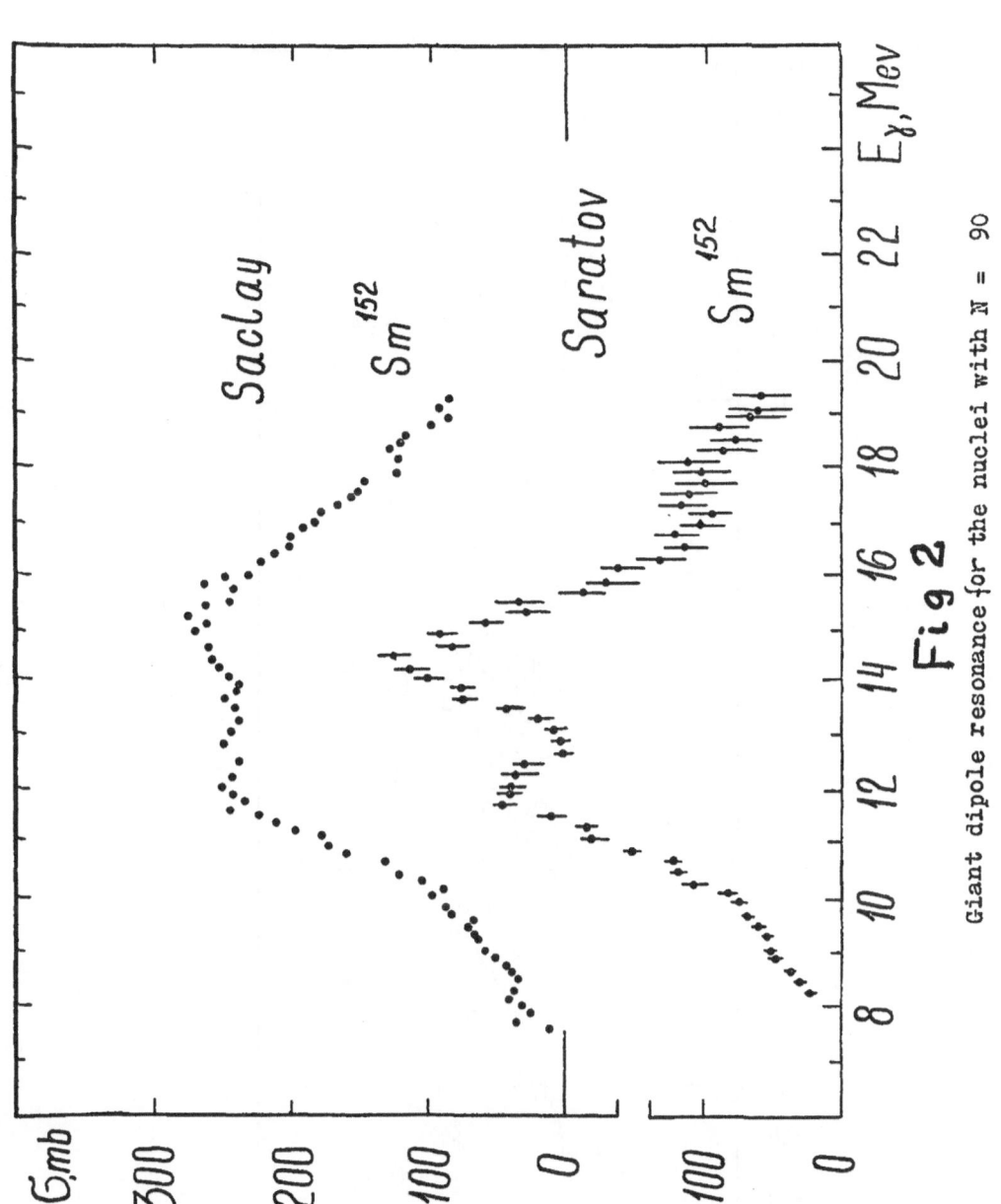

**Fig 2**

Giant dipole resonance for the nuclei with N = 90

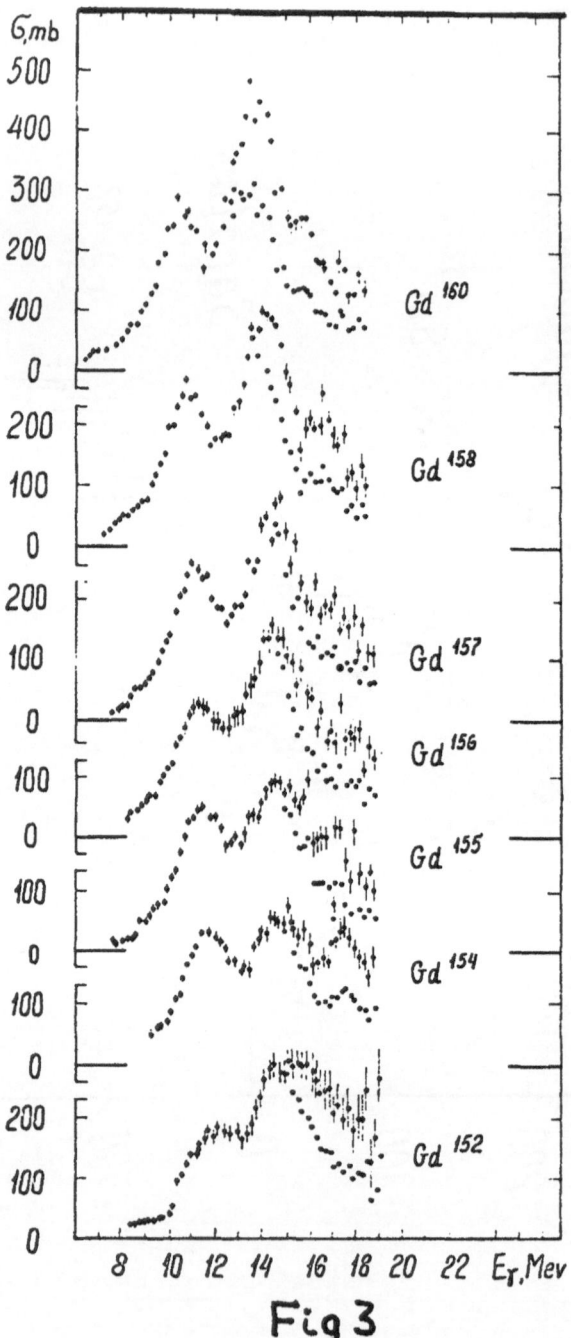

## Fig 3

Giant dipole resonance for Gd isotopes.

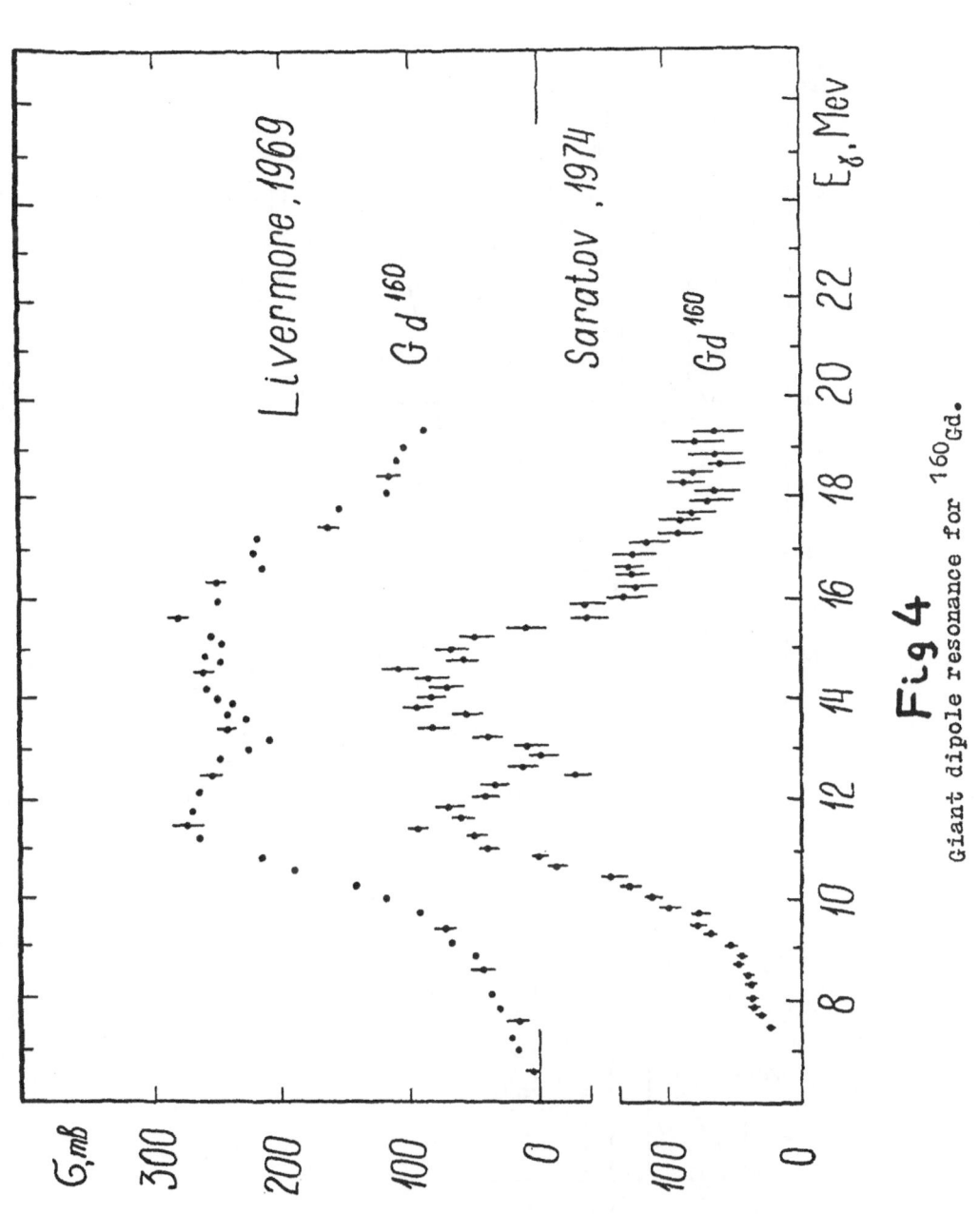

Fig 4

Giant dipole resonance for $^{160}$Gd.

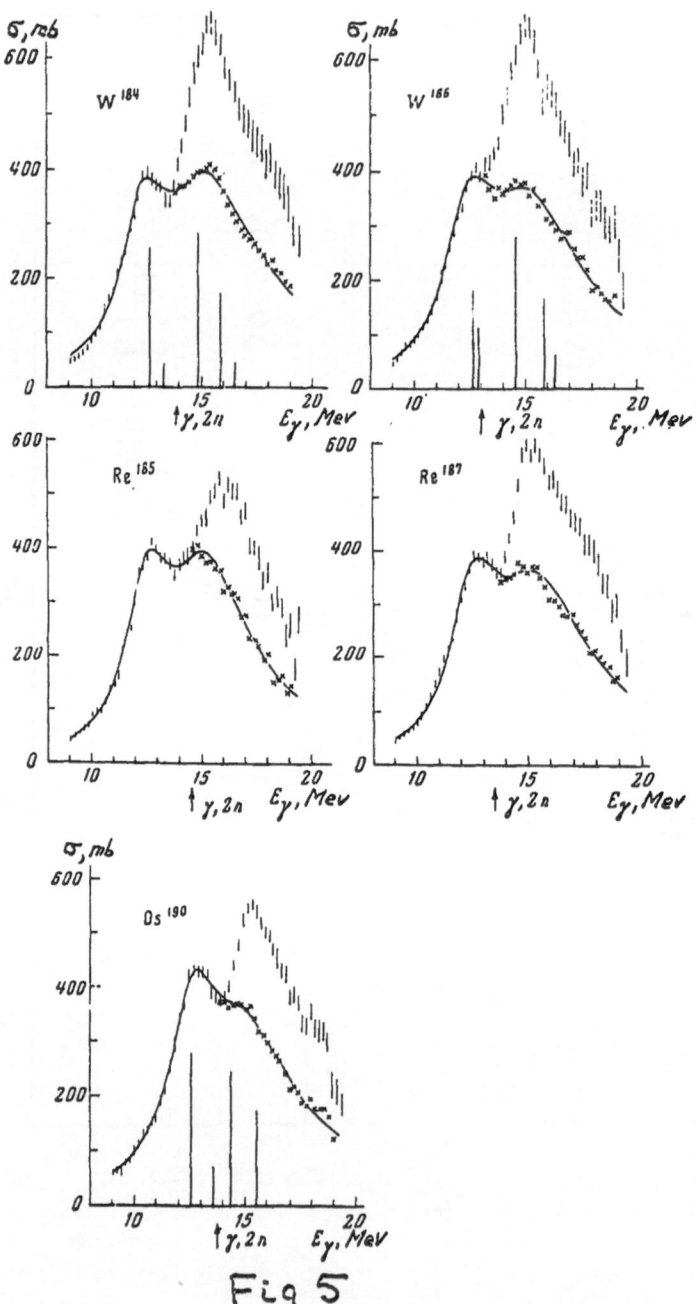

Fig 5

Giant dipole resonance for the nuclei with A ~ 190

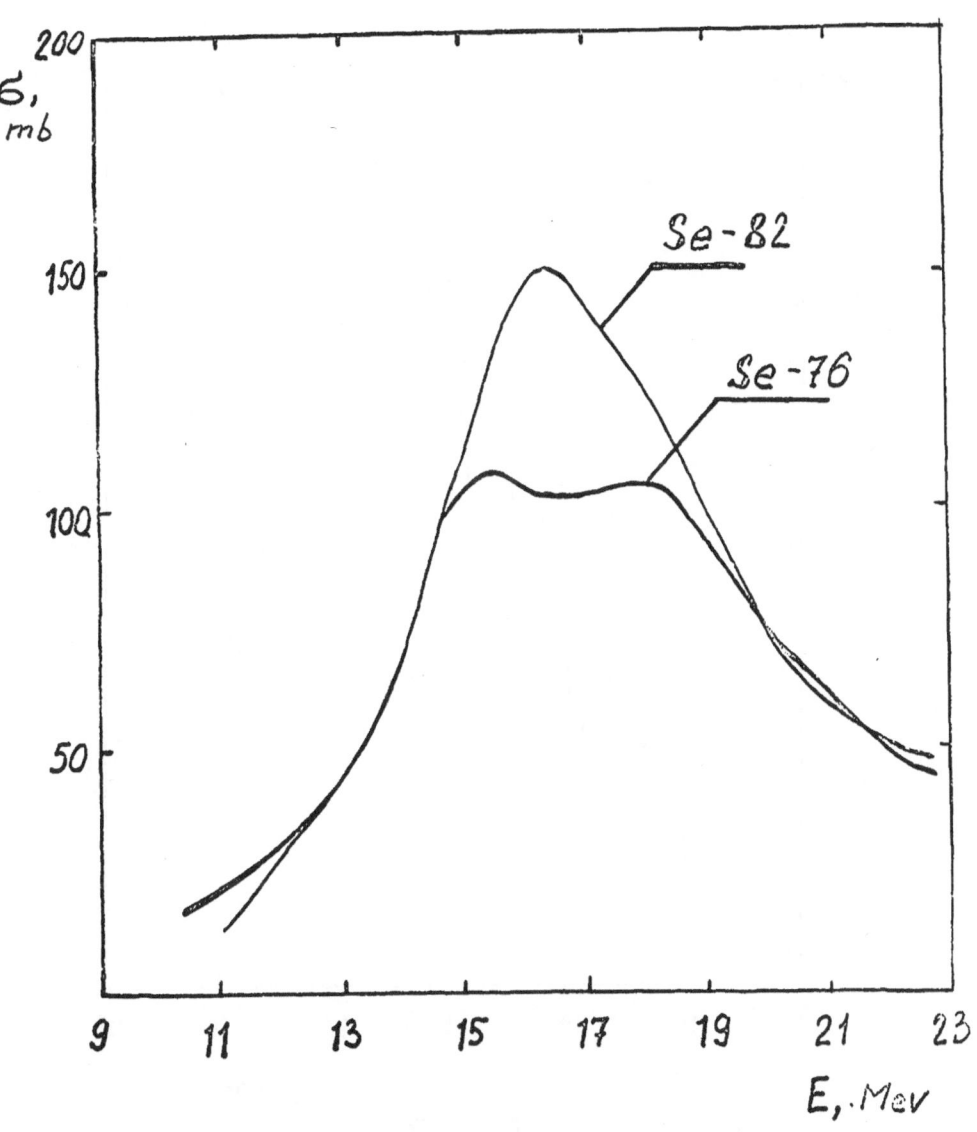

## Fig 6

Photoneutron cross sections for $^{76}$Se and $^{82}$Se.

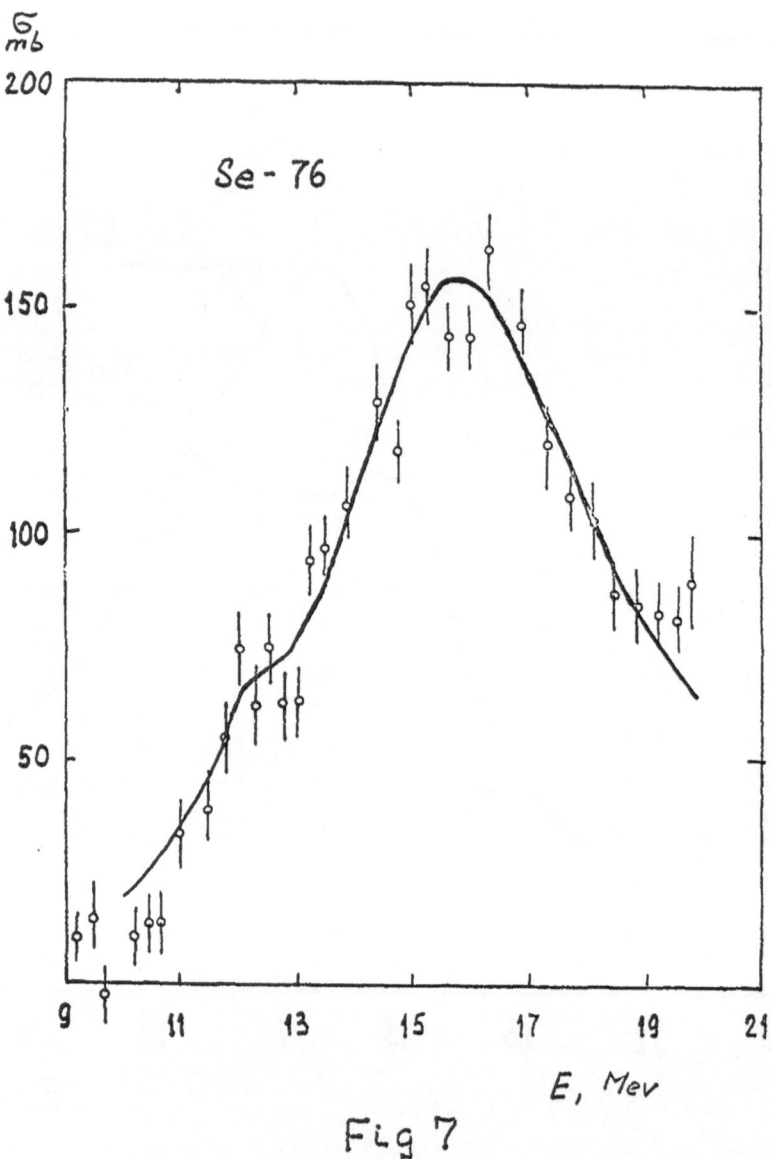

Fig 7

Total photoabsorption cross section for $^{76}$Se.

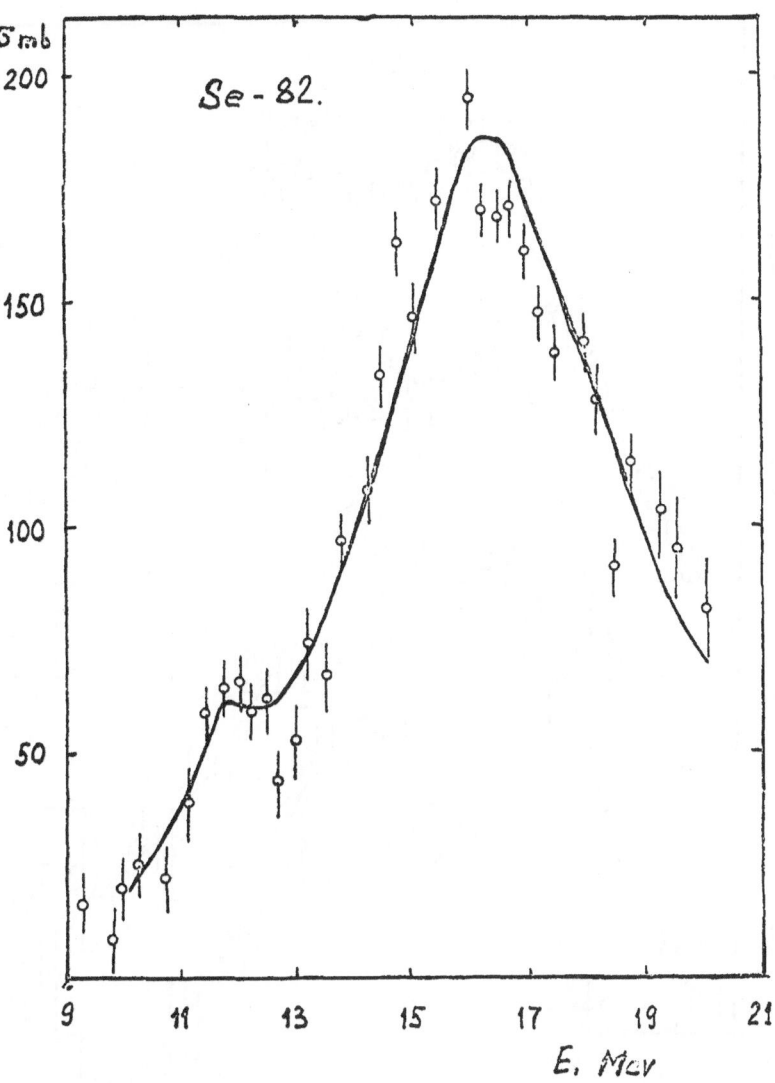

## Fig 8

Total photoabsorption cross section for $^{82}$Se.

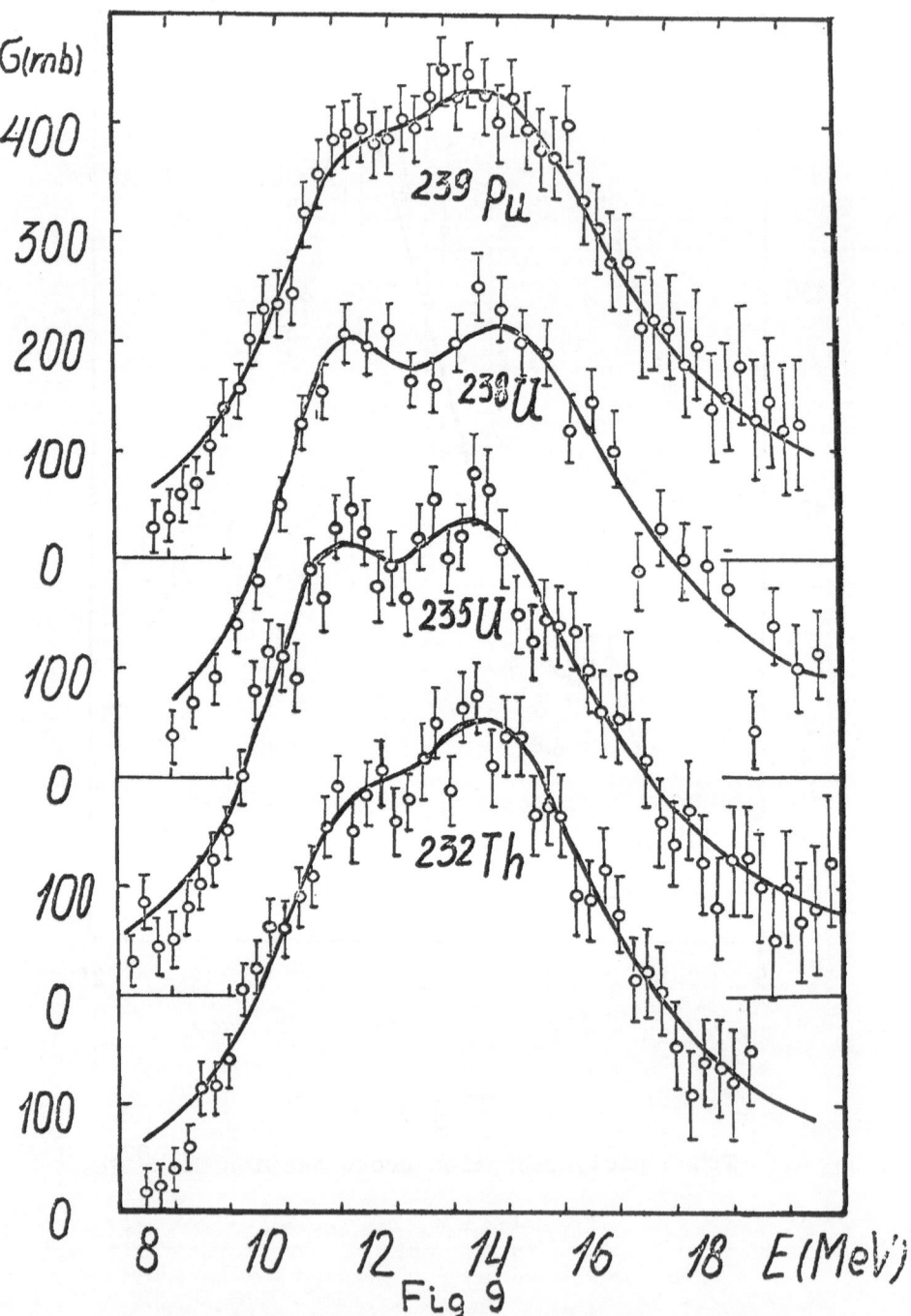

Fig 9

Total photoabsorption cross sections for $^{232}$Th, $^{235}$U, $^{238}$U and $^{23}$Pu and their respective best two Lorenz line fits.

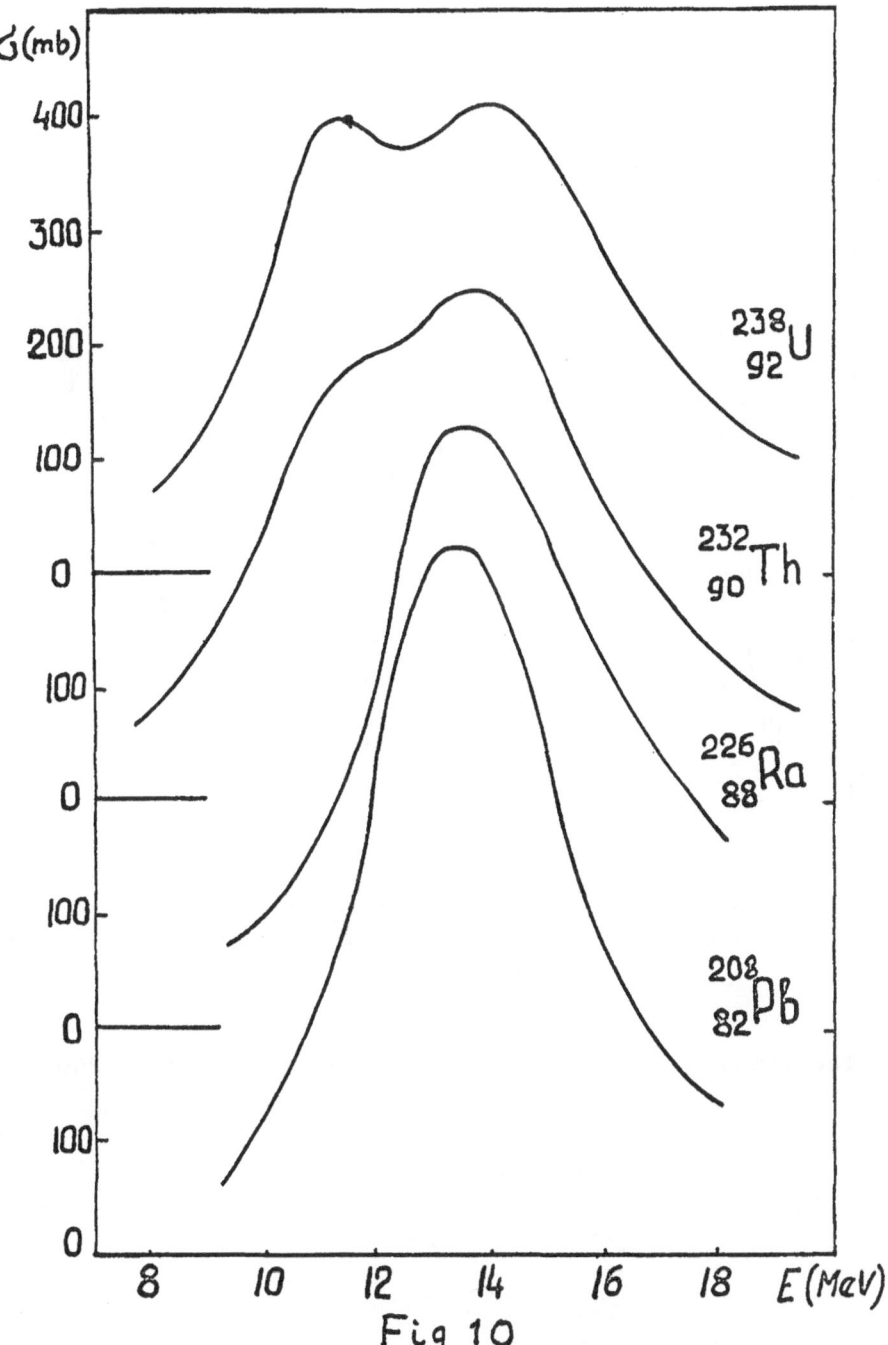

Fig 10

Cross section shape evolution for Z = 88-92.

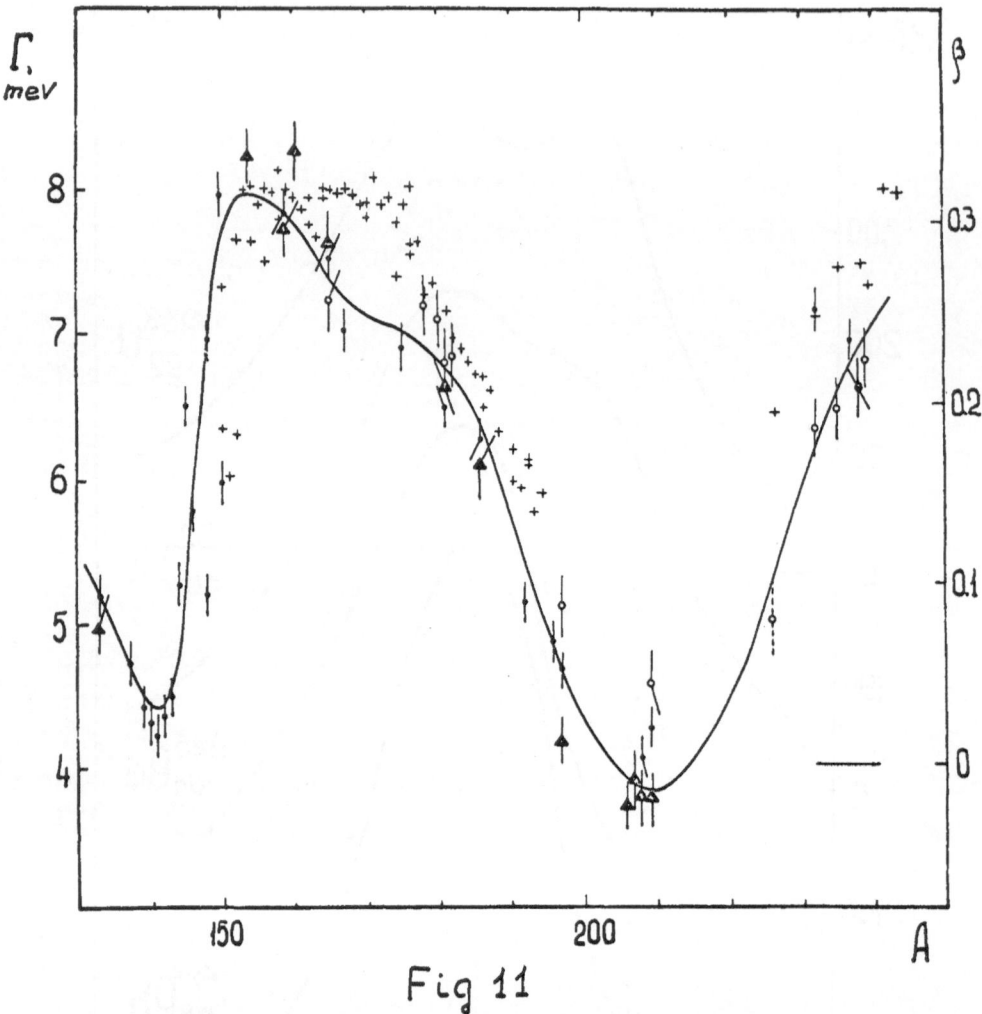

Fig 11

The width of giant dipole resonance for $165 \leqslant A \leqslant 209$ nuclei

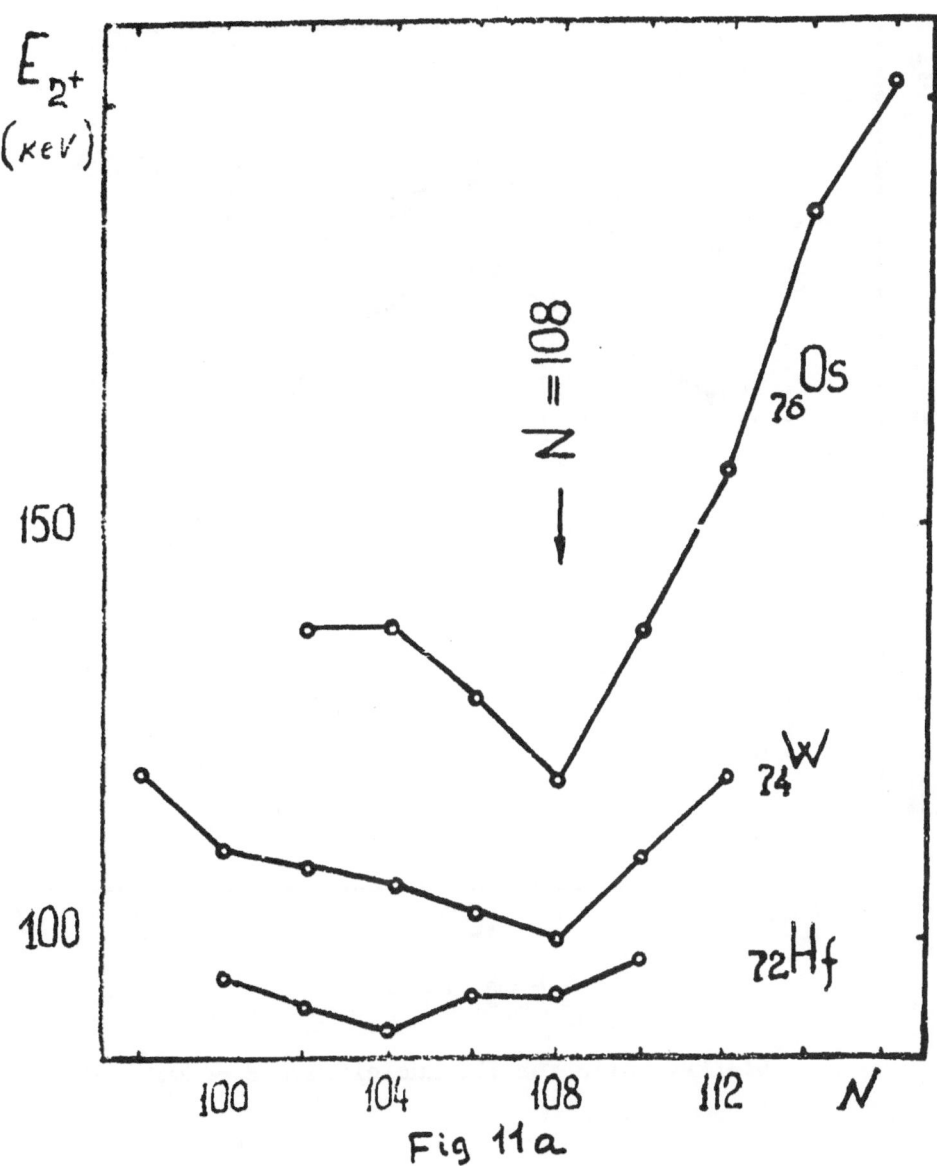

Fig 11a

E2+ values for the nuclei with N ~ 108

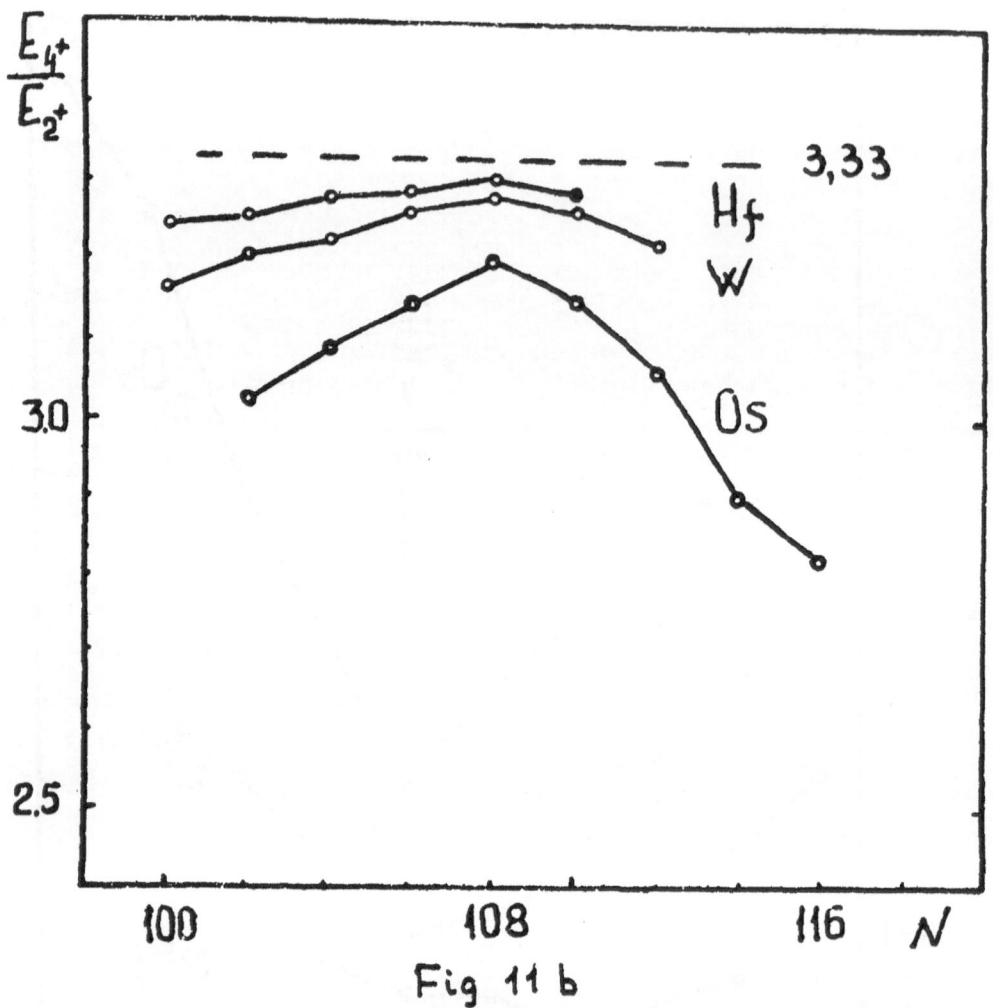

Fig 11 b

E4+/E2+ ratio for the nuclei with N ~ 108

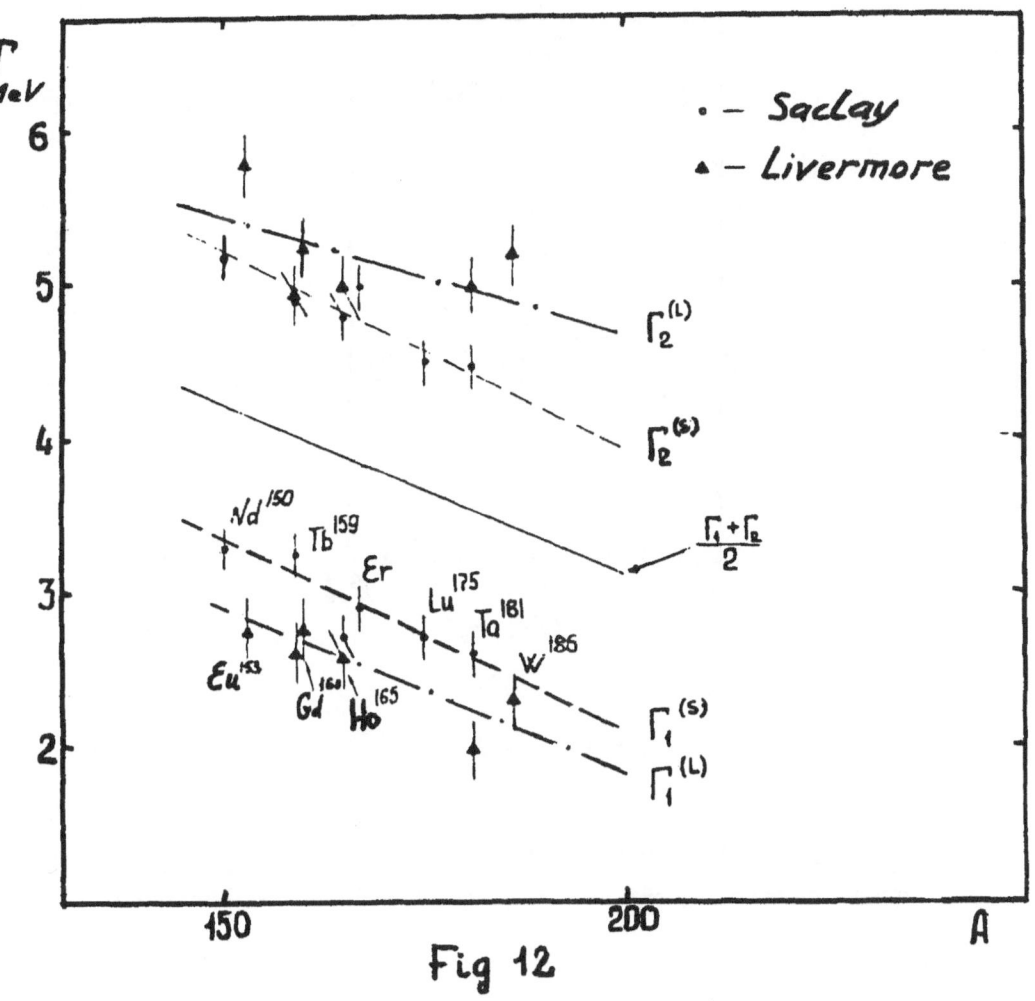

**Fig 12**

Evolution of the Lorentz line widths

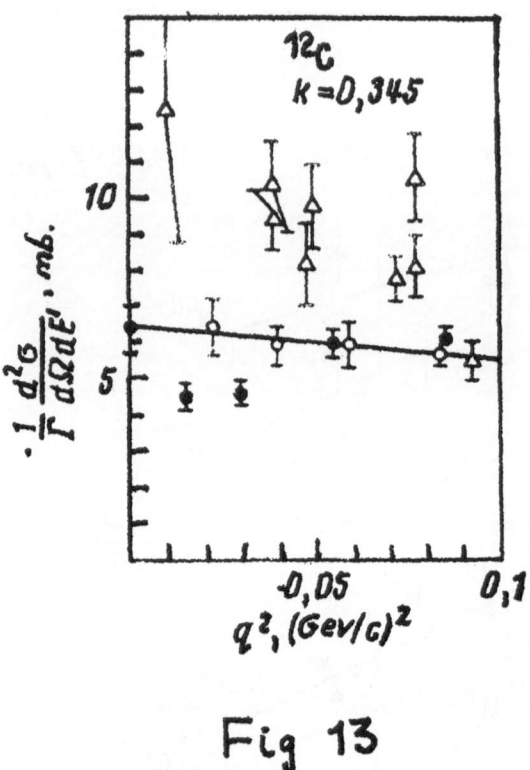

$$\cdot \frac{1}{\Gamma}\frac{d^2\sigma}{d\Omega dE'}, mb.$$

## Fig 13

$\sigma_{\gamma A}$  cross section for $^{12}C$ nuclei.

crear circles- /29/, dark circles - data for hydrogen.

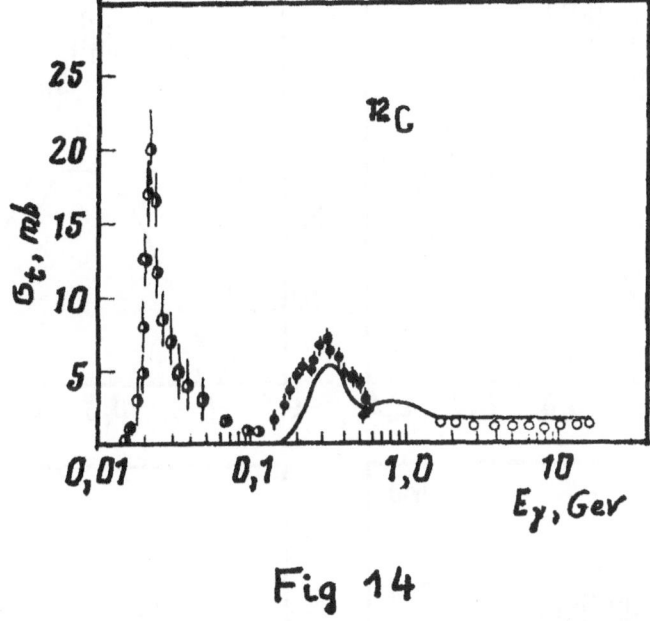

## Fig 14

Total hadronic photoabsorption cross section for $^{12}C$.

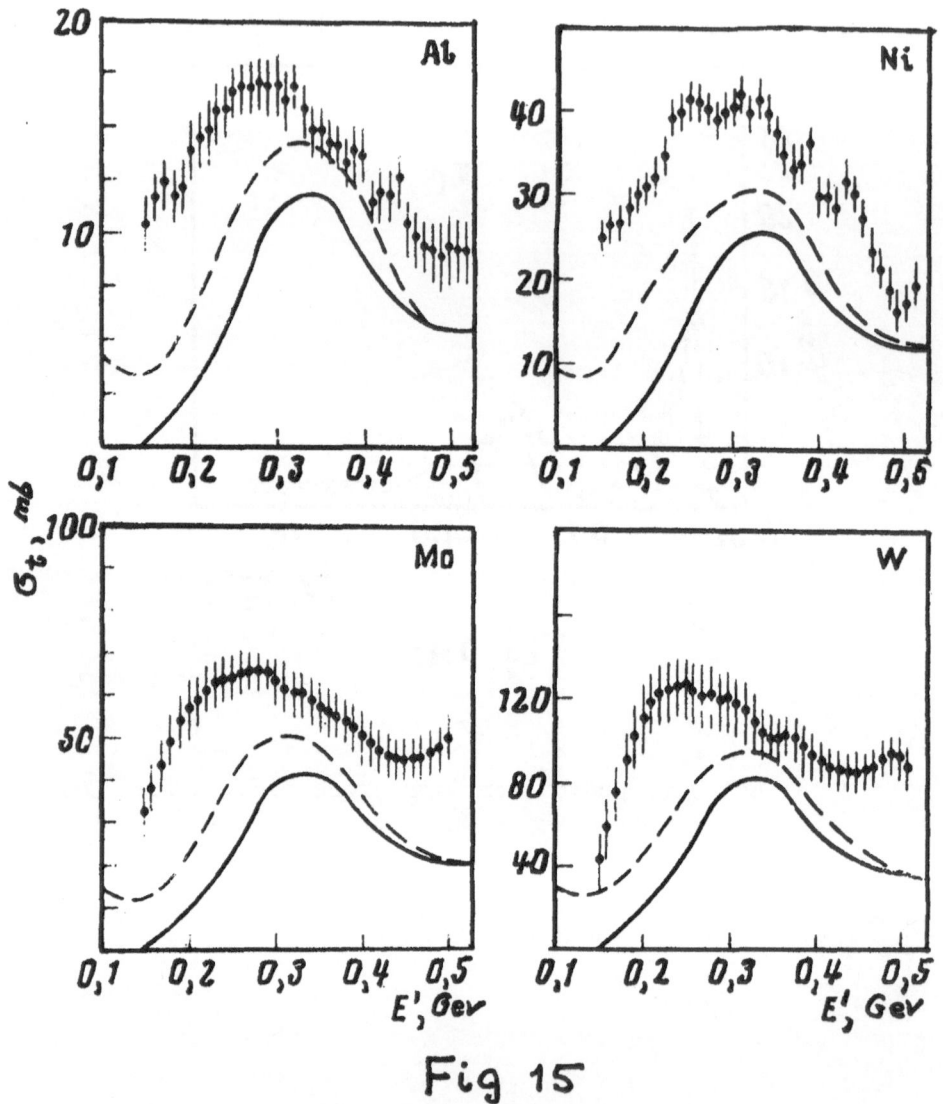

## Fig 15

Total hadronic photoabsorption cross sections for some nuclei

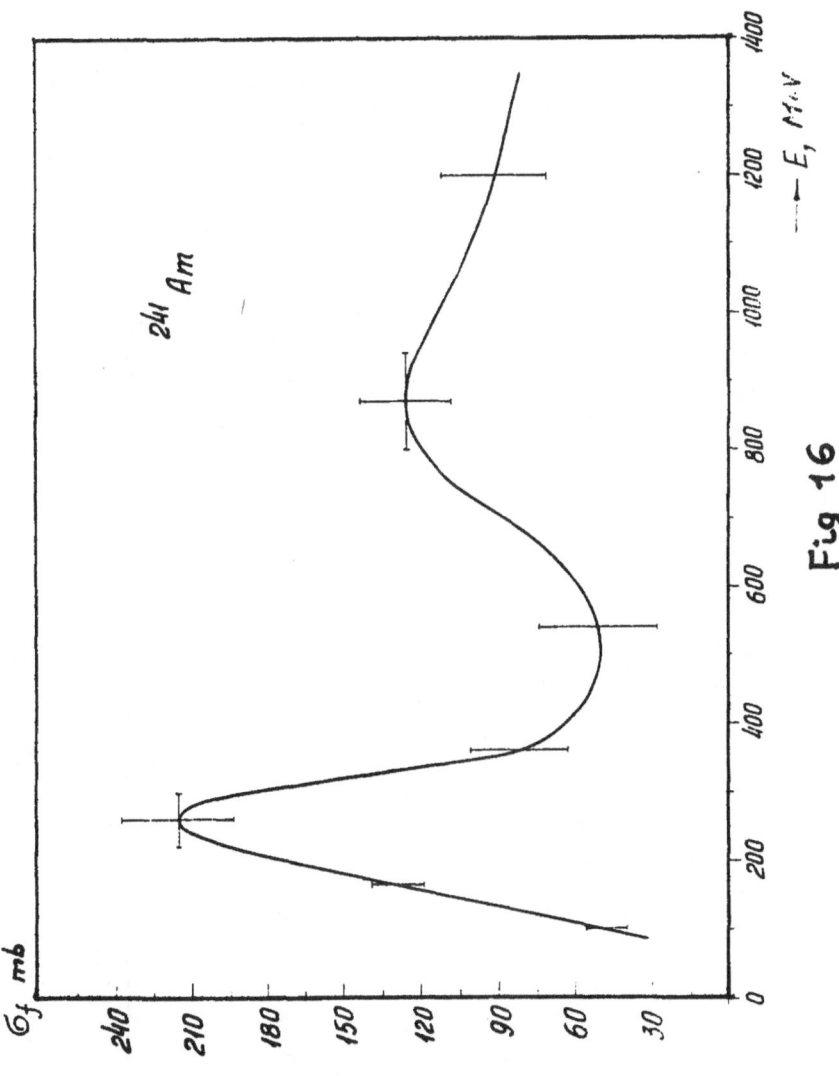

Fig 16

Photofission cross section for $^{241}$ Am

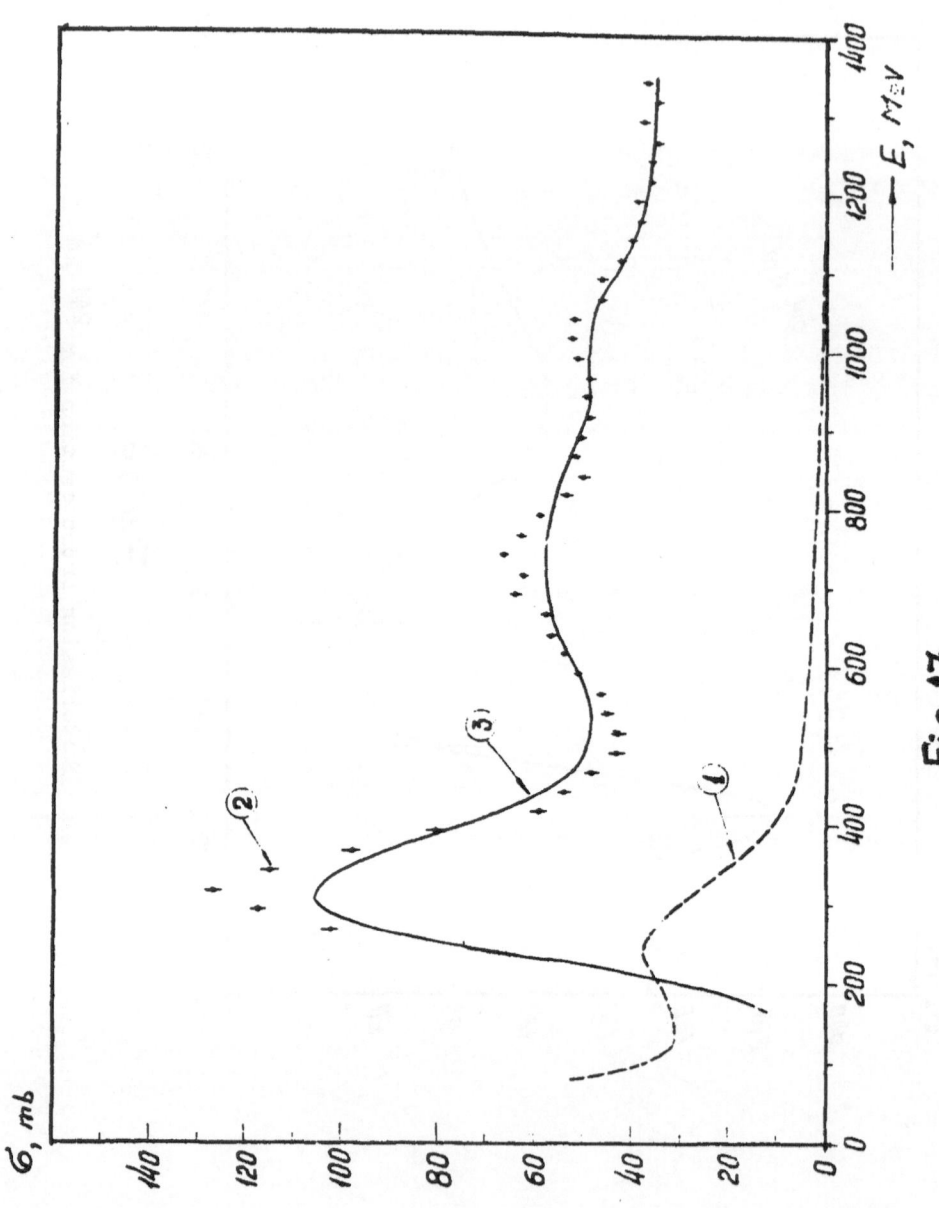

Fig 17

Total hadronic photoabsorption cross section for $^{241}$Am.

Table I   Lorentz line parameters for $^{76}$Se and $^{82}$Se

| Nucleus | $B_n$ Mev | $B_p$ Mev | $E_1$ Mev | $E_2$ Mev | $\sigma_1$ mb | $\sigma_2$ mb | $\Gamma_1$ Mev | $\Gamma_2$ Mev | $\int \sigma\, dE$ Mev·mb | $60\frac{NZ}{A}$ Mev mb |
|---|---|---|---|---|---|---|---|---|---|---|
| $^{76}$Se | 11,2 | 9,5 | $15,9\pm0,17$ (16,5) | 12,0 | $156\pm7$ (100) | 22 | $5,9\pm0,6$ (8,9) | 1,5 | 935 | 1130 |
| $^{82}$Se | 9,3 | 11,2 | $16,3\pm0,12$ (16,5) | 11,7 | $185\pm8$ (150) | 27 | $5,1\pm0,5$ (5,9) | 1,5 | 996 | 1190 |

Table 2   Lorentz line parameters for $^{232}$Th, $^{235}$U, $^{238}$U and $^{239}$Pu nuclei.

| Nucleus | $\sigma_1$, mb | $\Gamma_1$, Mev | $E_1$, Mev | $\sigma_2$, mb | $\Gamma_2$, Mev | $E_2$, Mev |
|---|---|---|---|---|---|---|
| $^{232}$Th | 247±26 | 3,90±0,4 | 10,99±0,16 | 362±26 | 4,67±0,38 | 13,9±0,13 |
| $^{235}$U | 283±39 | 3,23±0,55 | 10,74±0,18 | 354±33 | 4,92±0,58 | 13,77±0,23 |
| $^{238}$U | 286±30 | 2,99±0,48 | 10,37±0,13 | 351±25 | 5,10±0,63 | 14,25±0,18 |
| $^{239}$Pu | 227±39 | 3,47±0,57 | 11,05±0,13 | 362±31 | 5,23±0,59 | 14,01±0,21 |

Table 3   Deformation parameters and quadrupole moments for the same nuclei.

| Nucleus | $\beta_{exp}$ | $\overline{\beta}$ | $Q_0$, $\delta_H$ | | |
|---|---|---|---|---|---|
| | | | | [ ] | [ ] |
| $^{232}$Th | 0,28 ±0,03 | 0,274 | 10,0± 0,8 | 10,2 ± 1 | 9,65±0,1 |
| $^{235}$U | 0,30 ± 0,03 | 0,285 | 11,0±0,9 | 12,8±1,3 | 11,12±0,2 |
| $^{238}$U | 0,31 ±0,03 | 0,300 | 11,7±0,9 | 11 ± 1 | 11,3±0,1 |
| $^{239}$Pu | 0,29 ± 0,03 | 0,302 | 11,0±0,9 | | 11,02±0,3 |

Table 4     Surface motion characteristics for the Z 90 nuclei.

| Z | Nucleus | $E_g^{2+}$, kev | $E_g^{4+}$, kev | $E_g^{6+}$, kev | $E_\beta^{0+}$, kev | $E_\gamma^{2+}$, kev | $\mu$ | $\gamma$ | $\bar{\beta}$ |
|---|---|---|---|---|---|---|---|---|---|
| 88 | 226 Ra | 67,8 | 210 | 416 | | | | | 0,19 |
| 90 | 232 Th | 49,8 | 163 | 334 | 725 | 788 | 0,25 | 0,156 | 0,274 |
| 92 | 234 U | 43,5 | 143 | 296 | 812 | 921 | 0,23 | 0,143 | 0,285 |
| | 235 U | | | | | | | | 0,285 |
| | 238 U | 44,7 | 148 | 308 | 994 | 1062 | 0,20 | 0,140 | 0,300 |
| 94 | 238 Pu | 44,1 | 146 | 304 | 942 | 1028 | 0,21 | 0,136 | 0,302 |
| | 239 Pu | | | | | | | | 0,302 |
| | 240 Pu | 42,8 | 142 | 294 | 861 | 938 | 0,22 | 0,140 | 0,301 |

# ELECTROEXCITATION OF GIANT MULTIPOLE RESONANCES

Y. Torizuka

Laboratory of Nuclear Science

Tohoku University, Tomizawa, Sendai, Japan

## I   Introduction

In the last five years a lot of experimental evidences
for giant multipole resonances (GMR) has been obtained through
various experimental techniqes.  A survey of GMR has been made
by Satchler,[1] Hanna,[2] Borzov and Kamerdzhrev,[3] Bergere,[4] and
Bertrand.[5] Here I present rather limitted discussions on GMR
observed mainly by inelastic electron scattering.

The GMR have generally been observed as a bump riding on
a large background.  Hence, major uncertainties of the GMR data
arises from the approcimation used for the background.  Usually
the background consists of noises from the devices, radiation
tail, and it might be including a part of resonance states or
quasielastic scattering.

Fig.1 shows $^{208}$Pb spectrum obtained at Darmstadt[6] near the
giant dipole resonance (GDR) region measured at incident electron
energy of 50 MeV and scattering angle of 129°.  The GDR of this nucleus
was    observed at 13.5 MeV in the photonuclear reaction.  In
Fig.1 the peak appears at 14.1 MeV.  The solid line indicates

the background which arises mainly from the radiative effects. From the scale of the vertical axis the background is more than 90% of the total cross section. Fig.2 shows $^{208}$Pb(p,p') spectra at incident proton energy of 66 MeV obtained at Oak Ridge,[7] where we see that the GMR part, compared with the background (dashed line), is only a small fraction. Fig.3 shows the $^{208}$Pb(e,e') spectrum at 250 MeV and 25°, where the radiation tail caluculation is shown by the solid curve.[8] Using higher energy electrons the calculated radiation tail is rather flat and decreases to about 60% near the GDR.

II  Evidence for new giant resonances

Fig.4 shows the spectra for Ce obtained at Darmstadt.[9] The nuclear excitation part was determined phenomenologically assuming the size of the GDR and the overlapped peaks were decomposed into the El (15 MeV), E2 (12 MeV), and Ml (9MeV) giant resonances. The inelastic electron scattering spectra on $^{90}$Zr obtained at Sendai[10] is shown in Fig.5a. At the top of this figure the spectrum obtained in photonuclear reaction on $^{90}$Zr is shown. In the scattering spectra a broad peak at 16.7 MeV shifts to a peak at 14.0 MeV. Accordingly the bump riding on the smooth background is decomposed into the two peaks and obtained q-dependence as shown inFig.5a. The cross section for the 16.7 MeV peak is represented by the El theoretical curve. The q-

dependence of the 14.0 MeV peak is described by E2 (or E0) curve.
The transition strength for the 14.0 MeV peak exhausts about 56%
of the isoscalar E2 energy weighted sum rule (EWSR).

Electron scattering can excite strongly isoscalar (T=0) as
well as isovector (T=0) modes. While inelastic deutron or alpha-
particle scattering can excite only isoscalar mode. Fig.6 shows
the $(\alpha,\alpha')$ spectra on $^{27}$Al, $^{40}$Ca, $^{90}$Zr, and $^{208}$Pb obtained at Texas[11].
In contrast to the GDR systematics $E_x=80/A^{1/3}$MeV, the E2 peaks
are observed at $E_x=63/A^{1/3}$MeV and indicated to be isoscalar.
The resonance line shape, Excitation energy, and EWSR strength in
the $^{90}$Zr$(\alpha,\alpha')$ are in excellent agreement with the above electron
scattering result as seen in Fig.7 (Ref.12). Fig.8 shows the
$^{208}$Pb(e,e') spectra obtained at Sendai.[13] The GDR known from the
photoreaction is represented in the figure using the q-dependence
of the hydrodynamical model. We see a fine structure in the
excitation energy region corresponding to the isoscalar E2 giant
resonance and a broad bump at 22 MeV, which may correspond to
the isovector E2 giant resonance. The similar peaks are also
found in other nuclei, which follow the systematics $E_x=120/A^{1/3}$
MeV. The discrepancy between $^{208}$Pb$(\alpha,\alpha)$ and (e,e') spectra in
the region of 10 MeV will be discussed later. A major problem
is that although we have removed the radiation tail the unfolded
spectra contains non-resonant parts.

## III  Radiation tail

From 1974 to 1975 Fridrich came Sendai and investigated the radiation tail.  According to his suggestion we used the following tail function;[14]

$$\frac{d^2\sigma}{d\Omega dE_2} = [\{f_S(E_1-\omega,k)+f_{rad}(E_1-\omega,k,\tfrac{t}{2})+P_{col}(E_1-\omega,k,\tfrac{t}{2})\}$$

$$\times \frac{d\sigma}{d\Omega}(E_1,E_1-\omega)$$

$$+ \{f_S(E_1,k)+f_{rad}(E_1,k,\tfrac{t}{2})+P_{col}(E_1,k,\tfrac{t}{2})\}$$

$$\times \frac{d\sigma}{d\Omega}(E_2+\omega,E_2)]K^S(k)K^B(k)$$

where $f_S$ and $f_{rad}$ are the probabilities for one-photon emission. We can approximate the multiple-photon emission effect by multiplying the one-photon emission effect by the correction factor $K^S(k)K^B(k)$.  For the internal radiation term $f_S$ we used the peaking approximation formula given by Mo and Tsai.[15] This formula, compared with the non-peaking cross section given by Maximon and Isabelle[16], can be used up to 50 MeV within an error of 3%. For the bremsstrahlung and collision terms $f_{rad}$ and $P_{col}$ we used the formulas in Ref.14.

The spectra unfolded by the currently used radiation tail formula were negative in the region just below the elastic peak. This failure was considerably removed by using the new formula.

The raw and unfolded spectra at 250 MeV and 25° are shown in Fig.9. The cross section just below the elastic peak and the cross section at large energy losses shows to be reasonably zero. The same unfolding was applied to other spectra involving other nuclei. The results are quite satisfactory.

## IV  Comparison with the RPA

A series of the $^{208}$Pb spectra in Fig.8 was reanalyzed by the use of the new tail function and is displayed in Fig.10. In order to improve the statistical accuracy the three-point smoothing was applied to the range above 15 MeV. A broad bump at 22 MeV which may correspond to isovector E2 GMR is seen manifestly. The inelastic electron scattering cross section of the GDR normalized to the line shape and transition strength as derived from the photonuclear reaction on $^{208}$Pb is displayed in Fig. 10. To obtain this cross section, we used quite different models, i.e. the Goldhaber-Teller (GT)[17] which oscillates at the nuclear surface and Steinwedel-Jensen-Jensen model (SJJ)[18] which oscillate at the inside of the nucleus. For the same transition strength the cross section obtained by the SJJ model is much larger than that of the GT model. Also we notice that nontheless the accuracy of the tail subtraction was improved we must fit the GDR cross section to the spectrum together with the large background. The hatched area indicates the contribution from the transverse part which estimated from the experiment at the backward angle and the calculation from the Fermi-gas model.

In order to investigate the large background part we wish to compare the spectrum with the shell model calculation which predicts the strength distributions for GMR. Such calculations were perfomed with a large configuration in the random phase approximation (RPA) based on the Skyrme interaction,[19] for example Liu predicts the strength distributions for the isoscalar and isovector E0, E2, E3, and E4 excitations up to an excitation energy of 40 MeV. The RPA is in good agreement with experiment, i.e, the predicted strengths for the low-lying collective states are within 10% of the experimental values. The inelastic scattering cross section from the RPA strength distribution was caluculated using the q-dependence of the Tassie model[20] for the isoscalar states and the GT or SJJ model for the isovector states. In Fig. 11 the sum of the theoretical E0, E1, E2, E3, and E4 form factors is compared with the spectrum at 250 MeV and 25°. The form factor obtained from the GT model calculation is in good agreement with the measured spectrum. The SJJ model caluculation does not agree with the observed one indicating that the SJJ model form factor is too large. From the above comparison we may know that the observed spectra at low q and low excitation energy may be regarded as the sum of the GMR states. Then, if the q-dependence of the multipole states are known, a set of more than several spectra may be decomposed into these multipole states.

V  Multipole expansion of the nuclear continuum

A set of the five spectra in $^{208}$Pb (Fig.8) in the momentum transfer range from 0.447 to 0.714 fm$^{-1}$ was decomposed, channel by channel, into E1, E2. E3 and higher multipole cross sections[21]. For this purpose we have used the q-dependence of the Tassie model for the isoscalar states and GT or SJJ model for the isovector states. The E2 states above the GDR is presumably isovector and the other multipoles except for E1 were assumed to be isoscalar. The spectra were divided into successive bins of the equal intervals of 210 keV in the range below 15 MeV and 500 keV beyond 15 MeV. The contribution of the transverse form factor (the hatched area in Fig.8) was subtracted. The form factor for each bin is assumed to be the sum of the E1, E2, E3, and higher multipole components:

$$W_c^{(n)} = a_1^n |F_{E1}|^2 + a_2^n |F_{E2}|^2 + a_3^n |F_{E3}|^2 + a_4^n |F_\alpha|^2$$

where $a_i^n \geq 0$ and $a_i^n$ are determined by the least-square fitting. The higher multipole term was assumed to be E4 (set I) or the sum of E4 and E5 (set II) or the sum of E4,E5, and E6 (set III), where relative amplitudes were determined by normalizing to the corresponding EWSR. The E1 and E2 components obtained by set III is in good agreement with set II but do not agree with set I. The assumption in set I may not be realistic. The result

of the multipole expansion for the spectrum at 183 MeV and 35°
is displayed in Fig.12. The upper portion corresponds to the expansion using the q-dependence of the Tassie model for isoscalar
and the SJJ model for isovector (SJJ-model expansion). The
lower portion corresponds to the expansion using the Tassie
model for isoscalar and GT model for isovector (GT-model expn-
sion). The right hand scale in the figure indicates B(EL)/MeV,
which cannot be applied to the E2 strength of the SJJ expansion
above 17 MeV. The fitting errors were determined by the contur
enclosing $\chi^2 \leq 2\chi^2_{min}$. The percentages of the corresponding EWSR
for the specific excitation energies are tabulated in Table I.
The errors involve the uncertainty of the DWBA caluculation,
difference between set II and set III and fitting error.

The results obtained are discussed as follows.

A) Giant dipole resonance: The E1 strengths between 9.5 and 26
MeV extracted with the GT and SJJ models exhaust, respectively,
$136\pm^{21}_{33}$% and $128\pm^{17}_{29}$% of the E1 EWSR consistent with 117±8% obtained
from the $(\gamma,n)$ reaction[22]. The GT and SJJ form factors corresponding to a 100% WESR with a Breit-Wigner line profile with a width
of 4.05 MeV are shown together with experimental ones in Fig.11.
Both GT and SJJ curves reproduce nicely the experimental E1 form
factors, while an excess strength (about 20% of the E1 EWSR) is
in the region from 7.4 to 9.5 MeV.

B)  Giant quadrupole resonance:  The obtained E2 strengths are
concentrated in the regions centered at 11 and 22.5 MeV in both
GT and SJJ model expansions.  The deduced E2 strengths, however,
depend upon the model employed i.e. the sum of the E2 strengths
in the region 7.4∿15 MeV exhausts $120\pm^{25}_{12}$% and $42\pm^{15}_{8}$% of the isoscalar E2 EWSR for the expansion with the GT and SJJ models,
respectively.  The E2 form factor around 22.5 MeV with a width
of ∿5 MeV may be attributed to isovector and exhausts $95\pm^{40}_{13}$%
and $41\pm^{12}_{6}$% of the isovector EWSR for the GT and SJJ expansions,
respectively.

C)  Giant monopole resonance:  A difficulty arises because electron scattering cannot distinguish between E0 and E2 excitations.
The E0 strength may be  involved partly into the strength distribution of E2.  The residual strength subtracting the E2 part
may be identified to be E0.

The E2 strength obtained by the GT expansion shows a structure at 10.5 and 13.5 MeV.  The alpha particle scattering $^{208}$Pb
spectrum[11] indicates the presence of the corresponding E2 peak
at 10.5 MeV with a rather narrow width of ∿3 MeV, but no structure is seen at 13.5 MeV.  If one assumes the presence of an E0
excitation at 13,5 MeV instead of E2, this discrepancy may be
removed because electrons can excite strongly both E0 and E2
whereas hadrons excite E2 strongly, but E0 only weekly except for
specific angles.  The E0 strength between 12,5 and 15 MeV is

found to be $97\pm^{27}_{14}$% of the isoscalar E0 EWSR and the remaining E2 strength between 7.4 and 12.5 MeV exhausts about $76\pm^{14}_{8}$% of the isoscalar EWSR. The isoscalar E2 value in Table I includes the contribution from the bound sate (16%). If, however, we use the SJJ model the monopole strength at 13.5 MeV almost vanishes.

Evidence for E0 resonances has been presented at excitation energies; $E_{ex}= 80/A^{\frac{1}{3}}$ MeV from 80-MeV deutron inelastic scattering by Marty et al.[23] In $^{208}$Pb they have observed a bump at 13.5 MeV with a width of 2.5 MeV at the angles where an E0 excitation is favoured, which is in good agreement with the result obtained by the GT expansion.

D)  Giant octupole resonance

The giant octupole resonance states are made up from the 1 hω and 3hω shell model groups. Accordingly, the separation of the isoscalar and isovector states is very complicated. The decomposed octupole strengths distribute into peaks at 8.5 and 10.3 MeV and a broad bump centered at ∿16 MeV with a width of ∿6 MeV. The 16 MeV bump may be considered to be the 3hω isoscalar giant ocutapole resonance.

E)  Higher multipole resonance: The E4+E5+E6 excitations are rather flat lying the whole area of the spectrum and the amplitude approaches to that of the background used in the phenomenological analysis.

VI  Evidence for the giant monopole resonance in $^{90}$Zr

The same multipole expansion procedure was applied to the $^{90}$Zr spectra.[24] Here we used the Tassie model for the isoscalar and the GT model for the isovector states and decomposed into the E1, E2, E3, and E4+E5+E6 components.  The result for the spectrum at 183 MeV and 35° is shown in Fig.13.  The strength of the E1 component exhausts 114±25% of the EWSR which is consistent with the value deduced from the (γ,n) reaction.[25]  The E1 giant resonance, assuming a Breit-Wigner line profile normalized to the width of 4.0 MeV obtained from the photonuclear reaction, is shown in Fig.12.  This is in good agreement with the present experimental result indicating that the multipole expansion is reasonable.  The E2 peak at 14 MeV with a Breit-Wigner line of 4.5-MeV deduced from the inelastic alpha-particle scattering,[11] a broad E2 peak centered at 26 MeV with a width of 7 MeV, and a peak near 17 MeV assuming a resonance line with a width of 4.0 MeV were fitted simultaneously to the E2 spectrum in Fig.12. The peak at 14 MeV exhausts 84% of the isoscalar E2 ESWR, the peak centered 26 MeV exhaust 73% of the isovector E2 EWSR and if we assume the peak at 17 MeV to be E0 instead of E2 this peak exhausts 108% of the isoscalar EWSR.  Evidence for the giant monopole resonance at 17 MeV in $^{90}$Zr was presented by Marty et al.[23] from 80 MeV deutron inelastic scattering.

## VII  Magnetic Giant Multipole Resonances

The inelastic electron scattering spectra at backward
angles show large contribution from the transverse excitations.[26]
The transverse form factor calculated with the Fermi-gas model
reproduces the experimental form factor. A caluculation with
the single-particle model ( Woods-Saxon potential) is also in
good agreement with the experimental form factor.[27] Fig.14
shows the $^{90}$Zr(e,e') spectra at 155° corresponding to the various
momentum transfers.[28] The transverse part (data of points) was
separated from the total form factor (data of solid line) by a
Rosenbulth plot. The dashed and solid curves are the caluculation
from the Fermi-gas model. Bumps with a width of about 2.5 MeV
may be seen at 9 MeV. The area for these bumps were estimated
by assuming the background of the streight line. In Fig.15
the obtained form factors are compared with the sum of the M1,
M3, M5, and M7 calculations for the transition from $g_{9/2}$to $g_{7/2}$.
The experimental cross sections are much higher than this single
configuration. Fig. 16 shows the $^{208}$Pb(e,e') spectrum at 107
MeV and 155°. The separated transverse part is shown by the
hatching. A structure may be seen near 17 MeV. We need more
accuracy for the experiment and theoretical calculations for
the study of the magnetic giant multipole resonances.

VIII conclusion

Electron scattering can excite every electric and magnetic multipole states with isovector or isoscalar modes. On the other hand, in photonuclear reactions the observed cross section is mainly isovector GDR. The isoscalar giant quadrupole resonance (GQR) is excited selectively by inelastic $\alpha$-particle and deutron scattering. The GQR with isoscalar mode may be established by the combination of these reactions and inelastic electron scattering. Evidence for the isovector GQR, although electron scattering has observed has observed a rather narrow peak at 22.5 MeV in $^{208}$Pb, is few. We hope more data will be accumulated on this state. A method of extracting the monopole state is described here; if one can extract the cross section spectrum of E2-like structure from the inelastic electron scattering continuum the monopole part may be obtained by subtracting the E2 cross section known from other reactions. Then for $^{208}$Pb, $97 \pm^{27}_{14}$% or $10 \pm^{20}_{9}$% of the isoscalar monopole EWSR strength is obtained, depending on whether the GT of SJJ models, respectively, are used for the GDR. Evidence for the isoscalar giant octupole resonance may be suggested at $\sim$16 MeV in $^{208}$Pb. This cross section remains after the subtraction of the GDR and GQR from the inelastic electron scattering continuum. We need more investigation for magnetic giant multipole resonance.

# References

1) G.R. Satcheler, Physics Reports $\underline{14}$, 97 (1974)

2) S. Hanna, in Proceed. of Int. Conf. on Nucl. Structure and Spectroscopy, Amsterdam 1974, 249

3) I.N. Borzov and S.P. Kamerdzhiev, Report FEI 580, Obninsk, 1975

4) R. Bergere, Seminar on Electromagnetic Interaction of Nucl. at Low and Medium Energies, Moscow, 1975

5) F.E. Bertrand, to be published in Annual Review of Nuclear Science, 1976

6) T. Walcher, private communication

7) M.B. Lewis, F.E. Bertrand, and D.J. Horen, Phys. Rev. $\underline{C8}$, 398 (1973)

8) M. Sasao and Y. Torizuka, to be published

9) R. Pitthan, Z. Phys. $\underline{260}$, 283 (1973)

10) S. Fukuda and Y. Torizuka, Phys. Rev. Lett. $\underline{29}$, 1109 (1972)

11) J.M. Moss et al. Phys. Lett. $\underline{53B}$, 51 (1974)

12) D.H. Youngblood et al. Phys.Rev. $\underline{C13}$, 994 (1976)

13) M. Nagao and Y. Torizuka, Phys. Rev. Lett. $\underline{30}$, 1068 (1973)

14) J. Friedrich, Nucl. Instrum. Methods $\underline{129}$, 505 (1975)

15) L.W. Mo and Y. M. Tsai, Rev. of Mod. Phys. $\underline{41}$, 205 (1969)

16) L.C. Maximon and D.B. Isabelle, Phys. Rev. $\underline{136B}$, 674 (1964)

17) M. Goldhaber and E. Teller, Phys. Rev. $\underline{74}$, 1046 (1948)

18) H. Steinwedel, J.H.D. Jensen, and D. Jensen, Phys. Rev. $\underline{79}$, 1109 (1950)

19) S. Krewald and J. Speth, Phys. Lett. <u>52B</u>, 295 (1974),

      G. Bertsch and S.F. Tsai, Phys. Rep. <u>18C</u>, 125 (1975),

      K.E. Liu and G.E. Brown, preprint

20) L.J. Tassie, Australian J. Phys. <u>9</u>, 409 (1956)

21) M. Sasao and Y. Torizuka, to be published in Phys. Rev.

22) A. Veyssiere et al. Nucl. Phys. <u>A159</u>, 561 (1975)

23) N. Marty et al. preprint

24) S. Fukuda and Y. Torizuka, to be published in Phys. Lett.

25) B.L. Berman et al. Phys. Rev. <u>162</u>, 1098 (1967); A. Lepretre

      et al. Nucl. Phys. <u>A175</u>, 609 (1971)

26) K. Hosoyama and Y. Torizuka, Phys. Rev. Lett. <u>35</u>, 199 (1975)

27) Y. Kawazoe, private communication

Table I  Percentages of the EWSR in $^{208}$Pb below 26 MeV.  Contribution from the bound states (16% for isoscalar E2 and 20% for isoscalar E3) is included.

| Mutipole | Mode | $E_x$ (MeV) | GT expansion (%) | SJJ expansion (%) |
|----------|------|-------------|------------------|-------------------|
| E0 | T=0 | 12.5∿15 | 97 $^{+27}_{-14}$ | 10 $^{+20}_{-9}$ |
| E1 | T=1 | 7.4∿26 | 156 $^{+23}_{-35}$ | 145 $^{+18}_{-30}$ |
| E2 | T=0 | 7.4∿12.5 | 92 $^{+14}_{-8}$ | 52 $^{+12}_{-5}$ |
| E2 | T=1 | 15 ∿26 | 95 $^{+40}_{-13}$ | 41 $^{+12}_{-8}$ |
| E3 | T=0 | 7.4∿26 | 165 $^{+15}_{-71}$ | 94 $^{+14}_{-47}$ |

Figure Captions

Fig.1 The $^{208}$Pb spectrum obtained at Darmstadt in the giant
dipole resonance region measured by 50 MeV and 129° electron.
The background which arises mainly from the radiative
effects is indicated by the solid line.

Fig.2 The $^{208}$Pb(p,p') spectra by 66 MeV proton at Oak Ridge.
The excitation energies of the various peaks are noted
in MeV. The background is shown by the dashed line.

Fig.3 The $^{208}$Pb(e,e') spectrum at Sendai by 250 MeV and 25°
electron. The radiation tail calculation is shown by
the solid line.

Fig.4 Spectra of inelastic electrons scattered from Ce at various angles. The curves represent a decomposition into
overlapping giant resonances, predominantly E1 and E2 at
the two smaller angles, almost entirely M1 at the large
angle.

Fig.5 Excitation of $^{90}$Zr by electrons. The left side (a) shows
spectra of inelastic electrons for various momentum; at
top a photonuclear spectrum is shown for comparison.
A bump is resolved into two peaks at 16.7 (E1) and 14.0
(E2) MeV as a function of the momentum transfer. The
right side (b) shows the form factors corresponding to the
peaks at 14 and 16.7 MeV.

Fig.6   Spectra from the 96 MeV $(\alpha,\alpha')$ reaction on $^{27}$Al, $^{40}$Ca, $^{90}$Zr, and Pb (natural). The arrows are located at an excitation energy of $63/A^{1/3}$MeV.

Fig.7   The lower spectrum is $^{90}$Zr$(\alpha,\alpha')$ at 96 MeV. The upper is $^{90}$Zr(e,e') from Fig.5. The lines are the E2 (large peak) and E1 (small peak) components, respectively of the (e,e') data.

Fig.8   Spectra of inelastic electrons scattered from $^{208}$Pb with various momentum transfers. The solid curves represent estimates of the giant dipole resonance and the non-resonant background;the two curves indicate the uncertainties in these estimates.

Fig.9   The $^{208}$Pb(e,e') at 250 MeV and 25° before and after tail subtraction.

Fig.10  The $^{208}$Pb spectra re-analyzed by the subtraction of the radiation tail using a new formula. The dashed and solid curves represent the estimates of the giant dipole resonance, depending on the Goldhaber-Teller and Steinwedel-Jensen-Jensen models, respectively. The transverse contribution is shown by hatching.

Fig.11  The $^{208}$Pb spectrum in an excitation energy range about 100 MeV. The dot-dashed curve is the caluculation from the Fermi gas model with $k_f$=265 MeV, $M^*$=0.8M, and $\varepsilon_B$= 7.4 MeV. The sum of the cross sections from E0 to E4

strength distributions of the RPA was caluculated using
the q-dependence of the Tassie model for the isoscalar
mode and the Goldhaber-Teller model (full line) or the
Steinwedel-Jensen-Jensen model (dashed line) for the
isovectoe mode is shown.

Fig.12   The $^{208}$Pb spectrum at 183 MeV and 35° was decomposed
into each spectrum for El, E2, E3, and the sum of
E4, E5, and E6 using the q-dependence of the Tassie
model for the isoscalar mode and the isovector mode
the Goldhaber-Teller model (lower part) or the Steinwedel-
Jensen-Jensen model (upper part). The dashed curves
in the upper and lower El spectra are the caluculations
from the photonuclear data with the SJJ and GT models,
respectively. B(EL)/MeV at the right side cannot be
applied for the E2 component of the SJJ-model expansion
in the range higher than 17 MeV. The errors are those
from the fitting.

Fig.13   The $^{90}$Zr(e,e') differential total form factor $|W(q,E_x)|^2$
at 183 MeV 35°, transverse part $|W_T|^2(\frac{1}{2} + \tan^2\frac{\theta}{2})$, and
$|W_L(q,E_x)|^2$ corresponding to El, E2, E3, and E4+E5+E6.
The scales at the left side indicate form factors in
units of $10^{-4}$/MeV and at right side indicate B(EL) values
in units of $fm^{2L}$/MeV.

Fig.14 The $^{90}$Zr(e,e') differential form factors (data of solid line) at backward angles and transverse form factor $|W_T|^2(\frac{1}{2}+\tan^2\frac{\theta}{2})$ (data of points) separated by the Rosen-bulth plot. Solid and dashed lines are the caluculation from the Fermi-gas model.

Fig.15 The area for the 9 MeV bump in Fig.14 was estimated by assuming the background of the straight line. The obtained form factors are compared with the sum of the M1, M3, M5, and M7 calculations for the transition from $g_{9/2}$ to $g_{7/2}$.

Fig.16 The $^{208}$Pb(e,e')spectrum at 107 MeV and 155°. The transverse part separated is shown by the hatching.

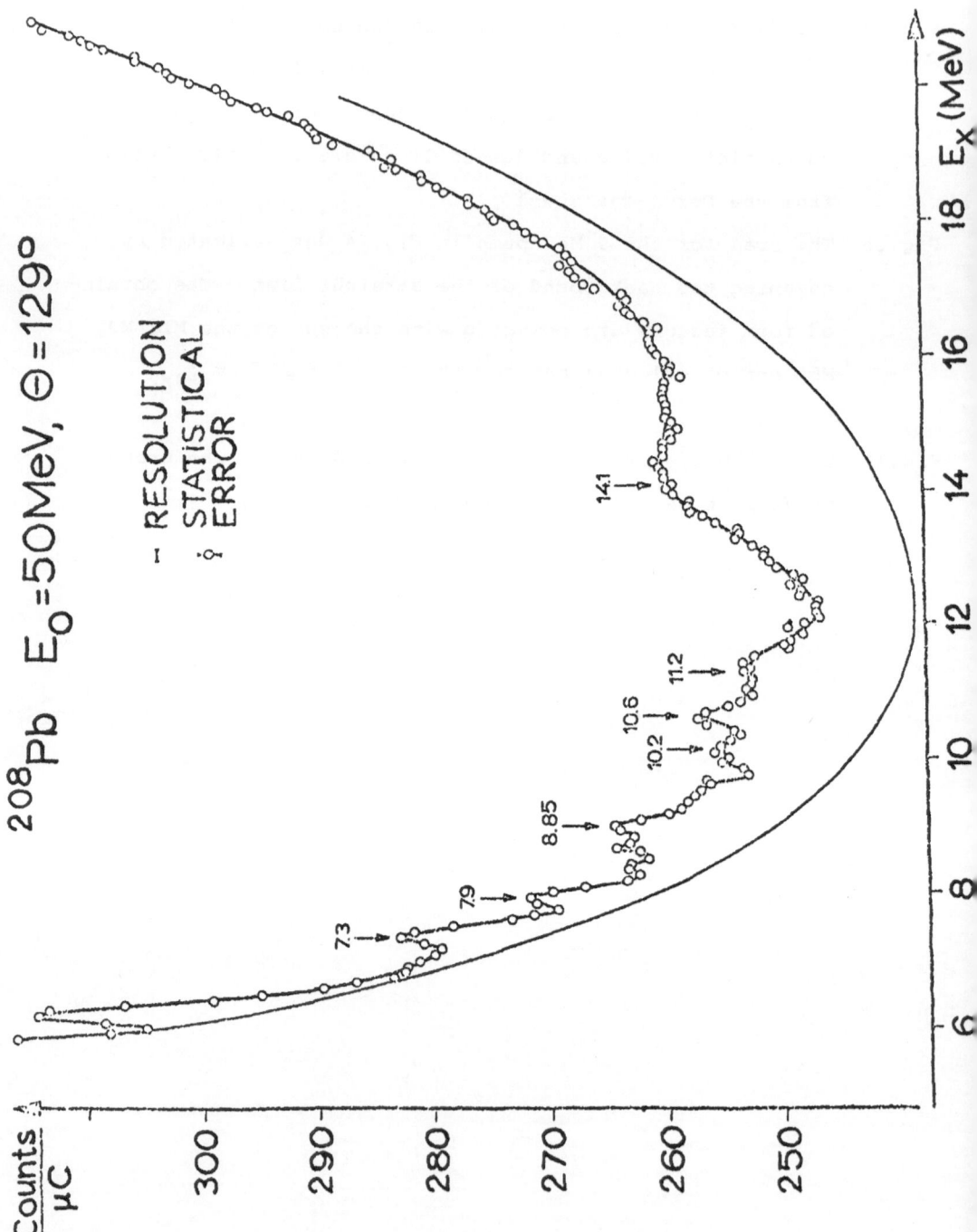

Fig.1

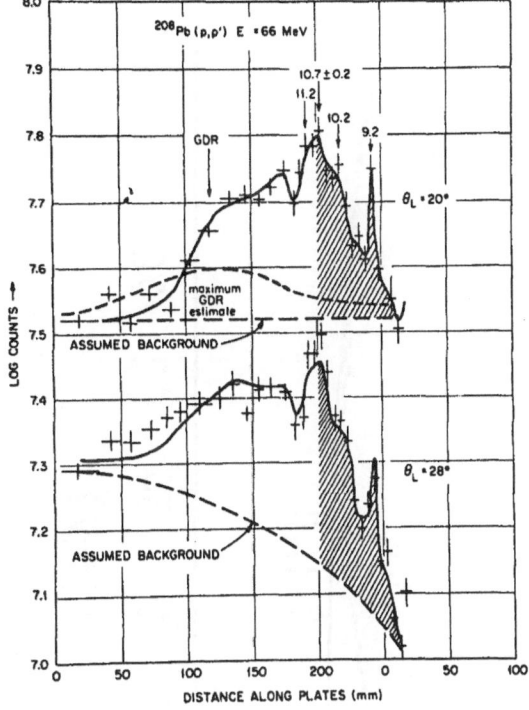

Fig.2

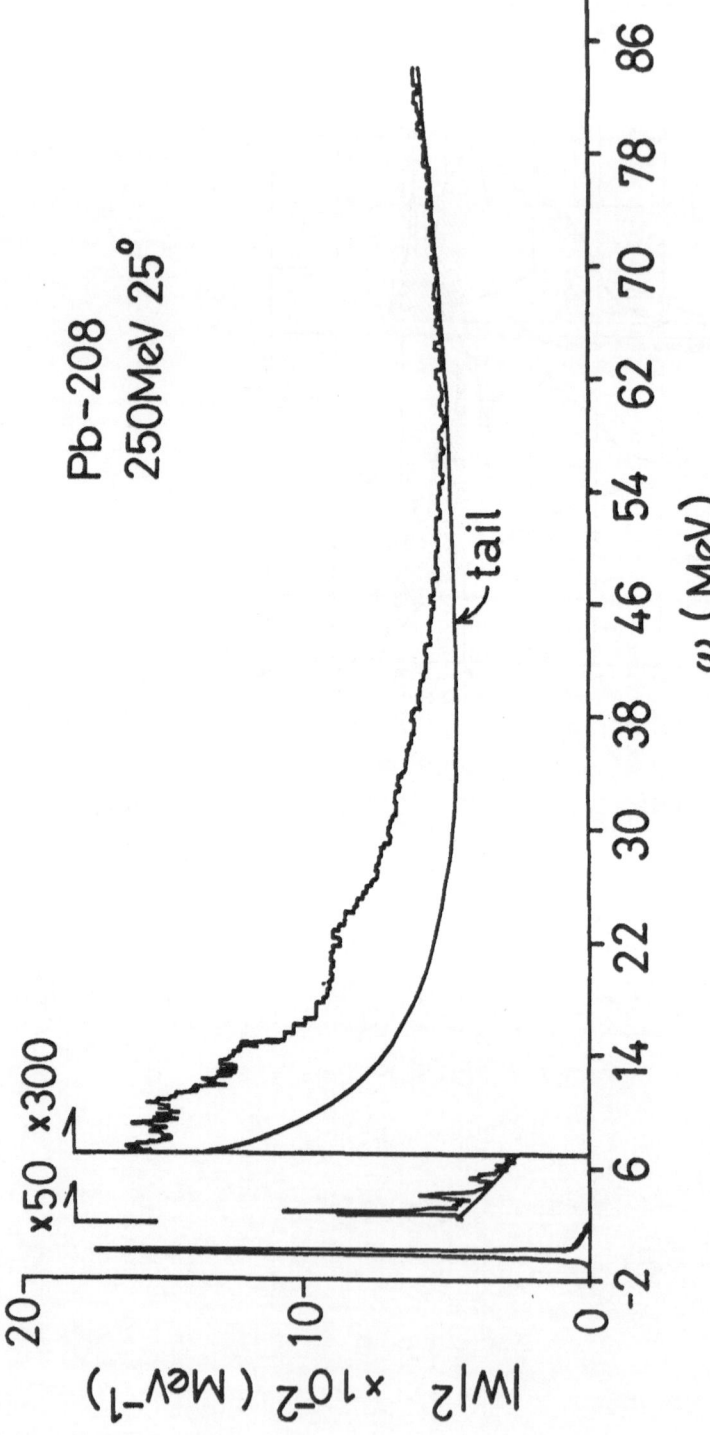

Fig.3

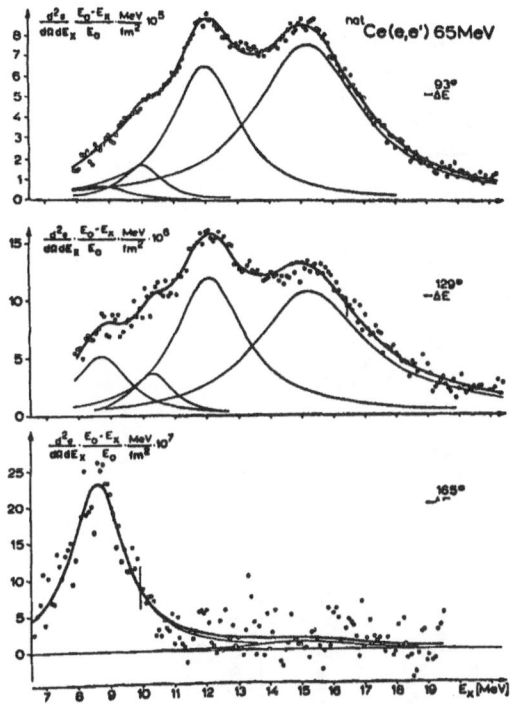

Fig.4

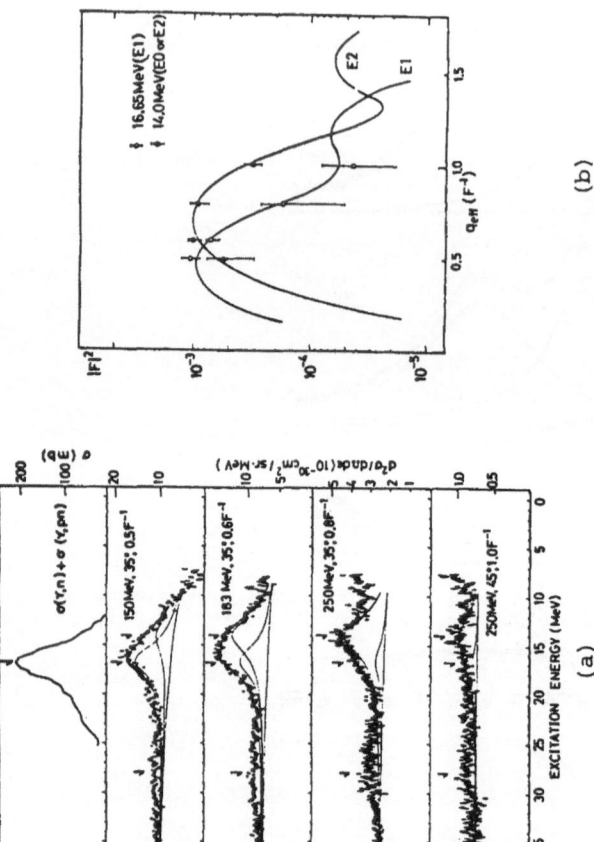

Fig.5

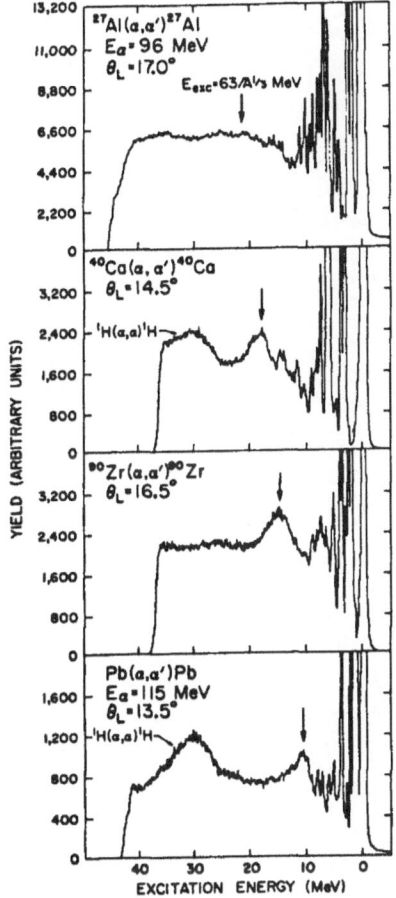

Fig.6

284

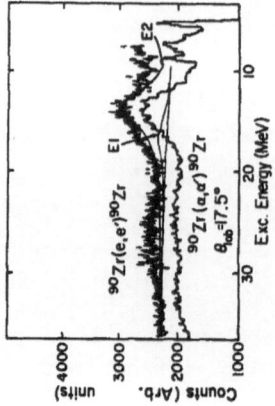

Fig.7

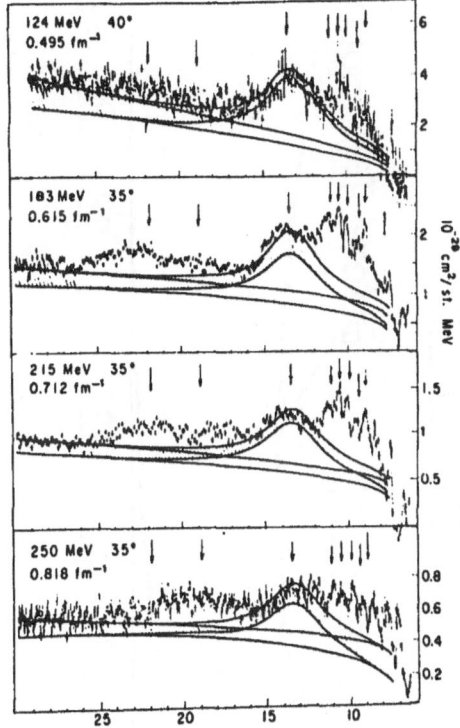

Fig. 8

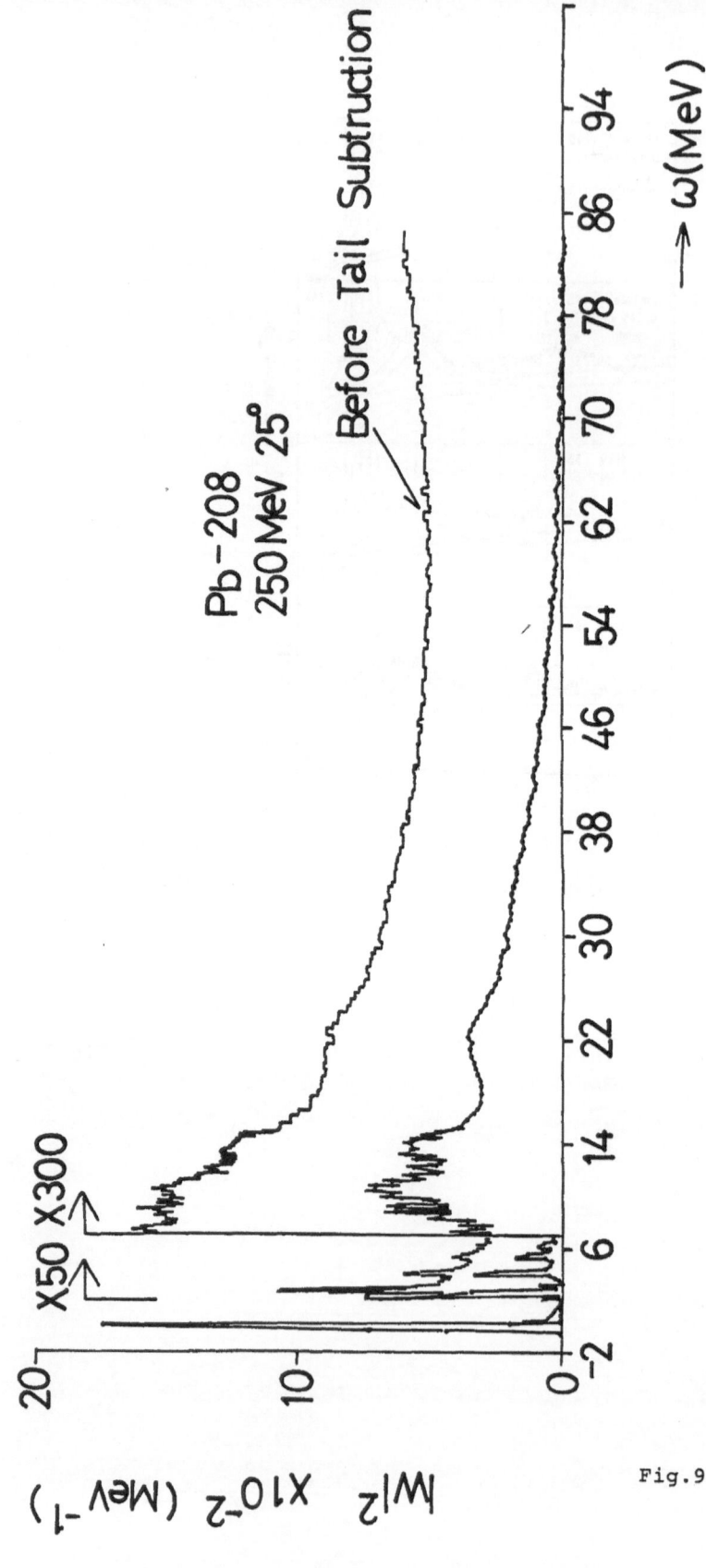

Fig.9

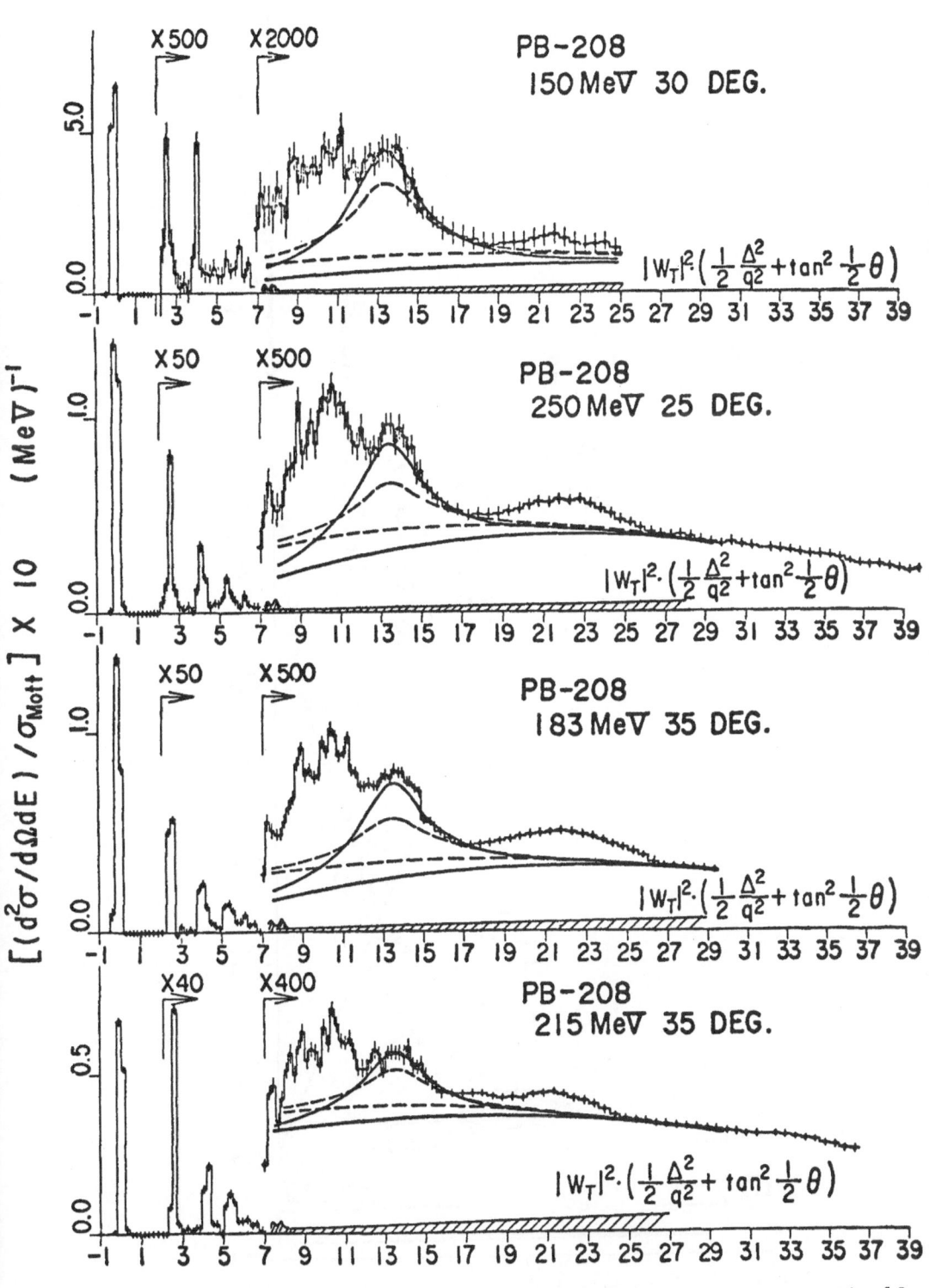

EXCITATION ENERGY (MeV)

Fig. 10

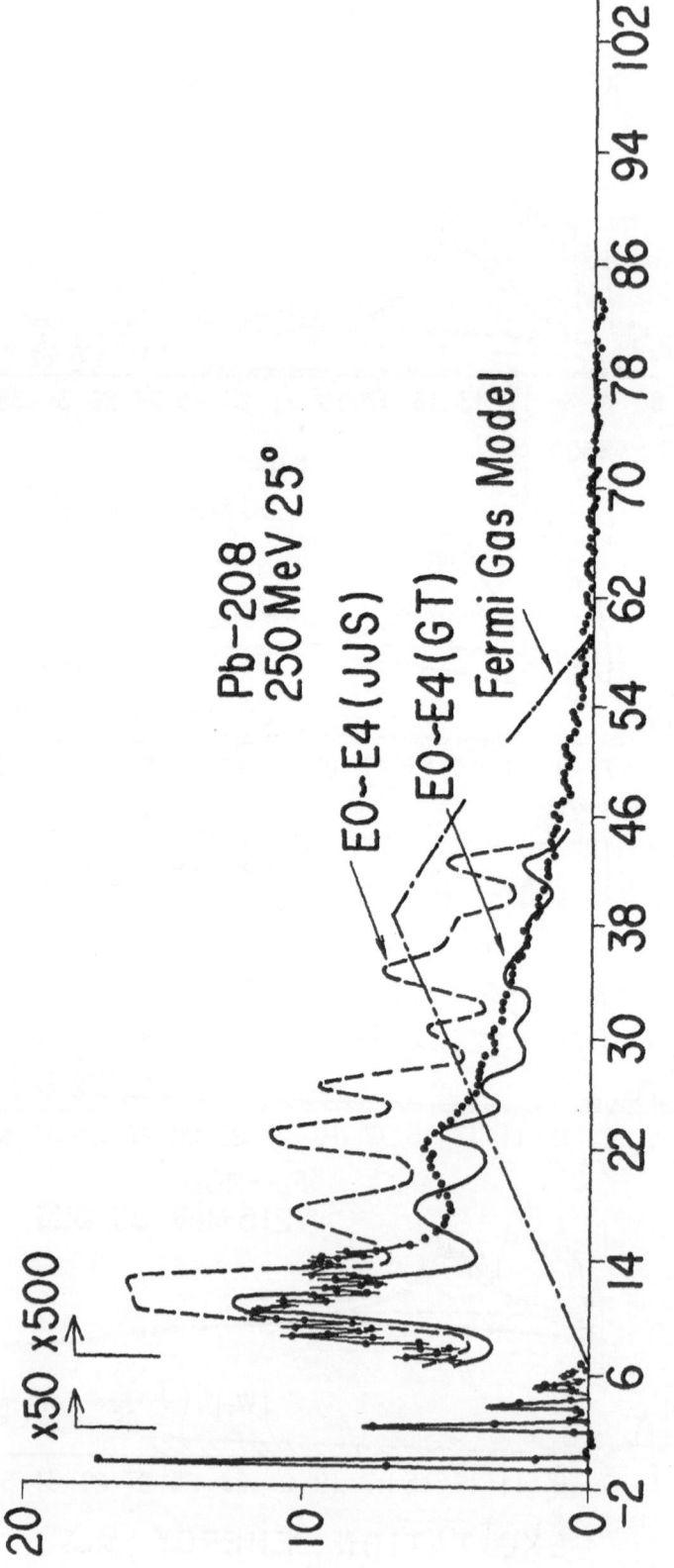

Fig.11

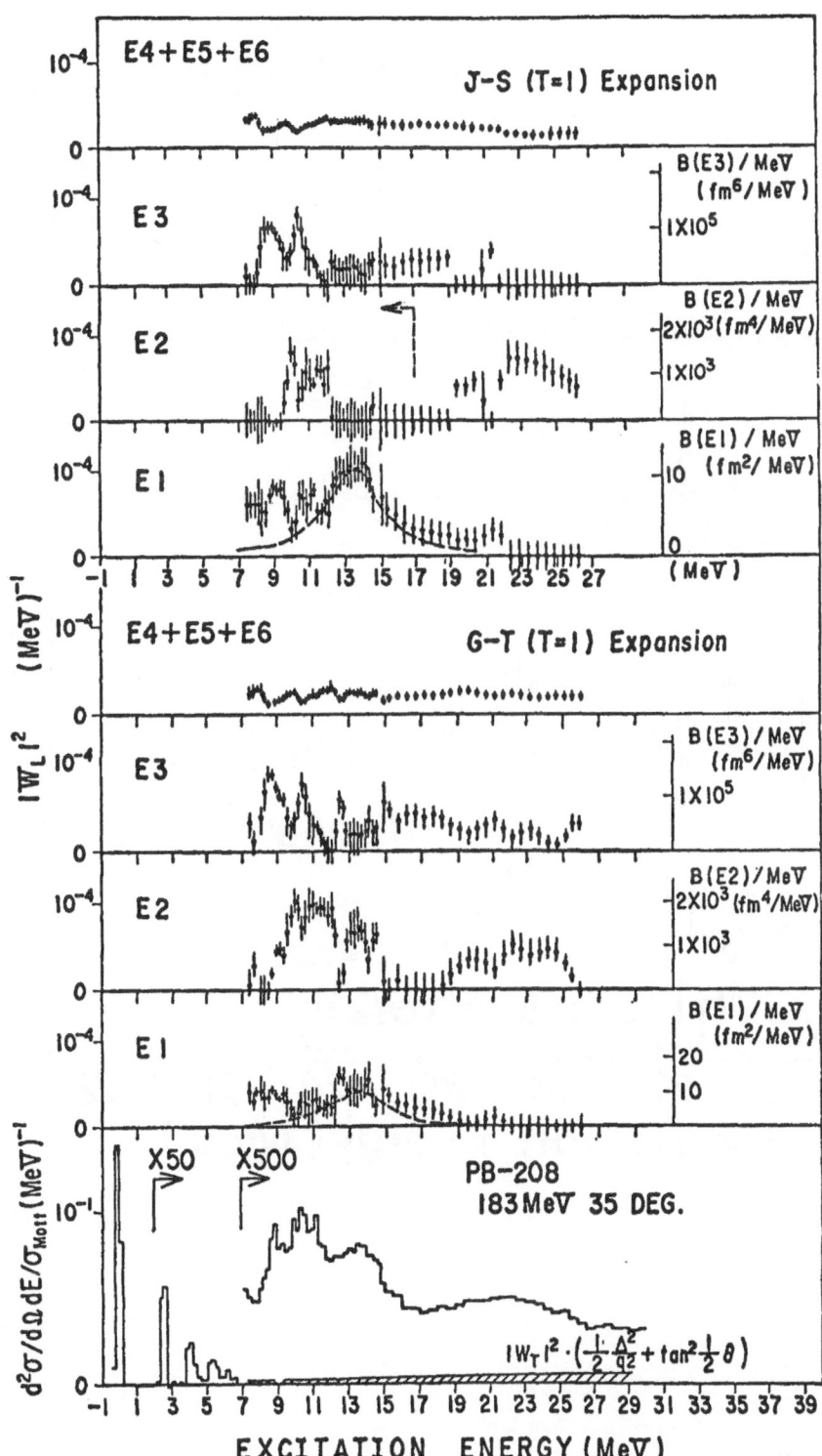

Fig.12

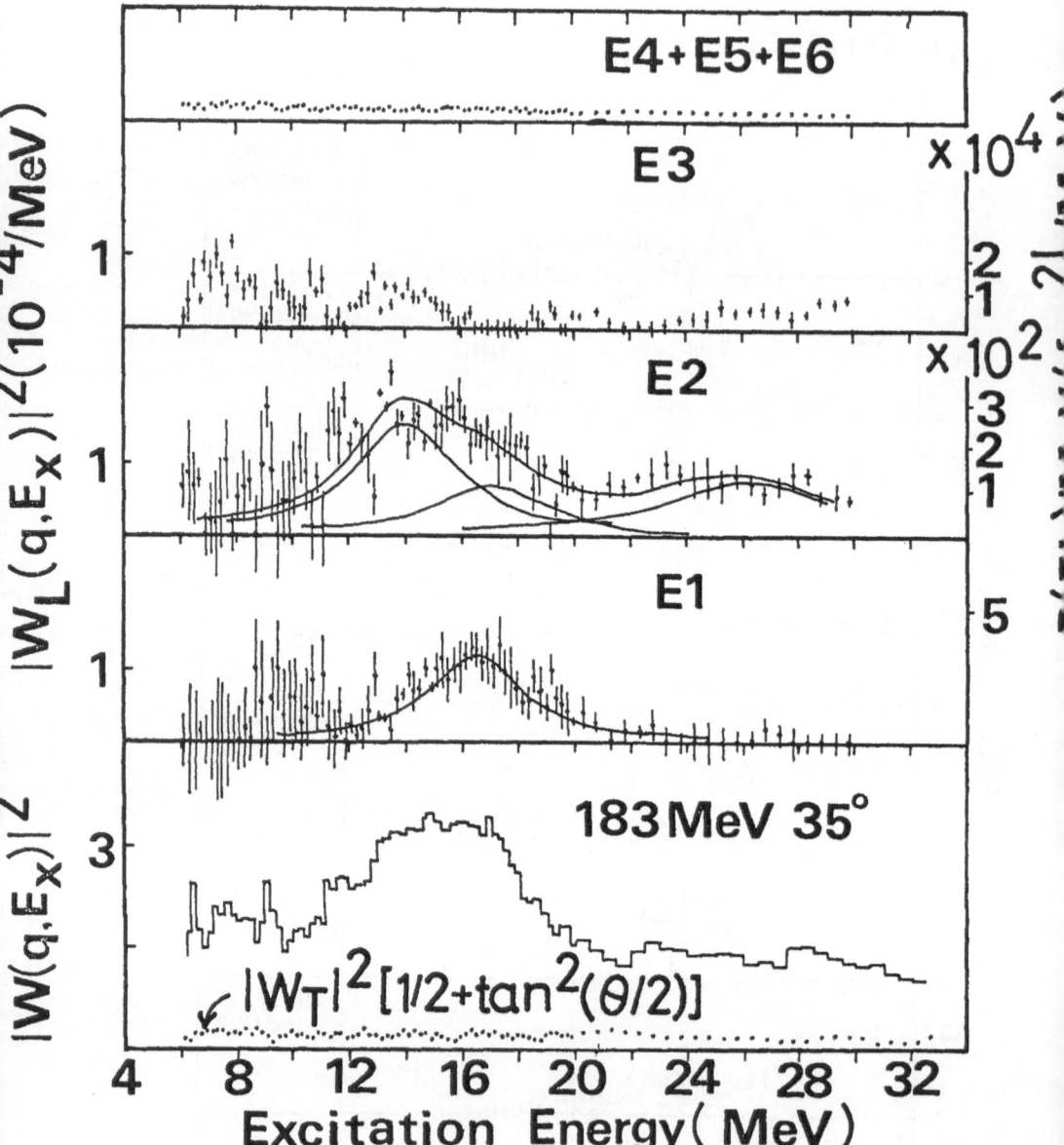

Fig.

$|W_L(q,E_x)|^2 (10^{-4}/MeV)$

E4+E5+E6

E3   $\times 10^4$

E2   $\times 10^2$

E1

$|W(q,E_x)|^2$

183 MeV 35°

$|W_T|^2 [1/2 + \tan^2(\theta/2)]$

Excitation Energy ( MeV )

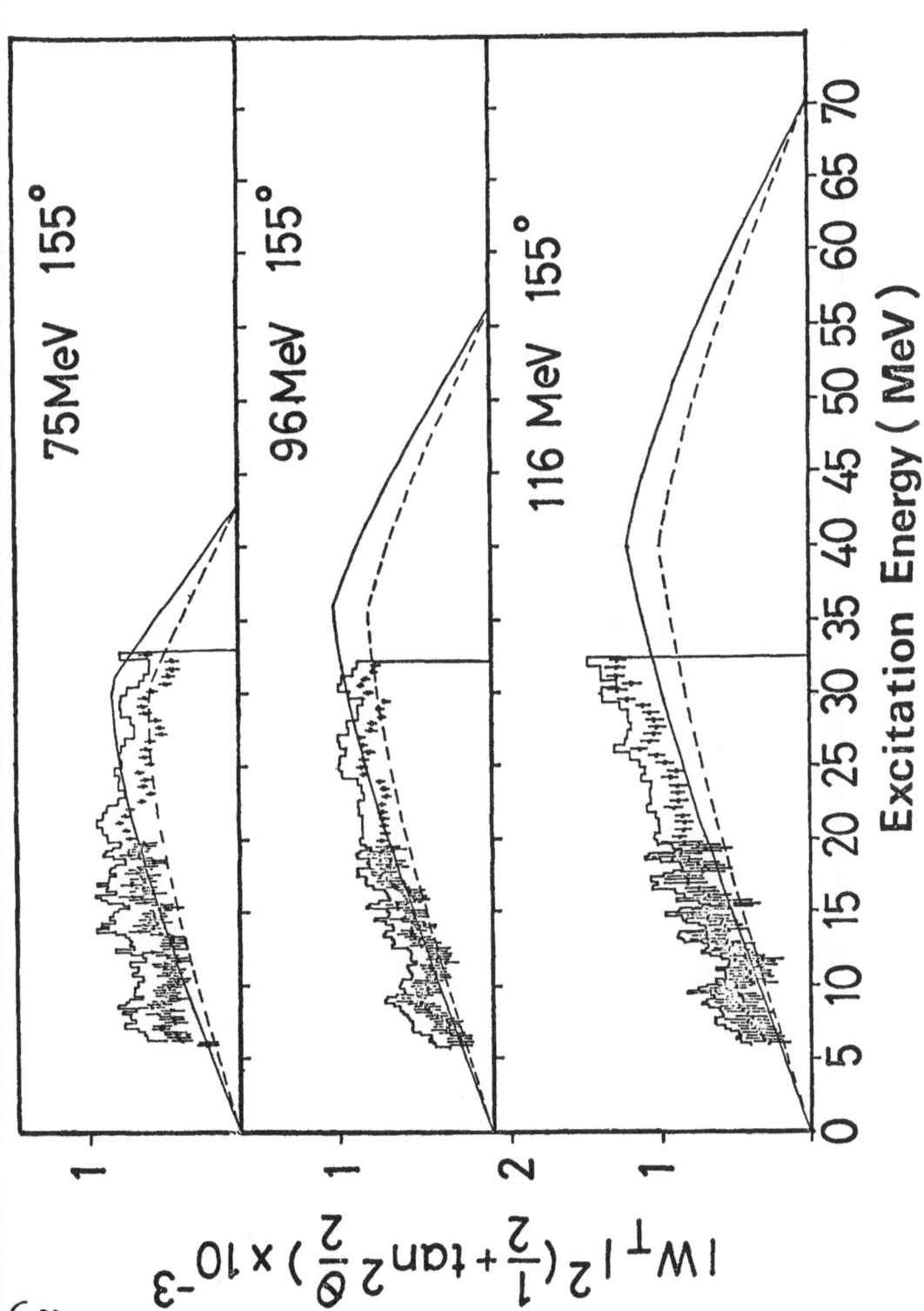

Fig.14

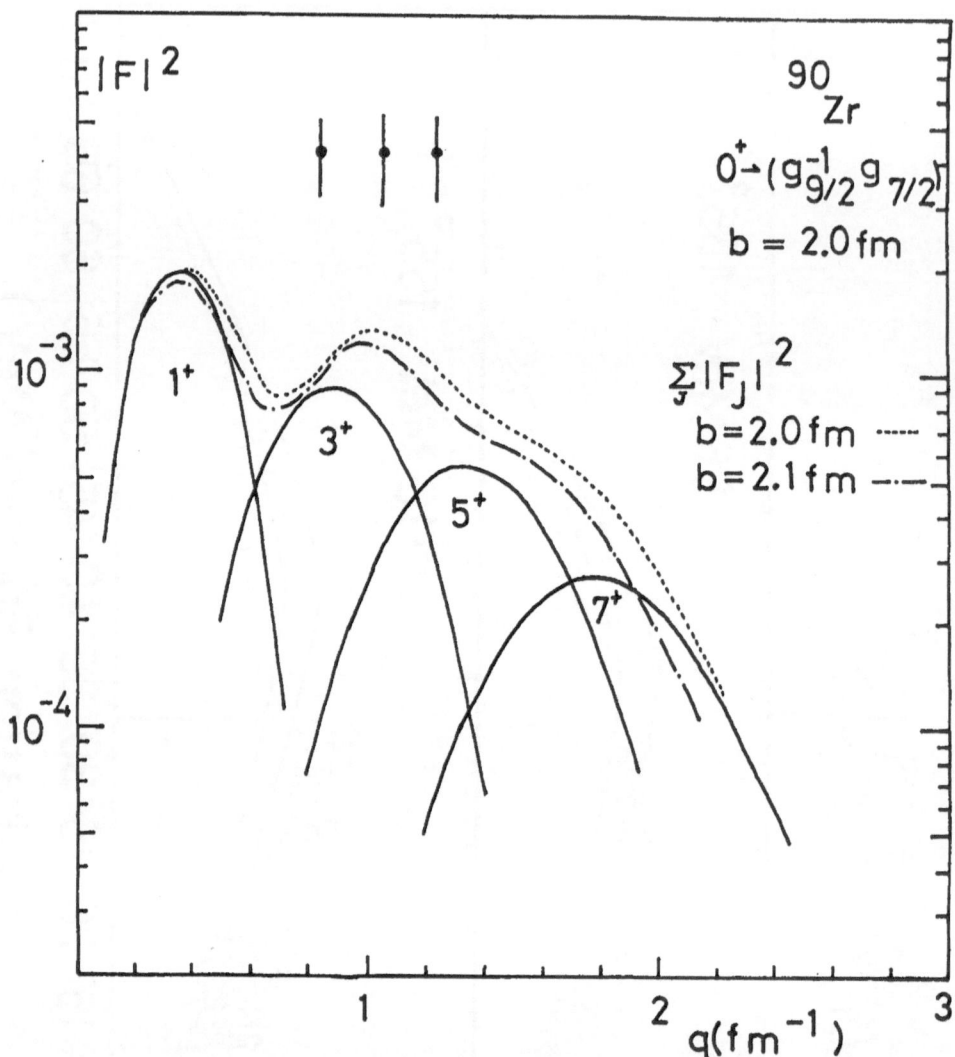

Fig.15

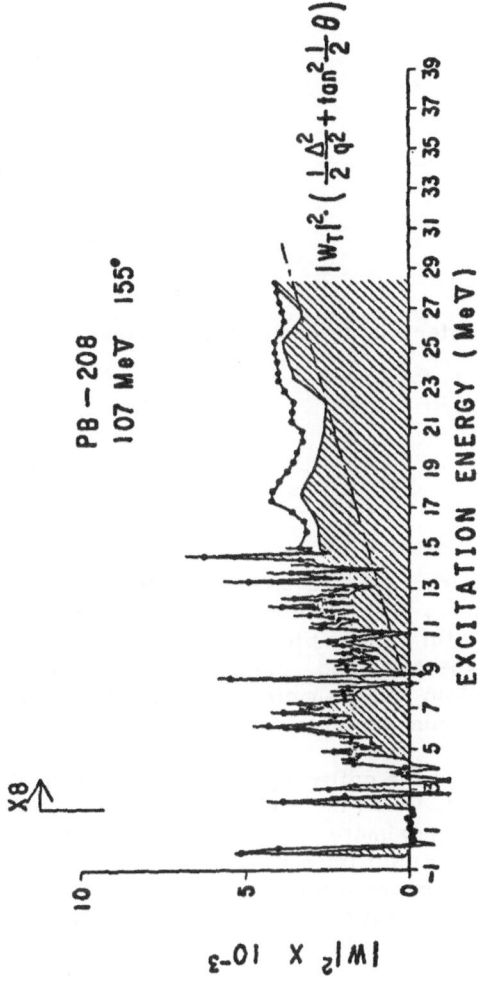

Fig. 16

Description of isoscalar giant resonances. A sum-rule approach.

O. Bohigas
Institut de Physique Nucléaire
Orsay

An approach to isoscalar resonances in terms of their energy moments was proposed. Besides the well-known moment containing the energy power plus one, moments with powers plus three and minus one are evaluated, in the selfconsistent R P A method. Values for the giant resonance energy, an upper limit on its width and a simple derivation of the quadrupole energy as $\sqrt{2}$ the oscillator quantum are obtained.

Relations between the collective energies from the G C M and R P A methods were studied. When the G C M approach is used with the usual Gaussian Overlap Approximation, the energy has a lower limit in terms of R P A sum-rules. If oscillator properties are assigned to the orbitals, the inequalities are converted into equalities. Some formulations of the G C M approach imply violation of sum-rules, so should be used with caution.

# The center of Mass Problem in Continuum Calculations of Nuclear Reaction Cross Sections ⊕

P.P. Delsanto°, A. Pompei, P. Quarati [+]

presented by P.P. Delsanto

In shell model calculations of nuclear reaction cross sections, spuriosities deriving from the center-of-mass motion can arise (1) and affect the results, particularly in the case of light nuclei.

In bound states calculations these spuriosities can be removed through the introduction of a strongly attractive harmonic oscillator potential. This procedure cannot be applied to continuum calculations, since an infinite nuclear potential is not compatible with the use of finite shell model potentials.

We have proposed (2) a numerical way of "tying" the c.m. to the origin, based on the recent treatment of the one particle continuum of Barret and Delsanto (3). In the B-D method the one-particle continuum is discretized through the separation of the configuration space into an external and an internal region, with a matching at the boundary of the external and internal wave functions, and its derivatives (natural boundary conditions). In the external region, where nuclear interactions are neglected, the w.f. is known but for the scattering phase shifts. If, at a given excitation energy, N channels are open, N-1 phase shifts can be kept fixed to an arbitrary value, while the remaining one is determined in such a way to satisfy the consistency requirement that one of the eigenvalues of the nuclear hamiltonian is equal to the excitation energy.

We have modified the B-D method to implement an interaction procedure that allows, through the dermination of the two phase shifts, to satisfy simultaneously the consistency requirement and the condition that the nuclear c.m. be in the origin. I.e. we impose that $\langle \psi | R^2 | \psi \rangle = 0$, at least in the limits of the numerical accuracy.

We also consider an intrinsic hamiltonian modified as in ref. 4.

As a preliminary test of our method we have computed $(\gamma, p)$ and $(\gamma, n)$ cross sections for $He_4$, a nucleus whose reaction states have been widely investigated in recent years (5-7). Our results are discussed and compared with those of recent measurements and other calculations.

---

⊕    Work supported in part by I.N.F.N.

°    On sabbatical leave of absence from the University of Puerto Rico at
                                       Mayaguez

+    Istituto di Fisica dell'Università di Cagliari and Istituto di Fisica del
    Politecnico di Torino

1.   J.P. Elliot and T.H.R. Skyrme: Proc. Roy. Soc. A232, 561 (1955)

2.   P.P. Delsanto and P. Quarati: to be published

3.  R.F. Barret and P.P. Delsanto: Phys. Rev. C10, 101 (1974)
4.  F. Palumbo and D. Prosperi: Nucl. Phys. A115, 296 (1968)
5.  S. Fiarman and W.E. Meyerhof: Nucl. Phys. A206, 1 (1973)
6.  A.H. Chung, R.G. Johnson and T.W. Donnelly: Nucl. Phys. A235, 151974
7.  S. Ramavataram, C.L. Rao and K. Ramavataram: Nucl. Phys. A226, 173 (1974)

# A simple model for resonance shifts

Mauro Giannini
Istituto di Scienze Fisiche
Università di Genova

Istituto Nazionale di Fisica
Nucleare- Sezione di Genova

In some recent pion photo-production experiments (1) a shifted and narrowed $\Delta$ -peak has been observed. A mechanism, which might give rise to such an effect, is that of one pion jumping from nucleon to another standing nearby. In this picture, which is similar to the $H_2$+ molecule description of the electronic motion, each nucleon is alternatively excited to a $\Delta$ -resonance and then de-excited. $\pi$ N system can be built up (2), whose parameters are fixed by fitting to the total $\pi$N cross section in the neighbourhood of the $\Delta$ resonance. The scattering amplitude $\pi$NN, with the two nucleons at fixed relative position, can be evaluated and the positions and widths of the resonant state estimated. The results depend on total spin, isospin and relative position of the nucleons; in some case a downward shift of~60 MeV and a narrowing of~35 MeV is achieved.

(1)  P.Argan et al. Physics Rev. Lett. __29__ (1972) 1191

(2)  R.Cenni, G.Dillon, and M.M. Giannini, preprint Genova 1975.

# NUCLEON POLARIZABILITIES AND DEEP
# INELASTIC ELECTRON SCATTERING

G. Matone and D. Prosperi
INFN, Laboratori Nazionali di Frascati, Frascati, Italy

Presented by D. Prosperi

During the past 15 years, a great deal of effort has been spent to investigate the low energy Compton scattering where the nucleon polarizabilities can be
extracted from. The present experimental evidence indicates for the proton the
following determinations :

$$\alpha_p \text{ (electric polarizability)} = (12.4 \pm 0.6) \times 10^{-4} \text{ fm}^3 ,$$

$$\beta_p \text{ (magnetic polarizability)} = (1.8 \pm 0.9) \times 10^{-4} \text{ fm}^3 ,$$

obtained with a fit of all the present available data[1].

The result $\alpha_p > \beta_p$ has to be considered as an extremely surprising conclusion[2]. In fact one knows from experiments that all photoabsorption processes
on nucleons are dominated by the $\Delta(1236)$ resonance and this excitation is of magnetic nature. So, one could expect the nucleon to be essentially a good parama-
gnetic object characterized by $\beta > \alpha$.

In order to inquire more deeply into this puzzling situation, the most general expression for the spin averaged Compton amplitude of off-mass shell photons has been rediscussed and the terms which, in the low energy limit and for
$q^2 \to 0$, contribute to the electric ($\alpha$) and magnetic ($\beta$) polarizabilities have been
individuated[1].

Furthermore it has been shown that these amplitudes are connected to the
structure fucntions usually defined in the deep inelastic scattering and moreover
the following sum rules for the polarizabilities have been deduced :

$$a + \beta = \lim_{q^2 \to 0} \frac{1}{2\pi^2} \int_{\nu\text{th}}^{\infty} \frac{\sigma_T(q^2, \nu')}{\nu'^2} d\nu' \; ,$$

$$a = \frac{(\lambda\hat{e})^2}{4M^3} + \lim_{q^2 \to 0} \frac{1}{2\pi^2} \int_{\nu\text{th}}^{\infty} \sigma_T(q^2, \nu') \frac{R(q^2, \nu')}{q^2} d\nu' \; ,$$

$$a - \beta = \frac{(z\hat{e})^2}{2M^3} + \frac{1}{2\pi^2} \int_{\nu\text{th}}^{\infty} \frac{\tilde{\sigma}(\nu')}{\nu'^2} (1 + \frac{\nu'}{M}) d\nu' + 4\hat{e} \; \Phi_1(M^2, -1) \; .$$

Besides the usual definitions, the other quantities are defined as follows:

$\sigma_T(q^2, \nu')$ = total photoabsorption cross section,

$$\tilde{\sigma}(\nu) = \sigma^{AV}(\Delta\pi - \text{yes}) - \sigma^{AV}(\Delta\pi - \text{no}) \; , \qquad R(q^2, \nu) = \frac{\sigma_1(q^2, \nu)}{\sigma_t(q^2, \nu)} \; ,$$

where $\sigma(\Delta\pi$-yes$)$ $(\sigma(\Delta\pi$ - no$))$ is the spin averaged cross section corresponding to parity flip multipoles (no parity flip multipoles) and $\Phi_1(s, \cos\theta)$ is a complicated function defined in terms of the absorbitive part of the amplitude in the channel t. While the first is the very well known forward dispersion relation yielding the result

$$(a + \beta) = (14.2 \pm 0.3) \times 10^{-4} \text{ fm}^3$$

the second one furnishes an independent evaluation of $a$ expressed through the experimental determination of $R(q^2, \nu)$. Data from SLAC on the deep inelastic electron scattering allows to obtain the following estimate[1]:

$$a_p = (9.3 \pm 2.0) \times 10^{-4} \text{ fm}^3 \; ,$$

which is in substantial agreement with the experimental value quoted previously.

Similarly a backward sum rule for $(a - \beta)$ is proposed even though its numerical evaluation is strongly dependent on the t-channel contribution. Nevertheless, within the present experimental knowledge, this rule gives values compatible with the previous evaluation $a_p > \beta_p$.

A big experimental achievement could be reached with a better determination of $\beta_p$, which at the present time is the most uncertain one. In

connection with this, the polarization of the incoming photon beam could pro
vide the possibility to separate with great accuracy the contribution of the
electric and magnetic polarizability to the Compton cross section. Therefore the LADON project, whose characteristics have been presented at
this School, appear to be among the most qualified tools to improve the
present knowledge of these fundamental structure parameters.

REFERENCES. -

(1) - G. Matone and D. Prosperi, Nucleon polarizabilities and deep inelastic
electron scattering, Frascati report LNF-76/38 (1976); submitted to
Nuovo Cimento.

(2) - T. E. O. Ericson and J. Hüfner, Nuclear Phys. B57, 604 (1973).

# Exchange effects in photon scattering on nuclei

Paolo Christillin
Niels Bohr Institute
Copenhagen

Marco Rosa-Clot
CERN
Geneva

Exchange effects manifest in the photon nucleus interaction as an enhancement of the total photoabsorption cross section and as a direct contribution to the scattering amplitude above giant dipole resonance.

Dispersion relation can be used to connect these two aspects of the problem relating the integrated photoabsorption cross section to the scattering amplitude.

To do this, first we construct the scattering amplitude up to the pion threshold, exploiting gauge invariance, and then we compare it with experimental data on photoabsorption in the light nuclei region. Furthermore, we focus our attention on deformed nuclei where we perform a phenomenological analysis and postulate the presence of an anisotropy of the exchange forces.

Experimental data in the giant dipole resonance region seem to support this analysis which is then used to predict effects in the high energy scattering amplitude.

# Selected Issues from
# Lecture Notes in Mathematics

Vol. 431: Séminaire Bourbaki – vol. 1973/74. Exposés 436–452. IV, 347 pages. 1975.

Vol. 433: W. G. Faris, Self-Adjoint Operators. VII, 115 pages. 1975.

Vol. 434: P. Brenner, V. Thomée, and L. B. Wahlbin, Besov Spaces and Applications to Difference Methods for Initial Value Problems. II, 154 pages. 1975.

Vol. 440: R. K. Getoor, Markov Processes: Ray Processes and Right Processes. V, 118 pages. 1975.

Vol. 442: C. H. Wilcox, Scattering Theory for the d'Alembert Equation in Exterior Domains. III, 184 pages. 1975.

Vol. 446: Partial Differential Equations and Related Topics. Proceedings 1974. Edited by J. A. Goldstein. IV, 389 pages. 1975.

Vol. 448: Spectral Theory and Differential Equations. Proceedings 1974. Edited by W. N. Everitt. XII, 321 pages. 1975.

Vol. 449: Hyperfunctions and Theoretical Physics. Proceedings 1973. Edited by F. Pham. IV, 218 pages. 1975.

Vol. 458: P. Walters, Ergodic Theory – Introductory Lectures. VI, 198 pages. 1975.

Vol. 459: Fourier Integral Operators and Partial Differential Equations. Proceedings 1974. Edited by J. Chazarain. VI, 372 pages. 1975.

Vol. 461: Computational Mechanics. Proceedings 1974. Edited by J. T. Oden. VII, 328 pages. 1975.

Vol. 463: H.-H. Kuo, Gaussian Measures in Banach Spaces. VI, 224 pages. 1975.

Vol. 464: C. Rockland, Hypoellipticity and Eigenvalue Asymptotics. III, 171 pages. 1975.

Vol. 468: Dynamical Systems – Warwick 1974. Proceedings 1973/74. Edited by A. Manning. X, 405 pages. 1975.

Vol. 470: R. Bowen, Equilibrium States and the Ergodic Theory of Anosov Diffeomorphisms. III, 108 pages. 1975.

Vol. 474: Séminaire Pierre Lelong (Analyse) Année 1973/74. Edité par P. Lelong. VI, 182 pages. 1975.

Vol. 484: Differential Topology and Geometry. Proceedings 1974. Edited by G. P. Joubert, R. P. Moussu, and R. H. Roussarie. IX, 287 pages. 1975.

Vol. 487: H. M. Reimann und T. Rychener, Funktionen beschränkter mittlerer Oszillation. VI, 141 Seiten. 1975.

Vol. 489: J. Bair and R. Fourneau, Etude Géométrique des Espaces Vectoriels. Une Introduction. VII, 185 pages. 1975.

Vol. 490: The Geometry of Metric and Linear Spaces. Proceedings 1974. Edited by L. M. Kelly. X, 244 pages. 1975.

Vol. 503: Applications of Methods of Functional Analysis to Problems in Mechanics. Proceedings 1975. Edited by P. Germain and B. Nayroles. XIX, 531 pages. 1976.

Vol. 507: M. C. Reed, Abstract Non-Linear Wave Equations. VI, 128 pages. 1976.

Vol. 509: D. E. Blair, Contact Manifolds in Riemannian Geometry. VI, 146 pages. 1976.

Vol. 515: Bäcklund Transformations. Nashville, Tennessee 1974. Proceedings. Edited by R. M. Miura. VIII, 295 pages. 1976.

Vol. 516: M. L. Silverstein, Boundary Theory for Symmetric Markov Processes. XVI, 314 pages. 1976.

Vol. 518: Séminaire de Théorie du Potentiel, Proceedings Paris 1972–1974. Edité par F. Hirsch et G. Mokobodzki. VI, 275 pages. 1976.

Vol. 522: C. O. Bloom and N. D. Kazarinoff, Short Wave Radiation Problems in Inhomogeneous Media: Asymptotic Solutions. V. 104 pages. 1976.

Vol. 523: S. A. Albeverio and R. J. Høegh-Krohn, Mathematical Theory of Feynman Path Integrals. IV, 139 pages. 1976.

Vol. 524: Séminaire Pierre Lelong (Analyse) Année 1974/75. Edité par P. Lelong. V, 222 pages. 1976.

Vol. 525: Structural Stability, the Theory of Catastrophes, and Applications in the Sciences. Proceedings 1975. Edited by P. Hilton. VI, 408 pages. 1976.

Vol. 526: Probability in Banach Spaces. Proceedings 1975. Edited by A. Beck. VI, 290 pages. 1976.

Vol. 527: M. Denker, Ch. Grillenberger, and K. Sigmund, Ergodic Theory on Compact Spaces. IV, 360 pages. 1976.

Vol. 532: Théorie Ergodique. Proceedings 1973/1974. Edité par J.-P. Conze and M. S. Keane. VIII, 227 pages. 1976.

Vol. 538: G. Fischer, Complex Analytic Geometry. VII, 201 pages. 1976.

Vol. 543: Nonlinear Operators and the Calculus of Variations, Bruxelles 1975. Edited by J. P. Gossez, E. J. Lami Dozo, J. Mawhin, and L. Waelbroeck, VII, 237 pages. 1976.

Vol. 552: C. G. Gibson, K. Wirthmüller, A. A. du Plessis and E. J. N. Looijenga. Topological Stability of Smooth Mappings. V, 155 pages. 1976.

Vol. 556: Approximation Theory. Bonn 1976. Proceedings. Edited by R. Schaback and K. Scherer. VII, 466 pages. 1976.

Vol. 559: J.-P. Caubet, Le Mouvement Brownien Relativiste. IX, 212 pages. 1976.

Vol. 561: Function Theoretic Methods for Partial Differential Equations. Darmstadt 1976. Proceedings. Edited by V. E. Meister, N. Weck and W. L. Wendland. XVIII, 520 pages. 1976.

Vol. 564: Ordinary and Partial Differential Equations, Dundee 1976. Proceedings. Edited by W. N. Everitt and B. D. Sleeman. XVIII, 551 pages. 1976.

Vol. 565: Turbulence and Navier Stokes Equations. Proceedings 1975. Edited by R. Temam. IX, 194 pages. 1976.

Vol. 566: Empirical Distributions and Processes. Oberwolfach 1976. Proceedings. Edited by P. Gaenssler and P. Révész. VII, 146 pages. 1976.

Vol. 570: Differential Geometrical Methods in Mathematical Physics. Bonn 1975. Proceedings. Edited by K. Bleuler and A. Reetz. VIII, 576 pages. 1977.

Vol. 572: Sparse Matrix Techniques, Copenhagen 1976. Edited by V. A. Barker. V, 184 pages. 1977.

# Lecture Notes in Physics

Vol. 44: R. A. Breuer, Gravitational Perturbation Theory and Synchrotron Radiation. VI, 196 pages. 1975.

Vol. 45: Dynamical Concepts on Scaling Violation and the New Resonances in $e^+e^-$ Annihilation. Edited by B. Humpert. VII, 248 pages. 1976.

Vol. 46: E. J. Flaherty, Hermitian and Kählerian Geometry in Relativity. VIII, 365 pages. 1976.

Vol. 47: Padé Approximants Method and Its Applications to Mechanics. Edited by H. Cabannes. XV, 267 pages. 1976.

Vol. 48: Interplanetary Dust and Zodiacal Light. Proceedings 1975. Edited by H. Elsässer and H. Fechtig. XII, 496 pages. 1976.

Vol. 49: W. G. Harter and C. W. Patterson, A Unitary Calculus for Electronic Orbitals. XII, 144 pages. 1976.

Vol. 50: Group Theoretical Methods in Physics. 4th International Colloquium. Nijmegen 1975. Edited by A. Janner, T. Janssen, and M. Boon. XIII, 629 pages. 1976.

Vol. 51: W. Nörenberg und H. A. Weidenmüller. Introduction to the Theory of Heavy-Ion Collisions. IX, 273 pages. 1976.

Vol. 52: M. Mladjenović, Development of Magnetic β-Ray Spectroscopy. X, 282 pages. 1976.

Vol. 53: D. J. Simms and N. M. J. Woodhouse, Lectures on Geometric Quantization. V, 166 pages. 1976.

Vol. 54: Critical Phenomena. Sitges International School on Statistical Mechanics, June 1976. Edited by J. Brey and R. B. Jones. XI, 383 pages. 1976.

Vol. 55: Nuclear Optical Model Potential. Proceedings 1976. Edited by S. Boffi and G. Passatore. VI, 221 pages. 1976.

Vol. 56: Current Induced Reactions. International Summer Institute, Hamburg 1975. Edited by J. G. Körner, G. Kramer, and D. Schildknecht. V, 553 pages. 1976.

Vol. 57: Physics of Highly Excited States in Solids. Proceedings 1975. Edited by M. Ueta and Y. Nishina. IX, 391 pages. 1976.

Vol. 58: Computing Methods in Applied Sciences. Proceedings 1975. Edited by R. Glowinski and J. L. Lions. VIII, 593 pages. 1976.

Vol. 59: Proceedings of the Fifth International Conference on Numerical Methods in Fluid Dynamics. 1976. Edited by A. I. van de Vooren and P. J. Zandbergen. VII, 459 pages. 1976.

Vol. 60: C. Gruber, A. Hintermann, and D. Merlini, Group Analysis of Classical Lattice Systems. XIV, 326 pages. 1977.

Vol. 61: International School on Electro and Photonuclear Reactions I. Edited by C. Schaerf. VIII, 650 pages. 1977.

Vol. 62: International School on Electro and Photonuclear Reactions II. Edited by C. Schaerf. VIII, 301 pages. 1977.